THE
ABANDONED
OCEAN

STUDIES IN MARITIME HISTORY
William N. Still, Jr., Series Editor

THE
ABANDONED
OCEAN

A HISTORY OF UNITED STATES MARITIME POLICY

Andrew Gibson

and

Arthur Donovan

University of South Carolina Press

UNIVERSITY OF SOUTH CAROLINA *BICENTENNIAL*

Published in Columbia, South Carolina, by the
University of South Carolina Press

Published 2000
First paperback edition 2001

Manufactured in the United States of America

05 04   5 4 3

The Library of Congress has cataloged the cloth edition as follows:

Gibson, Andrew, 1922–
    The abandoned ocean : a history of United States maritime
policy / Andrew Gibson and Arthur Donovan.
    Includes bibliographical references and index.
    ISBN 1-57003-319-6 (cloth : alk. paper)
    1. Merchant marine—United States—History. 2. Navigation—United
    States—History. I. Donovan, Arthur, 1938– II. Title. III. Series.
VK23 .G53 2000
387.5'0973—dc21   99-6133

ISBN 1-57003-427-3 (pbk.)

CONTENTS

ILLUSTRATIONS

Illustrations

T-2 tanker *Ideal X*
Portrait of Malcom McLean
Sea-Land vessel being loaded by a gantry crane
Superimposed silhouettes of a T-2 tanker and a Very Large Crude Carrier
Stern view of a large Roll-on/Roll-off (Ro-Ro) vessel
Lighter Aboard Ship (LASH) vessel
C-11 containership built for American President Lines

TABLES

Abbreviations

AMA	American Maritime Association
AMMI	American Merchant Marine Institute
CASL	Committee of American Steamship Lines
CCF	capital construction fund
CDS	construction differential subsidy
CINC	commander in chief
DOD	Department of Defense
DOT	Department of Transportation
DWT	deadweight tons
EFC	Emergency Fleet Corporation
FMC	Federal Maritime Commission
FOC	flag of convenience
JCS	Joint Chiefs of Staff
LASH	Lighter Aboard Ship
LNG	liquid natural gas
Lo-Lo	Lift-on/Lift-off
MARAD	Maritime Administration
MSP	Maritime Security Program
MEBA	Marine Engineers Beneficial Association
MSC	Military Sealift Command (successor to MSTS)
MSTS	Military Sea Transportation Service
MTMC	Military Traffic Management Command
NDRF	National Defense Reserve Fleet
NMU	National Maritime Union
OBO	ore-bulk-oil carrier
ODS	operating differential subsidy

OECD	Organization for Economic Cooperation and Development
OMB	Office of Management and Budget
OPA 90	Oil Pollution Act of 1990
Ro-Ro	Roll-on/Roll-off
RRF	Ready Reserve Force
SIU	Seafarers International Union
SOLAS	International Convention for Safety of Life at Sea
TEU	twenty-foot container equivalent unit
ULCC	Ultra Large Crude Carrier
UNCTAD	United Nations Conference on Trade and Development
USSB	U.S. Shipping Board
USTRANSCOM	U.S. Transportation Command
VLCC	Very Large Crude Carrier

ACKNOWLEDGMENTS

Like all books, this history has a history of its own. It first appeared as a doctoral dissertation that Gibson wrote while serving as a faculty member at the Naval War College in Newport, Rhode Island. The dissertation was completed in 1993, at which time Gibson received his Ph.D. from the University of Wales College of Cardiff. For their assistance at the Naval War College, Gibson wishes to thank Captain William M. Calhoun (U.S. Navy, retired), Dean of Academics; Captain Timothy Somes (U.S. Navy, retired), Professor of Force Planning; Frank Uhlig Jr., now Editor Emeritus of the *Naval War College Review;* and Alice K. Juda of the Reference Branch of the War College Library. Professor Richard O. Goss, who served as Gibson's supervisor at the University of Wales, deserves special thanks, as does Professor Alastair Couper, who greatly assisted in bringing the dissertation to a successful conclusion.

While completing his dissertation Gibson spent a year as a Visiting Professor at the U.S. Merchant Marine Academy in Kings Point, New York. He is especially grateful to RADM Paul L. Krinsky (USMS, retired), then Superintendent of the U.S. Merchant Marine Academy, for arranging this appointment and taking such a lively interest in his research. Thanks are also due to Dr. George Billy, Head Librarian at Kings Point, and the staff members of the library and the Department of Marine Transportation. It was also during this year that Gibson and Donovan, who was then serving as Head of Kings Point's Department of Humanities, got to know one another and decided to work together to turn the dissertation into a book. This became a bigger job than either of them imagined, but the scholarly benefits of collaboration have more than compensated for the time and effort expended. Donovan is particularly grateful for the constant support of two friends and colleagues at Kings Point, Dr. Warren F. Mazek, Academic Dean, and Dr. Jane P. Brickman, now Head of the Department of Humanities.

While writing this book our work has been supported in part by three grants. The Mormac Marine Group, Incorporated, provided one of these, and we are grateful to Vice Chairman James R. Barker for responding so promptly and generously to our request for funds. After beginning this project we were awarded two grants for a second collaborative project, a history of containerization. One of these came from the National Science

Foundation, the other from the Smithsonian Institution's Lemelson Center for the Study of American Innovation and Invention. Some of the time for research and writing provided by these grants, together with encouragement and assistance from generous colleagues, made completion of this book possible. Our wives, knowing us perhaps too well, shared our enthusiasms and complaints with bemused good will.

Frank Uhlig Jr. continued to make valuable suggestions for improving the manuscript and helped the authors obtain illustrations from the U.S. Naval Institute, for which we thank him. Frank Braynard, Curator at the American Merchant Marine Museum, and Norman J. Brouwer, Curator of Ships and Marine Historian at South Street Seaport Museum, were both extraordinarily generous in allowing the authors to copy and publish illustrations from their rich collections of maritime pictures. Martin Skrocki, Public Information Officer at the U.S. Merchant Marine Academy, contributed valuable photographic assistance.

The authors are grateful to Alexander Moore, Acquisitions Editor at the University of South Carolina Press, and Dr. William N. Still Jr., General Editor of the Press's distinguished Maritime History Series, for the enthusiasm with which they welcomed the manuscript. Many others also provided encouragement and assistance as we struggled to bring this work to completion and we thank them all. Of course the authors are responsible for any errors that remain.

Andrew E. Gibson
Advanced Research Fellow
Naval War College
Newport, Rhode Island

Arthur Donovan
Department of Humanities
U.S. Merchant Marine Academy
Kings Point, New York

THE
ABANDONED
OCEAN

INTRODUCTION

This book is about U.S. maritime policy, a subject far more extensive and complex than one might expect. Federal maritime policy is in fact an accumulation of legislative statutes, administrative programs, and judicial decisions that have appeared in piecemeal fashion over the years as the national government has responded to specific problems concerning its inland waters and surrounding oceans. To engage the subject in all its fullness one would have to give due consideration to maritime transportation, trade policy, fishing policy, environmental policy, and a host of similar topics. We do not address all facets of maritime policy, however, for we are not attempting to write a comprehensive treatment of the subject.

We concentrate instead on federal policies that promote, regulate, protect, and subsidize U.S. shipping in coastal and foreign trade. Since the national government, not long after its creation, required that ships sailing under the U.S. flag be U.S. built, we are concerned with the shipbuilding industry as well. And since the activities of the merchant and military navies have been intertwined since before the American Revolution, we also address naval history when it has a bearing on maritime policy. But always at the center of attention is our motivating and focusing question: Why, since the middle of the nineteenth century, has U.S. maritime policy been so strikingly unsuccessful in achieving its stated goal of promoting a commercially viable merchant marine engaged in foreign trade?

While our subject is policy, we write as historians. We follow our subject as it has evolved in time, our method being to explain how the main constituents of federal maritime policy were forged in the specific circumstances that forced and shaped federal action. We have done our best to employ appropriate analytic categories and tell informative and convincing stories, our goal being to render a complex and contentious subject intelligible to an interested general audience.

Although rendering the past intelligible is an appropriate goal for historians, we also hope our study will be of use to those charged with formu-

1

lating new maritime policies in the years ahead. Indeed, we are persuaded that anyone who grapples with a subject as venerable, complex, and troublesome as federal maritime policy can only succeed by first becoming thoroughly familiar with its long history and the forces that have shaped it. This study was thus motivated in part by the belief that illuminating the past can be helpful to those who must devise policies for the future.

Historians study challenge and response, the emergence of novelty and the ways in which those affected react to it. Maritime shipping is an ancient enterprise that dominated transportation long before industrialization led to the development of steam propulsion, railroads, motor trucks, and airplanes. During the more than two centuries that the United States has been an independent nation, the technology and economics of transportation have changed dramatically. These developments, and the political and social changes that accompanied them, generated many of the challenges that the authors of federal maritime policies had to address. When describing the policies they created, we have tried to identify both the authors' intentions and the policies' unintended consequences. Those who made and implemented maritime policies were as influenced by circumstances and events as those governed by their policies. Policies and their consequences are historical artifacts and we have done our best to treat them as such.

Ours is certainly not the first historical study of American shipping policy; indeed we could not have written this book without the assistance of earlier works that have provided much of the information and analysis we cite. Nonetheless, the subject deserves reexamination, for it has been many years since it was last studied systematically and international shipping has been quite radically transformed in recent decades. The end of the cold war and the subsequent expansion of global trade, together with the worldwide adoption of containerization, have made the final years of the twentieth century a period of especially rapid change. Federal policy has been slow to respond, however, and this tension between new realities and unaltered policies provides much of the dynamic for our study.

In the two world wars that occupied the first half of the twentieth century the United States served as the "arsenal of democracy" as well as a combatant. During these years and on through the cold war the federal government committed itself to sustaining a U.S.-flag merchant marine capable of carrying a significant portion of U.S. foreign trade and providing the sealift needed to fight large-scale wars overseas. These two policy goals, one commercial and the other military, defined and justified federal maritime programs during most of the twentieth century. Previous students of maritime policy have largely evaluated federal policies in terms of their success in meeting these expectations. But at the end of the century, following the collapse of Communism in the West, the end of the cold war, and an enormous surge in global trade, these two policy commitments are of questionable appropriateness and adequacy.

Given the dramatic wave of deregulation that swept through the federal transportation agencies in the 1970s and '80s, one might reasonably expect that federal maritime policy would already have been fundamentally reconstructed. In fact, however, federal maritime programs have been only slightly altered by the change in attitude toward regulation that has freed the railroads, the trucking industry, and the airlines from stifling governmental control. Largely because of its strong linkage to military readiness, maritime policy, like federal support for the major providers of military hardware, has been governed by the dynamics of global politics and military preparedness rather than by commerce. But the end of the cold war marked the end of the special status accorded to many components of the defense enterprise. Commercial considerations must now be given primary consideration. While the military's need for sealift in time of emergency must still be met, the strategic assumption that the United States must be ready at all times to engage unilaterally in large-scale land war overseas against adversaries of comparable strength is no longer beyond question. The maritime policies that served the nation during a century of war no longer seem entirely relevant. If they are not revised in light of current realities, they will become increasingly dysfunctional and counterproductive.

Global trade has flourished in the post–cold war world. Ship sizes have increased enormously, propulsion systems are simpler and more efficient, the speed of cargo-handling systems has greatly increased and is less labor intensive, and those managing shipping services are far more sophisticated than their predecessors. Maritime policy remains an important subject because maritime shipping continues to play a central role in world commerce. We study the past to prepare for the future. It is time U.S. maritime policy moved beyond being a hostage to the past.

Is America a maritime nation? Alfred T. Mahan, the noted author of the doctrine of seapower, had his doubts. Writing a hundred years ago, he questioned "how far the national character of Americans is fitted to develop a great seapower." Like most Americans of his age, however, Mahan quickly cast aside any doubt, concluding, "It seems scarcely necessary, however, to do more than appeal to a not very distant past to prove that, if legislative hindrances be removed, and more remunerative fields of enterprise fill up, seapower will not long delay its appearance."[1] Mahan may well have been right, for had "legislative hindrances" been removed in the 1890s, there is a distinct possibility that a healthy shipping industry might have been reestablished as it had been in "a not very distant past." But these "hindrances" were not and have not been removed, and "more remunerative fields of enterprise" continue to attract the attention of American investors. Thus one is left wondering, as did Mahan, whether in these circumstances America "is fitted to develop a great seapower."

We have organized our narrative into three parts containing fifteen chapters; these sections are connected by a number of interpretive themes that recur throughout the book. Part 1, "Free Trade and American Enterprise," includes the first five chapters and reaches from the colonial period to the end of the nineteenth century. The first chapter describes the natural advantages the English colonists enjoyed within the British Empire and their success as shipbuilders and shipowners in the colonial era. The American Revolution is also described, as are the problems involved in establishing a strong central government once the war was won. When that government was in place, it passed the same kinds of restrictive and tariff laws that Great Britain and the other European imperial powers had long used to protect trade within their empires so as to encourage shipbuilding and participation in the carrying trade. America had fought for freedom from the British Crown and, to an extent, for free trade; but it was prepared to adopt European mercantilist practices in order to engage in competitive trade as an equal.

The various wars and naval actions that occupied Americans during the first fifteen years of the nineteenth century are described in chapter 2, as are the treaties of reciprocity that the United States made with many trading partners once the British victory over Napoleon had ushered in a century of peace. Chapter 3 gives an account of the golden age of American maritime enterprise, that period between 1830 and 1860 in which American ships were the finest in the world and Yankee traders competed successfully around the globe. Even in these years, however, clouds were gathering on the horizon. Steam-driven and iron-hulled ships were beginning to appear. The Americans' success with wooden and wind-driven vessels made them reluctant to adopt the new technology. By the middle of the nineteenth century the British iron industry was producing great quantities of cheap iron plate; but the American iron industry, operating behind a high tariff wall, continued to produce small quantities of relatively expensive iron for its home market.

In 1849 the British repealed their Navigation Acts, a deregulation that allowed British shipowners to buy their vessels wherever they wished and register them in Great Britain. The repeal also opened the British coastal and imperial trades to ships of other nations. This was a bold step which the United States was unwilling to follow and the U.S. registry laws continued to require that all ships sailing under the American ensign be U.S. built. This restriction, when coupled to the shift from sail to steam, made much U.S. shipping relatively expensive and pitched the foreign-trading U.S. merchant marine into an irreversible decline.

Chapter 4 surveys the Civil War at sea and the subsequent industrialization of the North. A detailed account of the Lynch committee hearings held after the war reveals that the causes of the decline of the American merchant marine were well understood but that Congress was adamant in not allowing ships built abroad to be registered in the United States. It became clear for the first time that the shipbuilders represented a far more powerful lobby than the shipowners. Chapter 5 carries the story down to the end of the nineteenth century with accounts of early subsidy programs, competition between coastal shipping and the railroads, the turn from maritime enterprise to continental development, and attempts to consolidate maritime industries. It also points out that while in the final decades of the nineteenth century there was a naval revival, as the nation sought to hold its own in the frenzied imperialism of the Great Powers, the decline of the American merchant marine continued unabated.

Part 2, "War-Impelled Industries," spans the period from the beginning of World War I to the beginning of the war in Vietnam. Chapter 6 describes Woodrow Wilson's maritime strategy and the landmark Merchant Marine Acts of 1916, 1920, and 1928. It was in this period that the federal government adopted a forthright policy of building and maintaining a merchant marine capable of carrying a significant portion of the nation's foreign trade and of providing adequate auxiliary services for the armed forces in time of emergency. This was also the period in which a federal bureaucracy charged with fulfilling this mandate was first created.

Chapter 7 surveys the period between the crash in 1929 and the beginning of World War II. It begins with a detailed account of the Black committee, which investigated the management of the mail subsidy system established in 1928, and goes on to describe the Merchant Marine Act of 1936, which remains today the organic act for federal maritime policy. Chapter 8 carries the story through World War II and the construction of the second great war-built fleet to be launched within a thirty-year period.

Chapter 9 surveys the postwar period from 1945 to 1960; it focuses on the growing dependence of American-flag shipping on government subsidies and protection, and on the rise and growing divisiveness of the American maritime labor unions.

The continued long decline, which had been interrupted by two periods of heroic wartime effort, is the subject of part 3, "The Approaching End." Since part 3 addresses topics from the more recent past, the evidence employed differs somewhat from that used in the first two parts. Andrew Gibson, the senior author, was directly involved in many different aspects of the industry during the period covered in part 3, and the notes, newspaper clippings, and magazine articles he has collected throughout his career provide much of the information included in these chapters. When Gibson served as maritime administrator, his speeches, along with published accounts of his activities and those of the agency, were collected by MARAD's public affairs staff. His testimony before Congress was a matter of public record. Gibson was given this material when he left government service, and much of it has also been used as source material in part 3. Rather than burdening the text with lengthy references to these records, an autobiographical appendix in which Gibson describes his activities while in federal service has been included.

Discussion of the early years of the container revolution has also been informed by an unusual set of recollections. While completing this book the authors were engaged in interviewing shipping executives who played significant roles in the introduction of cargo containerization. This project was supported by the Smithsonian Institution's Lemelson Center for the Study of American Innovation and Invention. The tapes and transcripts of these interviews have been deposited in the Archives Center of the National Museum of American History, where they are available to the general public. While there are many statements in part 3 for which no specific citations are given, the descriptions and interpretations offered are nonetheless based on a wide variety of reliable sources.

Chapter 10 focuses on the Merchant Marine Act of 1970, which sought to stimulate American shipbuilding at a point when ships built during World War II had to be replaced. Chapter 11 surveys several changes that profoundly altered the commercial shipping industry in the postwar period. One of these was the beginning of modern containerized shipping and its integration into comprehensive systems of intermodal transportation. This fundamental change in the way general cargo is packed and loaded has in one generation utterly transformed the organization and operation of the carrying industry worldwide. Another change examined in this chapter is the increasing practice of registering ships under flags of convenience rather than the flags of the nations in which the companies that own the ships are located.

Chapter 12 takes up the troubled relations between the transportation needs of the military and the sealift capabilities of the merchant marine. It also provides a close look at how logistic support was provided during the Gulf War against Iraq. Chapter 13 summarizes presidential attempts to create new subsidy programs for shipbuilding and ship operations following the 1980 decision not to issue any new subsidies under the 1936 act. In Chapter 14 the story is brought down to the election of the 104th Congress in November 1994 and the first part of President Clinton's second term, a time in which the most recent maritime legislation, called the Maritime Security Act, became law. A final chapter reviews the nation's existing maritime policies in the light of the new political realities and questions whether existing policies, most of which were established more than a hundred years ago, are appropriate or even viable as we advance toward a new century. The array of restrictive U.S. legislation, at variance with that which exists in most advanced maritime nations, effectively inhibits the private investment required to reestablish a healthy industry.

PART I

FREE TRADE AND AMERICAN ENTERPRISE

1

FROM COLONY TO NEW NATION,
1600–1790

The British colonies in North America were settled and governed as part of an expanding British Empire. By the 1660s, when England began displacing Holland as the world's leading maritime power, European exploration, colonization, and overseas trade had been developing in close conjunction for over two hundred years. Mercantilist colonial theory had been largely worked out by then. Ideally, colonies would supply products of great intrinsic value, such as gold, silver, spices, and gems that would enrich the mother country when sold on the world market. Colonies lacking these resources could still prove valuable if they produced agricultural commodities such as sugar and tobacco that were in great demand and could be easily preserved and transported. Colonies incapable of producing any of these goods, as was the case with the New England colonies, were obliged to exchange the commodities they did produce, such as wood and fish, with other nations or their colonies, the goal still being to accumulate wealth through trade.

The expectations of colonial merchants in North America changed dramatically during the eighteenth century. At the outset they largely accepted their traditional colonial status and sought to make their way by exploiting their rights to trade within the British Empire. But as the century progressed their skills as shipbuilders, which enabled them to make good use of the new continent's abundant trees and harbors, and their success as merchants, especially in trading in the Caribbean, led them to think of themselves not merely as colonials, but as men of commerce who deserved the same political rights as their cousins in Great Britain. This change in their estimation of their own worth transformed the colonists' perception of the Navigation Acts that governed colonial trade. The restrictions these laws imposed on colonial merchants came to be seen as burdensome and eventually as one of the causes that led Americans to take up arms against their king.

The colonists' opposition to the commercial aspects of colonial policy and their insistence on political autonomy were strongly supported by many liberals in Great Britain. It is notable that both Thomas Jefferson's Declaration of Independence and Adam Smith's *Wealth of Nations* were published in 1776. America's advocates in Great Britain participated in the political debate occasioned by the colonial rebellion and incorporated the American critique of royal governance into their attacks on the "absolutist" practice of granting commercial monopolies to privileged corporations. This rather fortuitous conjunction of political arguments in favor of peaceful devolution and economic arguments against state-sanctioned monopolies made it appear to some that the American war for political independence was inseparably linked to the liberal campaign for free trade. Yet once they had won their independence, the Americans quickly demonstrated that while they would gladly trade freely when it was to their advantage, they were also prepared to use their newly won political autonomy to write navigation laws of their own whenever doing so appeared advantageous.

COLONIAL SHIPPING AND SHIPBUILDING

The New England colonies were linked to each other and to the larger world by the sea. It would be many decades before any substantial inland penetration took place. Hardy fisherman and seafarers were familiar with the western Atlantic. The risks it presented were readily accepted in return for the rewards it offered. The shallow soil that had built up on top of glacial rock in New England provided only marginal farmland, and in the northern climate harvests were always uncertain. The land of this area could not be relied upon to produce sufficient crops to sustain the first inhabitants. Only the coastal waters and the offshore fishing grounds provided an assured source of food. The only other resource that the area possessed in abundance was the timber that grew to the water's edge. The ready availability of this material, especially the white oak and pine, when coupled with an energetic, hard-working people possessing the necessary skills, established the first major industry in the new world—shipbuilding.

Most of the early vessels built in America were small craft for fishing and coastal transport. Before long, however, oceangoing vessels were being produced. The first of these was "a faire pinnace of thirty tons," christened *Virginia* and launched in 1607 on the Kennebec River in what is now Maine; it was built to return the survivors of the ill-fated Popham Colony to England.[1] Two other vessels are also mentioned in most shipbuilding histories of the period. One was built in the small Dutch settlement of New Amsterdam by an adventurous fur trader, Adrian Block. Block's original ship had burned during the previous winter, and in the summer of 1614 he launched a fully decked vessel, *Onrust*, on the Hudson River and explored Long Island Sound on his way home to Holland. Block Island, situated off the eastern entrance to the sound, bears his name to this day.

A more famous ship, *Blessing of the Bay*, was built in the Massachusetts Bay Colony in 1631 under the orders of the new colony's governor, John Winthrop. This vessel is generally considered to mark the beginning of the American shipbuilding industry that was to flourish for more than two hundred years and provide many of the early fortunes. A few years after the ship was launched this early industry was considered so essential that shipwrights in Massachusetts were exempted from the colonial militia, so that they might spend full time at their trade.[2]

Given the natural advantages they enjoyed and the lack of alternatives for profitable enterprise, many New England coastal towns began building boats and ships not only for their own use but increasingly for sale to others. It was not uncommon for an entire town to invest in the production of a ship, some of the people providing timber, some hardware and cordage, others sails, and still others labor. In time some of these same people would provide cargoes from the nearby farms, forests, and ocean for trade to distant shores.

While New England dominated early ship construction, building yards also appeared in New York, on the Delaware, and in many parts of Chesapeake Bay. By the beginning of the eighteenth century shipyards were also active on the coasts of the southern colonies. This area had the advantage of vast supplies of naval stores, such as the tar, pitch, and turpentine obtained from the local pine. This straight-grained tree also provided excellent deck planking, and the live oak of the area surpassed the white oak of New England in its strength and toughness. Yet, in spite of possessing the natural resources for shipbuilding, the South never developed a thriving shipbuilding industry. The fertile land and moderate climate provided much greater opportunities. Commerce and shipping could best be left to others, while the plantation owners concentrated their efforts on obtaining greater wealth from agriculture.

By the early eighteenth century American shipbuilding had become so competitive that British shipbuilders were beginning to feel threatened. It was reported in 1724 that the Thames shipwrights attempted, without success, to have legislation enacted to limit the import of ships built in the colonies. A record of American shipbuilding, prepared in 1769 from Lord

Colony	Number of Vessels	Tonnage
New Hampshire	45	2,452
Massachusetts	137	8,013
Rhode Island	39	1,428
Connecticut	50	1,542
New York	19	955
New Jersey	4	83
Pennsylvania	22	1,469
Maryland	20	1,344
Virginia	27	1,269
North Carolina	12	607
South Carolina	12	789
Georgia	2	50

Table 1.1: Colonial Shipbuilding, 1769.
Source: Data from Winthrop L. Marvin, *The American Merchant Marine: Its History and Romance from 1620 to 1902* (New York: Scribner's, 1902), 8.

Sheffield's *Review of American Commerce*, shows that the colonies produced some twenty thousand tons of shipping that year.[3]

While the North American colonies were growing and beginning to prosper, other British colonies in the West Indies were also beginning to produce wealth of a different kind. For these colonies, established on the islands of Barbados and Jamaica, the principal crop was sugar. The conditions for growing cane on the islands were ideal. The intensive labor required was provided by slaves, and sugar became the cash crop at the expense of all others. Sugar was considered so valuable that little land or labor was allocated to producing food for those who worked the fields. As additional land was cleared, the limited amounts of available timber disappeared. Thus a trade developed between the northern colonies and the Caribbean plantations, a trade that was based on the economic advantages enjoyed by each and produced wealth for both. For the North Americans, the import of sugar promoted new industries. Soon New England merchants built refineries for sugar and distilleries to make rum. The ships carrying sugar northbound returned to the islands with full cargoes of salt cod, timber, and barrel staves. Many of these descendants of pious Puritans quickly saw the profit to be made in the transport of slaves, something for which their sleek, fast ships were ideally suited. This eventually resulted in a triangular trade of sugar to New England, rum to the Benin Bight of West Africa, and slaves to the sugar plantations of the Caribbean.

The Yankee traders did not restrict themselves to the British colonies. By the eighteenth century trade with the French, Dutch, and Spanish Caribbean islands was well developed. While this was illegal under English law, it does not appear to have caused undue problems for the colonists. English merchants were generally aware that increased commerce was the only practical means by which the New England colonists could offset their potentially unfavorable trade balance with the homeland. If they were to continue to buy British goods, new sources of revenue had to be established. The enterprise of these early traders enriched not only Boston, Salem, and Newport, but merchants in London, Bristol, and later Liverpool as well.

Much of the commerce was centered in the towns of Massachusetts. Newport, Rhode Island, with its fine deep harbor on Narragansett Bay, also grew to be one of the principal trading centers of the Atlantic coast and would so continue until the War of Independence. New York at this time had only begun to develop into the great port that it would become in the next century. Philadelphia, situated on the Delaware River, exported wheat and flour as well as produce to the Caribbean Islands in exchange for sugar. Ships from that port also ventured to Central America. The mahogany brought back from Honduras as well as Santo Domingo formed the basis for the manufacture of the fine furniture for which Philadelphia became famous.

In the southern colonies of Virginia, the Carolinas, and Georgia, a strong reciprocal trade had developed early with Great Britain. Unlike the New England colonies, their land and climatic conditions readily produced a variety of crops for which the motherland was a natural consumer. In return for British manufactured goods these colonies exported large quantities of tobacco from Virginia and North Carolina as well as rice, indigo, and naval stores from the Carolinas and Georgia.

THE NAVIGATION ACTS AND COLONIAL TRADE

Whatever long-term effects the British Navigation Acts may have had on American shipping policy, there can be little doubt that the early American colonists profited substantially from the protection provided by these laws.[4] British trade and shipping restrictions had a long history. Some suggest that they went back to the time of King Alfred and the beginnings of the English navy. Certainly by the end of the sixteenth century Queen Elizabeth I had moved decisively to restrict the trading privileges that foreigners enjoyed with Great Britain. A major concern of the early laws that protected shipping and fishing was providing manpower for the growing

English navy as well as aiding infant industries. The principal laws that set the trading patterns for the American colonial period were first enacted in 1651 during the time of Oliver Cromwell, the Lord Protector. Further restrictions were added during the twenty-five years that followed.

These acts were designed to implement the economic philosophy generally known as mercantilism, and they were as simple as they were effective. Mercantilist practices were intended to protect home agriculture and industry from foreign imports, the goals being high employment and economic self-sufficiency. The Navigation Acts specifically protected English shipping from foreign competition in both international and domestic trade. The larger aim of these laws was to establish a favorable balance of trade and promote the accumulation of bullion. It was generally accepted at that time that a nation's wealth and power were directly proportional to the quantity of precious metal held by its Treasury. Mercantilists assumed that overseas colonies would serve as a source of raw materials for the mother country and would in turn provide a protected market for the manufactured goods produced by home industries. Such trade would in turn stimulate the growth of the merchant fleet. Seventeenth-century entrepreneurs confidently asserted that the industry and trade they engaged in were of great benefit to the nation, and they believed that the state had a responsibility to protect their enterprises. If a mercantilist economy was to continue providing profits for its beneficiaries, the state had to protect the nation's trade. Nowhere in this philosophy can be found the notion that a colony could in time become a competitor to the nation or empire that had nurtured it.

British navigation law stated that (1) no goods from Asia, Africa, or America could be imported into England except in British ships; (2) goods manufactured in Europe could only be imported into England or its overseas possessions by English ships or ships belonging to the producing country; (3) products grown or manufactured in foreign countries outside of Europe could not be transshipped via European ports to avoid the provision giving a transportation monopoly to British ships. A further restriction, included later, added that ships covered by the 1651 act, in addition to being British owned and having a British master and crew, had to be British built.

Until the mid-eighteenth century American colonists, being Englishmen, used these laws to their advantage. In so doing they often ignored or circumvented the intention of laws that were designed primarily to benefit the mother country. By combining the natural advantage of low-cost ships and crews with aggressive management, American commercial shipping continued to grow. While England could do little to curtail the expansion of American shipping, the growth of America's manufacturing capability threatened an increasing number of British industries; this could be and was curtailed.

Beginning in the latter part of the seventeenth century, laws were enacted to minimize or eliminate certain types of colonial competition. Woolen goods were not to be exported from the colonies to Great Britain, nor were they to be traded from one colony to another. While the export of pig iron to British manufacturers was encouraged, the use of any machinery for converting this material into finished products was prohibited. As restriction after restriction was enacted, a Bostonian complained in 1765, "[A] colonist cannot make a button, a horseshoe nor a hobnail, but some sooty iron monger or respectable button maker of Britain shall bawl and squall that his honor's worship is more egregiously being maltreated, injured, cheated and robbed by the rascally American Republicans."[5]

While the good citizen of Boston was venting his anger over British protectionism, the various colonies, particularly the one in which he resided, were enacting discriminating duties against their colonial neighbors. Most colonial charters allowed the erection of such barriers. Massachusetts, for example, imposed double duties, one according to what the item was and the other on its value, on all goods that were not imported directly from the place of origin. This particularly impacted the bordering colonies of New Hampshire, Rhode Island, and Connecticut and was clearly designed to encourage direct shipments to its own ports. On a smaller scale the colonies were copying the prohibition on transshipment contained in the 1651 act. While most of the duties imposed were generally intended to raise revenue for the colony, it was not uncommon to attempt to promote the welfare of their shipowners by imposing a lower level of duties on goods that were carried in ships of that colony.

THE WAR FOR INDEPENDENCE

When the colonies were young, their commerce and shipping generally benefited from the protection afforded by the Navigation Acts. They had the additional advantage of being able to evade or ignore the acts' restrictions when it suited them. During much of this early period, Great Britain was preoccupied with European wars and could devote few re-sources to enforcing colonial regulations on the far side of the Atlantic.

By defeating France in 1763 Britain gained control of all that had been French Canada. This victory was followed by a determination to draw the empire more closely together and strengthen the institutions of British merchant capitalism. The Sugar Act, passed before the end of the war, imposed a tax intended to recover from the colonies part of the cost of sending the British army to protect them in the French and Indian War. This act also sought to suppress the American sugar trade with the French islands, a trade which had developed extensively during the European war. The French plantation owners, when deprived of their home market by the British blockade, were willing to sell at prices substantially below those offered by the English growers in Barbados and Jamaica. The shrewd New England traders had taken full advantage of this situation, and now Parliament was determined to bring this trade to an end.

Despite the new tariff, many in the colonies attempted to sustain the French trade by devising stratagems to avoid the tax or by outright smuggling. One of their subterfuges was to obtain clearances for their ships from the British island of Anguilla by claiming that this was the origin of the cargo, even though the island hardly produced one full cargo of sugar a year.[6] The British navy was ordered to increase its efforts to stamp out such smuggling. Colonial courts were directed to issue general search warrants (called Writs of Assistance) to aid in the effort to identify cargoes imported illegally. The attempt to reimpose mercantilist restrictions caused deep resentment, especially in Boston, where the crackdown was felt most intensely. As early as 1761 James Otis, a prominent Boston merchant, vigorously attacked the recently imposed writs; John Adams later credited Otis with firing the "opening gun of the Revolution."[7]

Bitterness between the colonists and the British Parliament continued to grow. Every act of resistance was met with increasingly harsh measures designed to break the opposition. On several occasions the colonists organized boycotts of British goods. Although normal commerce was eventually reestablished, ill feeling remained, particularly in Massachusetts and Virginia, where it was kept alive by a growing group of revolutionary agitators.

In 1773 the East India Company was given what amounted to a monopoly on the import of tea to the colonies by being exempted from a tax that had been imposed. This had the effect of cutting out the local merchants and warehouse owners, who were determined to resist. When the first shipment arrived in Boston, the ship was boarded and the cargo dumped in the harbor, an event now known as the Boston Tea Party. The British retaliated by closing the port and forcibly quartering troops in the homes of citizens. Although few sought open rebellion, the fuse had been lit and the explosion that followed soon led to open warfare. The war that ensued was fought not only on land but on the far reaches of the ocean as well.

Open warfare between the rebellious colonists and Great Britain broke out in 1775. The new Continental navy improvised by Washington consisted mainly of small craft used to harass the enemy in harbors and coastal waters; it was no match for the British. American privateers, on the other hand, were a significant force. Commerce raiding by armed merchantmen or privateers was a well-established form of maritime warfare in the eighteenth century. Colonial shipowners had frequently been encouraged to engage in this type of warfare against French merchant ships. Colonial governors were authorized to issue the Letters of Marque and Reprisal that legitimized these assaults, which otherwise would have been considered outright piracy. The potential rewards of privateering for shipowners and ships' crews were considerable, and many entered into this activity with relish.

This practice had a long history. As early as the thirteenth century, when commerce began to flourish on both sides of the English Channel and the North Sea, it became apparent that some sort of maritime order had to be established. On the small island of Oleran, lying off the French coast, a previously established maritime court began to deliver verdicts that gained a growing acceptance. In time these became known as the Judgments of Oleran. These judgments covered many aspects of shipboard life, such as defining the relationship of master and crew. They codified the masters' and shipowners' responsibilities for the transport and delivery of cargo as well as for salvage, should that become necessary. They also provided that a prince could issue a license to a ship's master to attack his enemies, the theory being that anyone who suffered injury on the high seas was entitled to take reprisal.[8]

This type of warfare suited the Americans' prevailing free spirit and sense of adventure. When the British navy shut down most colonial commerce, there were plenty of ships and seamen available for privateering. By 1781 there were some 450 American privateers actively engaging the enemy. They raided from Canada to the Caribbean and were seen frequently in the English Channel and the Bay of Biscay. During the first two years of the war the commerce raiders had great success and captured some seven hundred British ships.[9] John Paul Jones, sailing as an officer of the Continental navy, also attacked merchant ships in the English Channel and along the Scottish coast in raids that became the substance of legends. His victory over HMS *Serapis* in 1779 also humiliated the British, even though the convoy *Serapis* was defending managed to escape. As the war progressed, however, the British navy was able to assert its formidable might and ranged with little interference along the entire American coastline.

Without the assistance of France, particularly its fleet, the Americans probably could not have won the war. By 1781 the colonies' financial resources had been exhausted by six years of warfare. Washington's few remaining troops were unpaid and half-starved. The decisive moment came in 1781 at the Battle of Yorktown, where the British forces under General Cornwallis surrendered in October. The victory was made possible by the French fleet's victory over the British fleet off the Chesapeake Capes. The French ships then put ashore troops and heavy weapons that fortified the besieging force, while the fleet under Admiral De Grasse imposed a blockade that kept the New York–based British fleet from aiding Cornwallis. Although peace did not come until the following year, Yorktown essentially ended British efforts to subdue its former colonies.

THE NEW NATION AND PROBLEMS OF STATEHOOD

When the peace treaty ending the Revolution was signed in 1783, the economy of the new Republic was in shambles, particularly in the northern states. The war had taken a huge toll, and the nation that shortly before had been a prosperous group of colonies now faced a troubled future. Much of its merchant marine had been destroyed and trade was at a standstill. The many influential merchants who remained loyal to the Crown had fled, and the former belligerents were still at swords' points.

In July 1783, Great Britain declared that trade with its West Indian colonies could only be carried by British ships. British shipowners were also prohibited from buying ships from the former colonies. While this might have been expected, it was a devastating blow, for prior to the war a substantial portion of the merchant ships flying the British flag was American built. These orders also severely restricted the importation of goods from the United States. As if to add insult to injury, the British Orders in Council stipulated that when the allowed goods were imported, they could only be carried on vessels owned by residents of the states in which the cargoes originated.

The Articles of Confederation, an early attempt to form a national government, failed to provide sufficient central authority for the new nation. Individual states, for instance, retained the right to impose additional tonnage duties on foreign vessels calling at their ports. Virginia set duties at one dollar per ton on British vessels, but only half this amount was collected from vessels flying the French or Dutch flag; no duty at all was imposed on American ships.[10]

Boston remained the hub of opposition to British commercial domination. Britain's decision not to purchase American ships left New England's shipyards with little to export. The virtual elimination of trade with the British West Indies left most of its ships idle. Boston therefore had every reason to retaliate against British restrictions. In April 1785, the merchants of Boston once again pledged not to buy any English imports or allow the use of any local facilities for the storage or sale of English merchandise. The boycott was so effective that by July no British merchantman was to be found in its harbor.

American reactions to British restrictions did little to promote American shipping. Since few American ships traded with Great Britain, British merchants were able to pass along to their American customers the full cost of the duties imposed by the states. The failure of the vessel tonnage duties led New York and Pennsylvania to institute a system of discriminating duties on the imports themselves, the goal being to encourage development of local industries and to raise revenue. Interstate rivalries blunted this effort, however, since states like New Jersey, having few ships and little manufacturing to protect, insisted on maintaining a system of free ports.

Having been effectively barred from trade with Great Britain and its colonies, American shipowners did what enterprising shipowners have always done, they went looking for new markets. Opportunity presented itself in the Far East, particularly in China, an area long considered the private preserve of the British East India Company. As early as 1784 an American ship, *Empress of China*, sailing out of New York, called at Canton. The following year *Grand Turk* out of Salem commenced a trade for which that Massachusetts town became justifiably famous. Before long ships from New England and New York were familiar sights in western Pacific and Indian ports.

In 1787 a Captain Kendrick, commanding the American ship *Columbia*, explored the northwest coast of the American continent, including the river that bears his ship's name. This was the beginning of a prosperous fur trade that found ready markets in the American East and in China as well. For many years thereafter American manufactured goods that had no market in Europe could be exchanged for the fur of the beaver and sea otter on the West Coast. The furs were then exchanged for silks and porcelain in China and tea in India.

One has to be impressed by the daring of these intrepid merchants and the crews that sailed the vessels carrying their cargoes. They sallied forth to these distant lands without a navy to protect them or local consuls to represent their interests. An American Pacific Squadron, charged with protecting American shipping, was not established until 1819. By that time a strong central government had been established under the Constitution and the growing number of New England whalers in the North Pacific were encountering problems that required government attention.

The government operating under the Articles of Confederation lacked the power needed to promote or regulate commerce. While foreign powers were interfering with and restricting American trade, sectional jealousies within the country were becoming increasingly apparent. The economic concerns of the largely agricultural South had little in common with the commercial objectives of their countrymen in the North. Such sectional rivalry, if not reconciled, threatened to tear apart the new nation. Attempts were made to resolve these problems. In April 1784 a congressional committee that included Thomas Jefferson issued a report saying in part: "Unless the United States in Congress assembled shall be vested with power competent to the protection of commerce, they can never command reciprocal advantages in trade; and without these our foreign commerce must decline and eventually be annihilated. Hence it is necessary that the States should be explicit and fix on some effectual mode by which foreign commerce not founded on the principles of equality may be restrained." The report concluded with a resolution: "That it be and is hereby recommended to the Legislatures of the several States to vest the United States in Congress assembled, for the term of fifteen years with the power to prohibit any goods, wares and merchandise from being imported into or exported from any of the States in vessels belonging to or navigated by the subjects of any power with whom these States shall not have formed treaties of commerce."[11] While the separate states prized the liberty they had wrested from Great Britain, those involved in foreign commerce realized a stronger central government was needed.

THE CONSTITUTION AND EARLY LEGISLATION
AFFECTING COMMERCE AND SHIPPING

The weakness of the government organized under the Articles of Confederation became increasingly apparent and advocates of a stronger union gained ascendancy. A Constitutional Convention was called to meet in Philadelphia in May of 1787. With the exception of Rhode Island, all states sent delegates, and George Washington was elected the convention's president.

Shipowners were prominent among those who wanted a strong central government. They understood that only countervailing laws could place them on a par with their foreign competition. They were joined by merchants opposed to barriers to interstate commerce, who needed a stable currency to conduct their business. Manufacturers also wanted a strong central government to protect them from foreign competition. The main opposition to these proposals came from agricultural interests favoring expanded overseas markets and continued cheap imports.

Another group campaigning for a new Constitution consisted of holders of Continental bonds, government certificates of indebtedness, and paper money. They understood that a strong central government was more likely to redeem these pledges than the existing Confederation. Much of this paper was held by speculators who stood to make handsome profits; some of these same men were delegates to the convention.[12]

The Constitution that was finally adopted significantly strengthened the power of the central government. Congress was given the power to regulate trade and protect industry, including shipping and shipbuilding. It could regulate commerce among the states and eliminate the obstructions that had been developing. Not surprisingly, Congress could impose taxes, establish credit, and redeem outstanding indebtedness.

When the new Congress met in April 1789, its first act was to levy a protective tariff on imported goods. Like earlier tariff bills passed by New York and Pennsylvania, this act encouraged use of U.S. ships by providing an across the board reduction of 10 percent for goods carried in American vessels. A special tax on tea could be halved if the tea was imported in American bottoms directly from the Far East. While this was an obvious effort to increase American trade, it also retaliated against the American merchants' principal trading rival.[13] The act further stipulated that only ships built in the United States could qualify for American registry. U.S. shipowners also received reductions in the port tonnage tax levied on all ships: on a U.S.-built, U.S.-owned ship, the tax was six cents per ton; for U.S.-built, foreign-owned ships, the tax was thirty cents per ton; for foreign-built, foreign-owned vessels, payment of fifty cents a ton was required. Furthermore, for U.S. vessels engaged in the coastal trade, the tonnage tax only had to be paid once a year instead of for each port call, as required of foreigners. Foreign vessels were not excluded from domestic trade at this time, but the discriminatory application of tonnage dues had the same effect in practice. It would be another twenty-eight years before a law was passed completely barring foreign ships from American coastal trading.

American shipping and trade began to expand almost immediately. Great Britain did not enact countervailing duties until 1797. The Americans had essentially beaten them at their own game and gone them one better. The British law pertaining to national registry prevented English owners from buying lower-cost American-built ships. Its effect was to put the

English shipowner at a distinct disadvantage when competing with Americans. There appears to have been little pressure on Parliament to respond. English merchants undoubtedly realized the American legislation would increase competition and lead to lower freight rates. The Americans had the significant advantage of a coastal trade that was effectively closed to outsiders. Shipowners engaged in foreign commerce used this advantage to improve vessel utilization. Americans could load a cargo in Bristol for Boston, New York, or Philadelphia, the major consuming areas. They could discharge that cargo and then load lumber, or local manufactured goods, produce, grain, et cetera, for Virginia, the Carolinas, or Georgia, where, after discharge, they could load tobacco, rice, naval stores, and later cotton for British or Continental ports. The Americans enjoyed another advantage when the Europeans went to war yet again in 1792. This war, begun in reaction to the revolution in France, continued with only brief interruptions until the defeat of Napoleon some twenty-three years later. British seapower promptly eliminated virtually all French commerce with its colonies in the West Indies, and enterprising Americans soon filled the void.

Rarely has any set of protective laws had such a dramatic and positive effect. In 1789 Americans carried only 23 percent of their imports and exports. By 1800 the number of U.S.-flag ships involved in international trade had expanded more than fivefold and American ships were carrying 89 percent of the country's imports and exports. This new-found market domination forced the British into grudging respect.

In 1794 President Washington sent Chief Justice John Jay to London as envoy extraordinary, his assignment being to demand that the British abide by the terms of the peace treaty and give up the forts they still held in the Northwest Territories. This was the "Old Northwest," that is to say the trans-Appalachian region of western Pennsylvania, Ohio, and other territories in the upper Mississippi basin. Some of the British commanders of these forts were openly encouraging Indians to attack newly arrived Americans who had begun to settle the area. While Jay was in London, an American force under the command of General Anthony Wayne eliminated this threat by decisively defeating a large group of Indians in an engagement known as the Battle of Fallen Timbers. Jay soon arranged an agreement for the evacuation of the forts. At the same time he negotiated the first commercial treaty between Great Britain and the United States. Although this treaty called for full trade reciprocity, little in fact changed and trade with the British West Indies remained virtually closed. The treaty awarded the United States some degree of commercial recognition, however, and marked a first step toward equal status with the principal trading nations of Europe.

Year	Shipping (in Tons)	Total Commerce (in Millions of Dollars)	% Carried in U.S. Ships
1789	123,893	–	23.6
1790	346,110	–	40.5
1792	411,438	52.3	64.0
1794	438,863	67.6	88.5
1796	576,733	140.0	92.0
1798	603,776	129.9	89.0
1800	667,107	162.2	89.0

Table 1.2: U.S. Tonnage and Foreign Trade, 1789–1800.
Source: Data from Winthrop L. Marvin, *The American Merchant Marine: Its History and Romance from 1620 to 1902* (New York: Scribner's, 1902), 50.

Free traders such as James Madison and Thomas Jefferson had no difficulty supporting protective duties for shipping and shipbuilding. When Jefferson was secretary of state in 1791 he argued strongly for the protection and promotion of U.S. shipping to avoid dependence on foreigners. He noted that if the United States was to become dependent on European nations for its essential transport, then when those nations were at war, the trade of this country would be seriously injured either by lack of service or by greatly increased freight rates. These early forms of protection were highly successful and cost the nation nothing, since the American producers of these goods (ships) and services (shipping) were highly competitive. In fact the American shipbuilders were so successful that, until prohibited by the enactment of Orders in Council in 1783, they were providing many of the ships purchased by British shipowners.

The maritime laws of the new nation forged a link between U.S. shipowning and shipbuilding that has up to recent times remained unbreakable. The requirement that all ships flying a nation's flag be built in that nation's shipyards was general practice at the time the first restrictions were enacted. The detrimental effects such a linkage might have on a country's shipping eventually became apparent to all nations except the United States.

2

MARITIME WARS AND RECIPROCITY, 1790–1830

The United States, having gained its political independence and established a central government, looked out with trepidation upon a hostile world. In Europe two Great Powers continued their prolonged contest for hegemony both at home and in those parts of the world where they traded or had colonies. Whether openly at war or replenishing their forces during brief periods of peace, France and England competed unrelentingly and with rising ferocity. Their struggle polarized international relations, creating a situation in which less-developed countries such as the new United States of America were expected to place themselves under the protection of one great power or the other. If a weak nation was to insist on neutrality, it would be asking for an indulgence that neither of the Great Powers was obliged or inclined to grant. Was it to insist on the right to trade without impediment, it would be denying the realities of power and its normal usages. The merchants of the United States forthrightly declared how they wanted other nations to behave toward them, and they did so with the self-righteousness and determination usually found among those who have recently achieved political freedom, but they hardly had the military means required to stand up to Great Britain or France. Indeed, the mighty ocean separating them was the best if not the only weapon they had.

And yet between 1790 and 1815 the United States fought strenuously to defend and expand its trade and avoid subordination to either of the warring nations. It was the Americans' good fortune that their struggles with the French and the British, when viewed from London and Paris, remained peripheral engagements of slight significance. Had the United States felt the full fury of either combatant, it would have had to pay dearly. Thus while war raged elsewhere, the United States was able to show a certain shrewdness and an often brazen stubbornness in defense of its independence and trade. In the end this determination was rewarded. When the European powers wearied of war, international relations became less tense and the interests of small nations could be more easily accommodated. At that point a new generation of self-confident Yankees quickly extended the nation's maritime trade and made the most of the many commercial opportunities that suddenly opened up to them. The United States had at all times been a bit player in the wars that raged across Europe between 1792 and 1815, but having survived intact, its right to negotiate commercial treaties and engage in trade on terms it found acceptable was acknowledged. While some cross-trading restrictions remained in effect after 1830, America had clearly achieved its commercial independence.

THE BARBARY PIRATES

The states occupying the Barbary Coast of North Africa (Morocco, Algiers, Tunis, and Tripoli) had long exacted tribute from the nations that traded along their shores. While Great Britain could have easily suppressed these seagoing bandits, it found it more convenient to enter into treaties that included relatively small annual payments for safe passage of its ships. This way of dealing with the problem protected British shipping while leaving the North African corsairs free to disrupt Britain's commercial rivals. Lord Sheffield summarized this strategy, and its implications for the Americans, in 1784: "It is not probable that the American States will have a very free trade in the Mediterranean. It will not be to the interest of any of the great maritime Powers to protect them from the Barbary States. If they know their interests, they will not encourage the Americans to be carriers. That the Barbary States are advantageous to maritime Powers is certain."[1] Sheffield knew whereof he spoke; as early as 1785 American ships were seized by Algerians near the Straits of Gibraltar.

The Moroccan problem was handled diplomatically. Thomas Jefferson noted that Morocco was the only North African state having Atlantic ports into which "the Algerians rarely come . . . and Tunis and Tripoli, never."[2] A Treaty of Amity and Commerce between Morocco and the United States was arranged in 1784. Jefferson considered this treaty a most important accomplishment; it turned out to be singularly long lived as well.

Algiers proved less amenable. Its corsairs captured a number of American ships and enslaved their crews pending payment of ransom. Congressional Republicans and Whigs alike, still savoring their own achievement of liberty, were outraged; yet the United States had neither the means nor the inclination to resort to force against even petty states. Urged on by Jefferson, President Washington's secretary of state, Congress agreed to authorize the construction of six frigates.

In 1791 the Senate hesitantly advised President Washington to pay the ransom demanded by Algiers, so that the seamen could be released. Two years later the American ambassador to Portugal initiated negotiations with Algiers, but little was accomplished. Negotiations were resumed in 1795 and after "a fair amount of bribe money [had] exchanged hands," a treaty with the Dey of Algiers was arranged.[3] The United States was to immediately pay $642,000 and shortly thereafter deliver an American-built frigate. Annual payments of $21,000 were to follow. Similar but slightly less expensive treaties were concluded with Tunis and Tripoli. The Senate ratified these agreements in 1796; shortly thereafter the Algerians began releasing the captured seamen and the Americas began building the frigate *Crescent* for Algiers.

This was not to be one of America's finest hours. The payment of blackmail never achieves its intended purpose, if that purpose is lasting peace. Undoubtedly there were those in the Congress who correctly maintained that this was something everyone else was doing. For them, this made the payments palatable. Others, more frugally minded, could point out that such payments, large as they were, were cheaper than building a navy. The value of a nation's honor being more difficult to quantify, it seems to have been ignored.

It is hardly surprising that soon after this agreement was reached demands were made for additional payments. But the situation had changed, for the United States now had several stout naval vessels and opinion was turning against paying further tribute. John Adams had been in favor of ransom, but after succeeding Washington as president he opposed additional payments. He instead sent a naval squadron to the Mediterranean to protect American shipping. By the time the four vessels arrived in May 1801, they found that the three North African states had declared war on the United States. By this act they elevated themselves from being outlaw "pirates" to legal "belligerents," a status recognized under existing international conventions. Richard Dale, the American commodore, attempted to blockade the Barbary ports while also protecting U.S. shipping; but he lacked the resources needed to do both, and little was accomplished. The following year a larger squadron arrived, but it, too, achieved little.

In the fall of 1803 Commodore Edward Preble finally gave the Americans something to cheer about when he led an action that British Admiral Nelson called "the most bold and daring of the age."[4] *Philadelphia*, one of Preble's two frigates, ran aground while chasing a pirate into the harbor at Tripoli and was captured. The captors soon refloated the ship and prepared to use it against Preble's squadron. To prevent this, Stephen Decatur, one of Preble's lieutenants, commandeered a captured Turkish vessel, entered the harbor at night, overpowered the crew, and set fire to *Philadelphia*, destroying it. The Americans returned from this raid without having lost a single man.

Preble then decided to level Tripoli. He augmented his squadron with borrowed Neapolitan gunboats and began a sustained bombardment. While this siege was underway, Commodore John Rodgers relieved Preble and the pasha reminded the new commodore that he was holding three hundred American hostages, including many from the *Philadelphia*. Rodgers's hand was forced and he agreed to ransom the captives. Nonetheless, this bold action along the shores of Tripoli demonstrated America's determination to defend its shipping, and there was little subsequent interference; it also added a memorable line to the Marine Corps hymn.

UNDECLARED WAR WITH FRANCE

While sparring with the Algerians, the United States was also engaged in a far more threatening struggle with France. This conflict is a case study in how strongly professed allegiances can lead to feelings of betrayal when circumstances change dramatically. France had provided crucial support to the colonies during their Revolution. The treaty these two allies entered into in 1778 pledged them to mutual support, a pledge that, following the achievement of American independence, seemed little more than an expression of enduring friendship. But the situation began to change in 1789 when the French launched their own political revolution. As this Revolution became increasingly radical, and as France's European neighbors grew increasingly alarmed, relations between France and America became less cordial. When France and its neighbors entered into open warfare in 1793, President Washington, eager to avoid entangling alliances, issued a declaration of neutrality. When the following year John Jay negotiated a commercial treaty with Great Britain, the French protested that the Americans had betrayed them. French reaction peaked shortly thereafter when Charles Pinckney, the American envoy to France, was threatened with arrest upon his arrival and summarily expelled. The French vowed they would not receive an American representative until the terms of the 1778 treaty were honored in full. The Americans obviously could not condone such ill-treatment, but there was also little they could do to oppose it.

The French in fact needed American assistance. Almost as soon as France went to war with Great Britain, it abandoned its navigation laws, a clear acknowledgment that it could not enforce them. France expected the

United States to step into the breech and carry much of its trade. But the British had anticipated this move; in 1756 they had issued an order stating that a country excluded from a trade in peacetime could not enter that trade in wartime. If the United States wanted the British to treat them as a neutral, they could not openly carry French trade. The French also expected that the Americans would be willing to assist them with loans to their government and bribes to appropriate officials, as they had done in North Africa. The French foreign minister, Charles-Maurice de Talleyrand-Perigord, let it be known he expected as much; but President Adams would have none of it.

Following Pinckney's expulsion, President Adams did what he could to salvage American honor. Congress was asked to legalize arming American merchantmen and to provide funds to complete three of the previously authorized frigates that were then under construction. These were *United States, Constitution,* and *Constellation,* all of which distinguished themselves in defense of the nation.

Adams also sent commissioners to France to see if the differences between the two former friends could be resolved. When Congress insisted on examining the minutes of the commissioners' meetings with their French counterparts, Adams would only identify the commissioners as X, Y, and Z. But once Congress had learned of French attempts to extort loans and bribes, their fury could hardly be contained. Commissioner Timothy Pickering's response to the French—"No, no, not a sixpence"— inspired the public war cry "Millions for defense, but not one cent for tribute."[5] And Congress did in fact provide more funds for shipbuilding and authorize the creation of a separate Department of the Navy, but they stopped short of actually declaring war on France.

The so-called quasi-war that followed was mostly fought in the Caribbean. America once again depended largely on privateers and armed merchantmen, but the navy's ships fought well, too. In 1799 the thirty-six-gun *Constellation* captured the French frigate *L'Insurgente* after heavy fighting; a year later it severely damaged the largest French frigate in the Caribbean, *La Vengeance,* but was itself badly battered as well. The twenty-nine-gun *Boston* distinguished itself by capturing the frigate *Berceau.* Naval vessels also protected convoys of merchant ships. This resistance began to discourage French raids on American shipping, which since 1794 had cost the new nation hundreds of ships. In 1799 and 1800 the French captured 159 U.S. ships, although many were later recaptured.[6] But France suffered losses too: the quasi-war cost it eighty-four privateers mounting some five hundred guns.[7] What France could not afford was war on another front while fully engaged in Europe. The Treaty of Montefontaine was therefore concluded and ratified by the Senate in 1801. It terminated the 1778 Agreement of Mutual Assistance and brought to an end, at least for some time, French raids on American merchant shipping. The successful conclusion of this conflict also gave Americans a welcome sense of their ability to manage their own affairs. Shortly after taking office the newly elected president Thomas Jefferson declared, "[W]e mean to rest the safety of our commerce on the resources of our own strength and . . . on every sea."[8]

JEFFERSON'S EMBARGO

Every nation interprets events to its own advantage. According to Jefferson, the quasi-war with France demonstrated that America could and would defend its maritime commerce; but in fact the British navy had carried the day. Britain's blockade of French ports effectively prevented France from deploying a naval force capable of threatening the U.S. mainland, as Britain was to do in the War of 1812. Stoddard, the American secretary of the navy, could therefore concentrate his small squadron in the West Indies, where it was able to stand up to the naval forces France had stationed there. In reality, the United States was still a militarily weak nation warily prowling the perimeter of a Great Power conflict.

Although the Treaty of Amiens brought a brief period of peace to Europe, the long war resumed in 1803 and America once again found itself the principal neutral carrier. Engaging in maritime commerce in such circumstances was hazardous business; but with luck and fortitude, money could be made. The hazards and necessities of war forced freight rates up, and those who succeeded in delivering cargoes were well paid. Many ships were lost, but ships could be quickly built in the numerous American shipyards and total U.S. tonnage actually grew steadily during the war years. Between 1782 and the outbreak of the War of 1812, American shipping prospered and the fortunes of many families involved in shipping swelled during these years.

Britain's challenge, in an age when propulsion depended on wind and communication did not extend beyond the horizon, was to enforce its command of the seas. The law that prohibited neutrals from carrying the trade of belligerents remained in effect, but Americans regularly evaded it by smuggling or subterfuge. One commonly used ruse was to load cargoes on American ships in French or Spanish colonies, carry them to the United States, offload them and pay customs duties, and then reload the cargoes, which were now ostensibly American, for European destinations. Of course the British were eager to suppress this trade. In 1805 they seized the

Year	Shipping Tons	Total Foreign Commerce (in Dollars)	% Carried in American Ships
1801	630,558	204,384,024	89.0
1802	557,760	148,290,477	86.5
1803	585,910	120,466,699	84.5
1804	660,514	162,699,074	88.5
1805	744,224	216,166,021	91.0
1806	798,507	230,946,963	91.0
1807	810,163	246,843,150	92.0
1808	765,252	79,420,960	90.5
1809	906,855	111,603,233	86.0
1810	981,019	152,157,970	91.5
1811	763,607	114,716,832	88.0
1812	758,636	115,557,236	82.5

Table 2.1: U.S. Tonnage and Foreign Trade, 1801–1812.
Source: Data from Winthrop L. Marvin, *The American Merchant Marine: Its History and Romance from 1620 to 1902* (New York: Scribner's, 1902), 131.

American ship *Essex* and hauled its master into admiralty court in Halifax. The court ruled that the vessel was in fact engaged in a continuous voyage and that the 1756 order banning neutral trade that would not have been permitted in peacetime applied. From the American point of view, however, such losses were simply part of the cost of doing business, and as the customs records reveal, this business was booming. In 1805 the duties paid on imported goods that were reexported totaled $8 million; two years later they were 25 percent higher, at $10 million.[9]

1805 was also the year in which Britain's command of the seas was confirmed by its victory at Trafalgar. In this historic engagement the British fleet, commanded by Vice Admiral Viscount Nelson, defeated and largely destroyed the combined French and Spanish fleets off Cape Trafalgar, near the Straits of Gibraltar. Having crushed its enemies at sea, Britain was able to enforce its command of the ocean much more aggressively. The *Essex* decision, for instance, was to be applied more thoroughly; Britain stationed ships outside American harbors to stop all vessels and check their manifests for this purpose. The result was that many American ships were seized and sent to Halifax for condemnation by the Prize Court. In 1806 the British extended their blockade of the European continent northward, so that ports from Brest to the Elbe were added to those already closed to commerce. Napoleon then made a rather empty gesture of retaliation by declaring a total blockade of the British Isles. What this meant for the Americans and other neutrals was that any vessel that fell into British or French hands had to be considered a total loss. The hazards of engaging in the carrying trade during wartime were increasing rapidly.

The British targeted American merchant ships for their crews as well as their cargoes. They were motivated by the need to find enough experienced sailors to man their warships. The Americans, not surprisingly, considered the British practice of impressing seamen from American ships an outrage to their national honor. Although Britain's needs did not elicit much sympathy in the United States, they were real enough. As its naval fleet expanded, Britain needed ever more men, and especially trained seamen, to man its warships. Since the days of Queen Elizabeth the British navy had drawn many of its sailors from the merchant fleet, which was protected and encouraged as a vital source of military manpower. But service in the navy was hard and in many cases involuntary, and many men did what they could to escape it. One option, after 1783, was to slip off to an American ship. The British regarded such sailors as deserters and were determined to return them to naval service. The United States, like all other nations, regarded its ships as national territory and were unwilling to submit them to British searches or sanction the kidnaping of crew members on the high seas. This was not a matter of a few scattered incidents; by 1807 U.S. government records indicated that up to six thousand American seamen were serving in the Royal Navy against their will.[10]

The British navy, driven by necessity and fortified by an unchallenged sense of supremacy, became increasingly assertive. In June 1807 the British warship *Leopard*, patrolling off Norfolk, Virginia, demanded that the U.S. frigate *Chesapeake* heave to and submit to a search for deserters. Commodore James Barron, the American captain, refused, and *Leopard* opened fire on the unprepared *Chesapeake*. Barron's ship was soon a wreck and he was forced to surrender. The British conducted their search and departed with four of his men, leaving the unfortunate commodore to limp back to port and face an eventual general court martial.[11] Americans were outraged by this incident. But what could they do? Jefferson knew better than to challenge the British directly and turned to diplomacy instead. James Monroe, the U.S. ambassador to the Court of St. James, was instructed to demand compensation for the damage to *Chesapeake* and for the deaths and injustices inflicted on its crew. He was also to insist that the British cease impressing American seamen. All these demands were rejected, adding insult to injury.

At that point Jefferson reasonably concluded that, since the United States could not defend its maritime commerce from British depredations, it had better keep its ships out of harm's way. He knew such a policy would be injurious to American maritime interests, but he hoped it would be equally damaging to the British. Great Britain depended on American ships to carry its trade with the United States; perhaps interrupting that trade would bring the British to their senses and persuade them to respect American neutrality. In December 1807 Jefferson convinced Congress to pass an Embargo Act implementing this strategy. It prohibited American ships from calling at any foreign port, leaving only the coastal trade available to them. American merchant ships and the naval squadron in the Mediterranean were called home, and every ship operator was required to post a bond for twice the value of his cargo to ensure that it would not reach a foreign port.

The consequences of Jefferson's embargo illustrate how difficult it is to anticipate the effects of legislation designed to bend commercial operations to national purpose. Embargoes, unlike military interventions, work slowly, and it is difficult to maintain the political commitment required to sustain them. Great Britain had its entire empire to draw on, and with American ships out of the picture, British shipowners moved in with more capacity than had been expected and reaped rich rewards. France, having been driven from the seas, was living on the resources of the European continent it had conquered. In America, thousands of seamen were thrown out of work, and foreign commerce, which had reached $246.8 million in 1807, dropped to $79.4 million the following year.[12]

Jefferson was a farmer at heart. He knew nothing about ships and like most Republicans at the time distrusted the navy. Maintaining it seemed a waste, and it was considered a peculiarly British institution, one that posed a danger to the new nation's fledgling democracy. At Jefferson's urging Congress agreed to allow him to stop all naval construction, sell off some naval vessels for use as merchant ships, and concentrate on building a fleet of small gunboats for harbor defense. These gunboats proved to be worthless when the British attacked Washington in 1814.

The New England states were hit hardest by the embargo, and they reacted most strongly. Forcing ships into layup threatened shipowners with bankruptcy and many turned to smuggling, even though this exposed their ships and goods to seizure by the British. Talk of revolt spread as the embargo choked off trade, and the political support required to sustain the embargo began to weaken. But certain regions in Great Britain were feeling its effects as well, especially the western ports. In April 1808 Liverpool merchants petitioned Parliament, pointing out that their city carried on three-quarters of the American trade and expressing their concern that it would be irreparably damaged if the policy of harassing American ships were not abandoned.[13] But Jefferson's second term was coming to an end, and early in 1809 he accepted the embargo's repeal.

MR. MADISON'S WAR

Since the problem that led to the embargo had not been eliminated, a new, more focused policy of isolation was tried. An Act of Non-Intercourse was passed, prohibiting trade with Great Britain and France alone. This law lasted only a year and was followed by an even more limited bill authored by Nathaniel Macon. This new Non-Intercourse Act said that if either Britain or France withdrew their decrees restricting neutral shipping, then the United States would prohibit trade with the other country. Napoleon promptly accepted this gift from the American people by withdrawing his Berlin and Milan decrees, which he had been unable to enforce. James Madison, who succeeded Jefferson as president, was warned that this was a ploy to force the United States into war with Great Britain; but Madison was itching for action and accepted the challenge. He declared that if Britain did not follow France in promising to respect American neutrality, he would prohibit all trade with Great Britain. In March 1811 he did so, but with no discernible effect. Great Britain ignored the proclamation and the French continued to seize American ships whenever they had the opportunity.

Madison's action set the stage for war; all that was needed was a triggering confrontation. It was not long in coming. In May 1811 the British frigate *Guerriere* stopped an American merchantman just outside New York Harbor and removed one of its crew. Madison, eschewing diplomacy, ordered the frigate *President* to take station off Sandy Hook near the entrance to the harbor to prevent further actions of this type. As *President* approached New York, it sighted a British warship and closed with it after nightfall. The American ship hailed the other ship and in reply received gunfire, which *President* returned. It turned out that the British ship was the sloop of war *Little Belt*. It was no match for the American ship and was quickly reduced to wreckage.[14] Here was the revenge for the attack on *Chesapeake* that Madison and his countrymen had longed for. Madison rejected all British protests and relations between the two nations, already bad, got worse. On June 1, 1812, Congress declared war on Great Britain.

Declaring war on Great Britain appealed to several factions. Madison, in his June address to Congress, stressed Britain's insupportable violations

35

of American maritime sovereignty—seizure of ships, impressment of seamen, blockade of ports, failure to respect its rights as a neutral. "Free trade and sailor's rights" soon became a war cry. Madison also drew attention to British encouragement of Indian attacks on settlements in the Northwest Territories. Going to war with Great Britain also pleased the war hawks from the South and West; their leaders, Henry Clay and John C. Calhoun, saw the war as an opportunity to seize Canada. The Louisiana Purchase had recently doubled the original size of the United States, and an appetite for further expansion was widespread, but hopes that Canada could be annexed did not bring about the war. It was maritime issues, and especially the odious Orders in Council and impressment, that led to war.

New England, which had long borne the cost of British depredations, dreaded the coming war and the inevitable ruination of its trade. In an ironic turn of events, the Massachusetts island of Nantucket, a major whaling center, went so far as to declare its neutrality.

Although the Americans did not know it, the British considered the prospect of war on another front equally unappealing. Napoleon had conquered the Continent and Great Britain was fighting on alone. The American Non-Intercourse Act was isolating Britain far more than Jefferson's embargo had. Indeed, the day before war was declared, the new British prime minister had withdrawn the Orders in Council that mandated Britain's denial of American neutrality. But the hounds of war had been let slip and a price had to be paid.

Canada proved to be less vulnerable than the war hawks had anticipated. The British ranged the Atlantic coast at will, raiding and burning where they pleased. The capital in Washington was put to the torch. With the exceptions of Harrison's engagements in Canada following Perry's naval victory on Lake Erie and the Battle of New Orleans, which took place after the war was officially over, the United States had little to show for its land campaigns. And surprisingly, given Britain's overwhelming naval preponderance, the United States fought most effectively on the water.

At sea the few ships in the U.S. Navy fought valiantly in single-ship engagements. *Constitution* defeated *Guerriere* and later *Java*, to the enormous embarrassment of the Royal Navy. Its durability in battle earned it the name *Old Ironsides*. *United States* defeated *Macedonian*, and the sloops of war *Wasp* and *Hornet* distinguished themselves as well. But the main force of the British navy could not be challenged, and it eventually established a blockade from Maine to the Mississippi. British cruisers were stationed off all major ports, and the trapped American naval vessels were immobilized and dispirited. *President*, the last American frigate that attempted to put to sea, was almost immediately captured by a trio of British warships.

A much more decisive series of naval engagements took place on the northern freshwater lakes. Commodore Oliver H. Perry built or acquired a small but effective fleet on Lake Erie. During September 1813 he gave battle to a British force of similar size and completely destroyed or captured all its ships. Soon after he sent General William H. Harrison, the U.S. Army commander with whom he had been cooperating, the famous report, "We have met the enemy and they are ours." This victory, together with Harrison's victory in the battle at the River Thames, eliminated Great Britain's Indian allies under Tecumseh as a fighting force and left the Northwest securely in American hands.

Two months later, in November 1813, the British notified the American ambassador, James Monroe, that they were willing to enter into talks with the objective of restoring peace. President Madison accepted the offer and soon sent envoys to Ghent, Belgium, where the negotiations were to take place. The British, however, had clearly indicated that they intended to conduct these negotiations on their own terms. Since they anticipated a significant military victory that would strengthen their bargaining position, they purposefully dragged out the negotiations.

The British strategy had long been to gain control of New York and New England by invading from Canada. The invasion route was to be via the Richelieu River from the Saint Lawrence, thence down Lake Champlain, and then, after a relatively short portage, to the navigable waters of the Hudson River somewhere above Albany. The key to this strategy was controlling Lake Champlain, and on August 31, 1814, the British launched an overland expedition to capture Plattsburg at the head of the lake. The American militia commander, Alexander Macomb, successfully delayed their advance; but by September 6 the British force had reached the Saranac River, where they awaited the arrival of the British squadron operating on the lake before pressing on with the final assault. The ships arrived a week later. All that stood in the way of victory was a thirty-year-old lieutenant, Thomas Macdonough, commanding *Saratoga* (twenty-six guns), three other ships, and ten galleys. During a two hour engagement that provided a brilliant display of seamanship, Macdonough destroyed the entire British squadron except for its shallow-draft gunboats.[15]

With the failure to capture Plattsburg, after having failed to capture Baltimore some months earlier, the British were denied their hoped-for victory. At the same time, pressures for peace were increasing at home.

Demands to resume normal trade relations, not only with the Continent but also with the United States, were mounting and could no longer be ignored. At the same time armed American merchantmen and privateers were mounting damaging attacks on British maritime power. Hundreds of these vessels sailed forth to harry British commerce. By the time negotiations began, this fleet, according to some estimates, had grown to over five hundred ships and had taken over a thousand prizes. They roamed from the North Atlantic to the West Indies, and like the naval officer John Paul Jones during the Revolution, they marauded along the English Channel and other British coasts. The privateers were so successful in disrupting trade that prices for basic commodities in Great Britain rose to siege levels. Their effectiveness, and the Royal Navy's inability to control them, are nicely captured in a resolution passed by a group of Glasgow merchants in September 1814:

> That the number of privateers with which our channels have been infested, the audacity with which they have approached our coasts, and the success with which their enterprise has been attended, have proved injurious to our commerce, humbling to our pride, and discreditable to the directors of the naval power of the British nation, whose flag, till of late, waved over every sea and triumphed over every rival. That there is reason to believe that in the short space of less than twenty-four months, above eight hundred vessels have been captured by that power whose maritime strength we have hitherto impolitically held in contempt. That at a time when we are at peace with all the world, when the maintenance of our marine costs so large a sum to the country, when the mercantile and shipping interests pay a tax for protection under the form of convoy duty, and when, in the plentitude of our power, we have declared the whole American coast under blockade, it is equally distressing and mortifying that our ships cannot with safety traverse our own channels, that insurance cannot be effected but at an excessive premium, and that a horde of American cruisers should be allowed, unresisted and unmolested, to take, burn, or sink our own vessels in our own inlets, and almost in sight of our own harbors.[16]

With the abdication of Napoleon in 1814 and the apparent end of the war, the British Parliament had to make a crucial decision. Would they expend British blood and the nation's wealth in forcing the United States to accept peace on their terms, or would they capitulate to the American president's insistence that they return all captured territory and restore all pre-war boundaries? The British had some forty thousand men under arms in Canada, many of them veterans of the Peninsular War. The Duke of Wellington was offered command of this force and accepted, but he offered little hope of success and decided not to leave Europe until the following spring. He noted that what the British needed was "not a General or General Officers and troops, but naval superiority on the Lakes." In light of Britain's recent failures to attain that objective, he concluded, "[Y]ou have no right . . . to demand any concession of territory from America."[17]

Wellington's position immeasurably strengthened the hand of those opposed to continuing the war, and the demand that previously seized territory be retained was dropped. Although negotiations continued for some time, the main obstacle to peace had been removed. On Christmas Eve 1814 the British and Americans signed the peace treaty. The causes of the war were not mentioned, and Britain did not renounce the right to impress fugitive sailors, but America had effectively demonstrated that it would not accept being treated with disdain. The end of warfare between the Great Powers helped ensure that no further outrages of the sort that caused the War of 1812 would occur. As peace descended upon Europe and its colonial dominions, Americans exhibited a growing sense of national pride and a determination to exercise what they were later to call their "manifest destiny."

TREATIES OF RECIPROCITY

The Treaty of Vienna, which ended the Napoleonic War, marked the beginning of a century in which a peaceful balance of power was maintained in Europe, an era known as Pax Britannica. But in 1815 no one knew what the immediate future held for war-ravaged Europe. France was exhausted but far from humbled by its revolutionary and imperial exertions, and Great Britain was stretched thin by its efforts to suppress political radicalism at home, manage a socially disruptive industrial revolution, and hold together and finance the alliance against France. When the diplomats succeeded in bringing the shooting war to a close, it remained to be seen who would win the commercial contest that would inevitably follow. Great Britain, victorious on the field of battle, had every intention of winning the trade war as well. The question for the United States in such circumstances was what policy would provide the best possible opportunities for the expansion of its trade?

Every nation has to locate its trade policy on a spectrum that runs from extreme mercantilist protection on the one hand to complete freedom of trade on the other. And of course each nation's choice is considerably constrained by the policies of its trading partners. In 1815 the United States was a largely undeveloped, peripheral state with a disproportionately large merchant marine looking for opportunities to carry the world's freight. Great Britain, on the other hand, had a worldwide empire, a navy without equal, and was determined to extend and consolidate its global reach. Could each benefit from a mutual reduction in restrictions on trade? Was freer trade good for everybody? Then as now, the question was hotly debated, the positions taken often reflecting the particular interests of the parties involved.

During the period between the end of the American Revolution and the fall of Napoleon, the tariff walls between the United States and Great Britain had steadily grown higher. With the end of hostilities, the two nations agreed to eliminate all discriminating duties in effect between them. Britain, however, insisted that trade with the British West Indies remain closed to the Americans. This exclusion soon led to the establishment of a triangular trade that was a mirror image of the triangular trade the colonists had conducted so profitably before the Revolution. British ships loaded cargoes in Great Britain for America; after discharging these cargoes in the United States, they loaded American lumber, fish, flour, and produce for the West Indies; in the Caribbean they loaded sugar and molasses, which they took to Great Britain or the United States. American shipowners complained bitterly about being excluded from this trade, and Congress was roundly criticized for ignoring their interests when they removed the restrictions that had previously reserved American trade for American ships.

The Canadians also made a play to capture their trade with the United States. Gypsum, a mineral used to make plaster, was a principal export brought from the Canadian maritime provinces of New Brunswick and Nova Scotia to the United States. This trade had been open to ships of both countries, but the Canadians decided to require that this commodity be carried only in their ships and imposed a discriminating duty to that effect.[18] The United States reacted by passing the Navigation Act of 1817, which included most of the restrictions on trade contained in the English Navigation Acts of the seventeenth century. But the desire for free trade was not entirely extinguished; the act also contained provisions that suspended its restrictions for nations that did not discriminate against U.S. shipping. One of its immediate effects, therefore, was to close off British trade between the United States and the British West Indies. This 1817 act also mandated the complete exclusion of foreign vessels from the U.S. coastwise trade, regardless of other countries' policies governing their own coastal commerce.

The 1817 Navigation Act demonstrated America's determination to protect its commerce against unfair trade discrimination, but it also authorized bilateral agreements to eliminate barriers to trade. In time many agreements of this sort were reached, with more than forty such treaties eventually being negotiated. Encouraged by this progress toward freer trade, Congress in 1828 repealed the tonnage tax on American vessels and offered the same treatment to the ships of all nations that agreed to reciprocate. Two years later Great Britain finally opened trade with its colonies in the West Indies. Thus by 1830 the United States found itself essentially once more in the advantageous trading position it had occupied before its Revolution. Its highly competitive merchant marine could once again carry goods to and from the ports of the British Empire that bordered the Atlantic Ocean. Of course the Americans had to compete with a British shipping industry undistracted by war, which meant that the American share of the nation's trade fell from the unnaturally high levels it had reached before 1815. But with the growth of trade and the continuation of peace, the stage was set for a golden age of American maritime commerce.

As we look back on the America of 1830 we see a nation enjoying certain natural advantages that would work to its benefit for several more decades. Peace in Europe meant the United States could concentrate on internal development and commerce. This peace would eventually be shattered not by renewed war in Europe, but by the disaster of the Civil War in America. Commercially, the United States still built the best and least expensive oceangoing vessels, and its captains and crews operated them with skill and economy. This would remain true so long as ships were made of wood and driven by wind; but from the middle of the nineteenth century onward, with the steady advance of iron and steam, the United States found itself increasingly relegated to the technological margins.

Territorially, in the west the United States had acquired vast lands that increasingly drew attention away from the ocean to the east. The Louisiana Purchase brought with it control of the Mississippi River, that great watery highway into the continental heartland, and by the middle of the nineteenth century the nation was as enthralled by smoke-belching steamboats racing along the inland rivers as it was by the great clipper ships racing to San Francisco and China. The acquisition of Florida in 1819 completed the consolidation of the Atlantic–Gulf of Mexico coastline, and coastal traffic increased steadily as settlement and the economy grew. Fully protected from non-U.S. competition by the 1817 act, this coastal trade employed a larger tonnage of ships than foreign trade by 1820.[19]

The U.S. population increased along with its territory; between 1776 and 1810 it grew from 3 million to 7 million. As ways were found around or through the Appalachian frontier, the nation began to feel a greater sense of integration and independence. The nation was looking westward and inward and was less troubled by European developments than it had been during its first decades of independence. Self-sufficient and self-confident, the United States nurtured a sense of exceptionalism. It was indeed the beginning of a golden age for the U.S. merchant marine.

3

THE GOLDEN AGE,
1830–1860

The years between 1830 and the beginning of the Civil War in 1861 are remembered as the golden age of America's merchant marine. The achievements of this period are nowhere more visible than in the size, beauty, and speed of the commercial sailing ships built in the United States and sailed around the world by Yankee captains. In these three decades ships doubled in length and tonnage and speeds increased up to an occasional daunting fifteen knots (a knot is one nautical mile an hour) as the great clippers set records on passages around Cape Horn and to the Far East. Like the soaring stone cathedrals of the Middle Ages, wooden sailing ships in this era were pushed to the ultimate possibilities of their form and material. The great clipper ships, and their cousins the down-easters, were marvels of human skill, functional design, and aesthetic delight. They were, again like the High Gothic cathedrals, products of traditional skills and craft rather than science and engineering. For a brief but glorious moment they dominated oceangoing maritime commerce. The ascendancy of the clipper ship marked both the culmination and the approaching end of America's golden age.

Anyone standing on the deck of a clipper ship making its way out of a busy American port in 1860 could have easily seen why the heyday of these great vessels was rapidly passing. A brief glance high and low would reveal that their complex rigs required large crews to sail them. Although after 1840 most of the officers were still American, more and more Britons and Scandinavians occupied the fo'c'sle. During the golden age, as before, the life of the sailor was hard, poorly paid, and frequently brutal, as Richard Henry Dana made clear in his classic *Two Years before the Mast*. What changed around the middle of the century were the opportunities available to ambitious young Americans elsewhere. Manufacturing wages began to outstrip seagoing wages. While factories offered none of the adventure associated with seafaring, workers could count on enjoying the pleasures of hearth and home. Opportunities to make one's fortune at sea also began to decline as the amount of capital required to launch a seagoing venture increased beyond the means of those who aspired to be masters and owners. Ambitious young men steeped in the ideology of freedom, democracy, and opportunity now looked elsewhere for advancement.

The clippers were also doomed by the advent of steam. By the middle of the century all busy harbors were traversed by a few steam tugs, local ferries, coastal steamers, and the occasional oceangoing sidewheeler. The displacement of sail by steam, and of wood by iron and then steel, took place over many decades; but its progress was inexorable. Had this technological challenge not appeared, there would have been no alternative to wind and wood and their costs and limitations would have been accepted as unavoidable. But just as surely as industrialization transformed warfare and brought to an end the gallant age of mounted cavalry and thin lines of infantry, so too did the advent of steam and iron bring to a close America's golden age of sail. It was an age of heroic achievement, and once gone it became a source of endless nostalgic recollection. But by 1860 it belonged to the past.

THE RISE OF THE PACKETS

The year 1815 ushered in a era of great prosperity for American shipowners. The beginning of a prolonged period of international peace opened new markets and new possibilities. The United States had natural advantages, most notably an abundant supply of wood and harbors, and cultural advantages, especially its traditions of shipbuilding and commerce. American shipwrights, rising to the challenge, quickly moved beyond building bulky, slow-moving freighters according to traditional European designs and instead began to build larger, faster, and more sophisticated commercial packets and clippers. Out of this ferment emerged the world's first liner companies, enterprises that provided scheduled departures of general-cargo packets serving the North Atlantic trade.

The appearance of these "square riggers on schedule," as Robert Albion has aptly characterized this innovation, was driven by the increasing division of labor.[1] Adam Smith discusses this phenomenon in the first three chapters of his *Wealth of Nations*. Smith, writing before factories or transport were powered by steam, begins with the observation that "the greatest improvement in the productive powers of labour . . . seems to have been the effect of the division of labor."[2] He then develops his well-known example of the pin factory in which subdivision of the work and specialization by the

44

workforce enables a dozen laborers to produce thousands of pins in the time it would take an individual to make but a few. But Smith carefully notes that it is trade, not the technology of production, that gives rise to the division of labor; and it is the extent of the market that determines how far specialization can proceed: "As it is the power of exchanging that gives occasion to the division of labour, so the extent of this division must always be limited by the extent of that power, or, in other words, by the extent of the market."[3] Since the extent of the market is largely determined by the available means of transportation, one finds that the earliest industries to benefit from the division of labor lay along the coasts: "As by means of water-carriage a more extensive market is opened to every sort of industry . . . , so it is upon the sea-coast, and along the banks of navigable rivers, that industry of every kind naturally begins to subdivide and improve itself."[4] It was this dynamic, that is to say the more extensive markets and efficient transportation combining to promote an increased specialization of labor, that gave birth to packet liner service in New York.

The first successful packet service was inaugurated between New York and Liverpool in 1818. The ships of the first company, the Black Ball Line, displayed a large black circle on the topsail, the sail positioned above the fore course on the foremast. Packet ships carried any and all cargoes bound for their announced destinations, and they were committed to departing on schedule, whether full or not. Such "common carrier" liner service was a new kind of specialization, for previously most merchants engaged in foreign trade owned their own ships and dispatched them seasonally or when fully loaded. The beginning of packet service introduced a separation between ownership of the cargo and ownership of the ship. This specialization divided the risks entailed in a voyage and led to the development of new forms of management. Ship owners now had to be concerned primarily with attracting cargo and maintaining announced schedules, and they drove their ships and men hard to do so. In a very few years packet service had been extended to most of the principal U.S. ports, connecting them with the major ports of Great Britain and the Continent.

The British were never able to duplicate the speed and reliability of the American packets. They certainly were capable of meeting this challenge head on, but they lacked the incentive to do so. Since the British Navigation Acts barred the purchase of American-built ships, British shipbuilders contented themselves with serving their protected market while ignoring the American price and quality challenge. As is always the case, however, someone had to pay for this protection, and British shipowners and seamen were left to watch with dismay as local merchants regularly chose to send their goods abroad in American ships. Growing numbers of British passengers also chose to take passage on these same ships.

The success of the packets led to further divisions of labor and specialization. Syndicates were formed to expand the packet lines and provide new services. Owners soon found that to attract investment capital they had to demonstrate they could manage their business assets effectively and turn a profit. And whereas ship captains had previously represented the owner of both the vessel and the cargo, this was no longer possible on packet ships, which carried goods owned by many different parties. Port agents were therefore employed to obtain cargoes to be loaded and to look after cargoes that had been discharged. As ship captains gradually ceased to share in the ownership of their ships and cargoes, their traditional status as master of all aspects of the voyage declined.

In the 1820s New York City emerged as the foremost U.S. Atlantic port, a position it has maintained ever since. New York had a large harbor with easy access to major trade routes and a geographic position that made it the natural center for consolidating goods from inland regions and from other ports along the East Coast. Cotton and other commodities shipped from southern ports as far away as Galveston, Texas, were brought to New York and reloaded for shipment abroad. And after the Erie Canal was opened in 1825, grain and other commodities produced in the Old Northwest could be sent east along the Great Lakes and the Erie Canal and down the Hudson River to be loaded in New York for further distribution.

As the volume of trade increased, the size and number of packet ships grew as well. From 1836 to 1856 vessel capacity doubled, from five hundred to a thousand tons, and the number of packet ships increased from thirty-six to fifty-six.[5] Passenger traffic increased as well as more and more immigrants arrived from Europe—more than 1 million entered the United States between 1820 and the Civil War. While many of these immigrants came in packets, others found cheap accommodations in the empty upper deck spaces on westbound cotton freighters. No matter how they came, most immigrants found the crossing a miserable experience. Later on, both the United States and Great Britain would pass laws setting minimum conditions for ships carrying passengers.

THE CHALLENGE OF STEAM

Technological changes that thoroughly transform the way we live appear in retrospect so inevitable that it is very difficult to reconstruct how they were developed and how they displaced older ways of doing things. We now take for granted the nearly universal availability of electric power, automobiles, telephones, and airplanes. But it took time to develop the mature forms of these technologies, and no one could have successfully predicted in advance what those forms would be. At the beginning of the nineteenth century no great insight was needed to predict that steam-driven ships would one day dominate maritime commerce, but no one knew just how or when the new propulsion technology would achieve hegemony. How this transformation in fact took place is a complex and fascinating chapter in the history of industrialization.

In the twentieth century many types of technological development have been driven forward in roughly equal measure by military necessity and economic opportunity; in the nineteenth century the prolonged peace that began in 1815 allowed commercial factors to play a predominant role in setting the direction and rate of technological change. Before the naval arms race between Great Britain and Germany began in the final years of the nineteenth century, decisions about using steam power at sea were determined primarily by commercial opportunity and entrepreneurial initiative. This is not to say that national governments did not take an interest in fostering the use of steam. However, rather than investing heavily in military applications of steam power, they chose to provide state funds in modest amounts to carrying companies willing to take the lead in building and operating technically advanced commercial ships. The subsidies they provided played only a supporting role, for in both Great Britain and the United States the shift from sail to steam was dominated by industrial and market considerations throughout the nineteenth century.

By the middle of the eighteenth century steam engines were being widely used in England to pump water out of coal mines; but they were huge, cumbersome, inefficient devices. Many improvements had to be made before engines driven by the power of steam could be adapted to provide locomotion for ships and trains and rotary power for factories. Thanks largely to a series of key inventions made by James Watt during the latter half of the eighteenth century, new applications began to be attempted. By 1787 John Fitch of Philadelphia was operating a steamboat on the Delaware River, and in 1807 Robert Fulton had a steam-propelled vessel named *Clermont* operating on the Hudson River between New York City and Albany.[6] Fulton's venture was protected by a twenty-year monopoly granted by the state of New York and was financed by the New York investor Robert Livingston. It operated at a modest profit until 1824, when aspiring competitors succeeded in having the monopoly overturned. The U.S. Supreme Court, citing the commerce clause of the U.S. Constitution, ruled that the states lacked jurisdiction over interstate navigation.

Steam-powered vessels were naturally first used where sailing vessels were least able to cope with local conditions. Steam tugs provided a solution to the problem of moving ponderous sailing ships in and out of busy harbors. And since early steam engines lacked condensers and required a steady supply of freshwater, they flourished first on freshwater lakes and rivers. America, with its many interconnected broad rivers and internal lakes, soon took the lead in developing steam power for freshwater navigation. The shallow-draft, paddle-wheeled river steamboat, its decks loaded with cordwood for fuel, was the workhorse for communication on the Mississippi and its tributaries years before a network of railroads had been constructed.[7]

The British showed far more interest in adapting the steam engine to seagoing use. Not only was Britain's internal trade conducted mainly along its seacoasts, but it had a worldwide empire to maintain as well. In 1830 the Englishman Samuel Hall invented the surface condenser, which condensed steam exhausted from the engine back to water and returned it to the boiler. Previously seawater had been used in ships' boilers, but the brine scale created by boiling seawater made it necessary to shut the engines down frequently so that the boilers could be scraped clean. By 1850 John Elder, another Britisher, had invented the compound engine, that is to say an engine with two interconnected cylinders of different sizes. High-pressure steam was introduced into the smaller cylinder, where half its expansive power was used up; the exhaust from the first cylinder was then passed to the second, larger cylinder, where the rest of the steam's expansive power was turned into mechanical motion. A few years later a three-cylinder, triple-expansion engine was developed and became the standard for many years, with even the World War II Liberty ships being powered by traditional triple-expansion steam engines. These compound engines were far more efficient than their predecessors and hence required less fuel for a given voyage. Since coal displaced cargo, the amount of fuel that had to be carried was crucially important.[8]

Not everyone was enthusiastic about the new steam technology. The British navy, having defeated the French and established world dominance under sail, was not eager to surrender its manifest advantages to a radically new and unproven propulsion system. No matter what the long-term prospects of steam might be, the navy's immediate concerns were reasonable. Steam propulsion was doubly expensive. Since ships could only carry a limited amount of coal, they had to be rigged with masts, spars, and sails as well as boilers and engines. The paddle wheels used to drive early ocean-going steamships were also extremely vulnerable to damage by gunfire.

Resistance based on such technical considerations was strongly reinforced by a perfectly understandable distaste for the offensiveness of coal. As a British historian has noted, coal "was dirty and smelly. It covered their immaculate sails with smut and their holystoned decks with coal dust. Above all it was inartistic. There was skill and artistry in sailing a square-rigged ship. In working a steam engine there was none."[9] In 1828 the navy's position was stated by the First Lord of the Admiralty in the following terms: "Their Lordships feel it their bounden duty to discourage to the utmost of their ability the employment of steam vessels, as they consider the introduction of steam is calculated to strike a fatal blow at the supremacy of the Empire."[10] During the next thirty years technical developments in steam propulsion, and especially the replacement of paddle wheels by submerged screw propellers, did much to mollify the navy's initial reaction to this new technology.

American resistance to oceangoing steam propulsion, and to building the iron ships that soon came to be associated with it, arose from various considerations. The more positive were that the United States still had plenty of wood with which to build sailing ships and that Americans were demonstrating through superior ship design and construction that the limits of this technology had not yet been reached. A more negative consideration was that the United States was in no position to compete with Great Britain in the industrial technologies of steam power and iron construction. Great Britain, the pioneer of the industrial revolution, had what appeared to be an insurmountable lead in the production of iron and machinery. British preeminence was most visible in the work of Isambard Kingdom Brunel, perhaps the greatest engineer of the age, who completed his pioneering iron-hulled, steam-powered *Great Britain* in 1846. If forced to compete in this arena, the United States would never become a first-rate maritime power. And so the U.S. maritime industry stuck to what it did best, building and operating wooden sailing ships, and for several more decades the traditional materials and crafts held their own.

American shipbuilders continued to design new hulls for new trades. Soon after the invention of the cotton gin in 1793 the cotton trade began to expand rapidly. Specialized ships with flat bottoms and shallow drafts were built to navigate the Mississippi River to New Orleans and serve smaller southern ports. The growth of the immigrant trade from Europe, along with the scheduled service offered by the packets, called for other ships that were larger and faster than older freighters. Size and speed were carried further in the famous extreme clippers, which were sharper hulled and more weatherly than the packets. The clippers cost more to build, carried less cargo than other ships their size, were built light and hence were shorter lived, and cost more to operate than other ships; but they were fast. In trades involving long distances, and particularly on passages to California and the Far East, they commanded much higher freight rates and captured much of the market from other ships. Their designers and builders are still recalled as titans in an age of heroes: J. W. Griffiths, George Steers, William Webb, and Donald McKay. Striking and still popular images of McKay's dramatically named ships continue to provide vivid reminders of this long-gone age of greatness: *Sovereign of the Seas, Lightning, Flying Cloud,* and *Great Republic.* So long as the clippers and packets held their shares of the markets they served, American shipbuilders and operators were not much interested in steam.

But of course there were a few dreamers. The most notable were several investors in Savannah, Georgia, who had the paddle-wheel steamship *Savannah* built in New York in 1819. When it sailed to Liverpool *Savannah* used its engines for only three of the twenty-nine days required to cross. A buyer was sought in Europe, but none could be found. When *Savannah* returned to the United States its engine was removed and it ended its days as a sailing vessel. *Savannah* was an American "first," but it was neither able to provide a reasonable return to its investors nor did it represent the main thrust of American shipping in its golden age.

Mail Contracts: The First Subsidies

The shift from wood and sail to iron and steam transformed the way ships were financed just as profoundly as it changed the ships themselves. Mechanically powered metal ships are far more expensive to build and operate than wooden, wind-driven ships of comparable size. Of course there were offsetting advantages, especially in reliability of scheduling; but it was by no means certain at the outset that such advantages would be sufficient to justify the increased investment and risk involved in building and operating steamships. A new technology beckoned, but like other new industrial technologies, it required an initial commitment that was several orders of magnitude greater than the older, proven technology. By the 1830s, however, the British government realized that if the potential benefits of this new technology were to be realized, it would have to help underwrite the initial costs.

The British government decided to subsidize the cost of oceangoing steam navigation for two reasons. The first was to ensure that the far-flung British Empire was served by the best possible system of communications, the second was to increase the market for the huge industrial investments that had been made to produce large quantities of inexpensive coal, iron, and machine tools. In the first half of the nineteenth century the United States did not share these concerns, but in the second half of the century, when its rail system was being pushed across the high plains and western mountains, the U.S. government provided heavy subsidies to the railroad companies for very similar reasons. In both nations the government followed where industrialists led.

In 1833 Samuel Cunard of Halifax, Nova Scotia, inaugurated the first regularly scheduled steam-driven service across the North Atlantic. He began with one ship, *Royal William*, traveling between Great Britain and Canada. Although it operated under steam throughout the voyage, its engines had to be shut down periodically, so that its boilers could be cleaned.[11] In 1834 the British government signed its first contract for the private carriage of the Royal mail; the contractor was the General Steam Navigation Company, which served Rotterdam and Hamburg. This contract indicated that the government had decided to use the device of mail subsidies as the means by which it would underwrite the costs of steam navigation. Shortly thereafter similar contracts were signed with the East India Company for mail service between Suez and Bombay and with the Peninsula and Oriental Steam Navigation Company (P&O) for service between England and Alexandria.[12] In 1839 Samuel Cunard signed a mail contract to help underwrite the cost of his North Atlantic service. This contract provided him with $437,000 per year for seven years and obliged him to build four steamers of 1,150 tons each and to provide twice-a-month service to Boston and Halifax from Liverpool. Cunard made good use of the funds provided and in 1843 added two large new vessels to his fleet.[13]

A decade after the British began providing mail subsidies for steamship service the United States began to worry that it might be missing the boat on the new technology. Americans had grown used to dominating the North Atlantic trade routes and were unwilling to abandon them to their British competitors. The only oceangoing commercial steamship built in the United States before 1845, other than the *Savannah*, was the 750-ton auxiliary-screw ship *Massachusetts*. It was not a commercial success, however, and was sold to the government for use as a transport during the war with Mexico.[14] In March 1845, Congress handed the problem to the U.S. postmaster general by authorizing him to contract for mail service between the United States and foreign countries. Edward Mills, a New York businessman, seized the opportunity and proposed to create the Ocean Steam Navigation Company for this purpose. He promised to build four ships at least as big and fast as Cunard's and to provide monthly sailings to Bremen and Le Havre. The postmaster agreed to pay him $350,000 per year for this service, but the venture stalled when Mills was unable to raise sufficient capital to go ahead.

Communications with Germany posed a special problem at this time. Cunard had established a monopoly on westbound transatlantic mail service, which meant that mail from the Continent had to pass through England before being put on a Cunard ship. Thus, while passage from Liverpool to New York took on average fourteen days, mail from Bremen took twice as long to get to New York. By 1845 thirty thousand Germans were emigrating annually to the United States; this provided a strong incentive to establish direct service from Germany.[15] The city of Bremen offered to exempt Mills's company from all port and tonnage duties as well as duties on coal purchased in its port. The Bremen Senate went further when Mills was unable to raise the necessary capital and, together with other German states, provided the funds needed to get the Ocean Steam Navigation Company started. Two ships, *Washington* and *Hermann*, were placed in the Bremen–New York run in 1847 and 1848, inaugurating the first American mail service contract.

Despite this promising start, the company did not flourish. The proposed additional ships were not constructed, and, although the government renewed the mail contract in 1852, it ceased providing funds in 1858. The Ocean Steam Navigation Company was therefore liquidated, but its supporters in Bremen nonetheless considered this first experiment a success. The port had expanded rapidly and become a major gateway for passengers in northern Europe. A few years later the defunct Ocean Steam Navigation Company was replaced by the wholly German company North German Lloyd.[16]

In 1848 Edward K. Collins, owner of the successful Dramatic Line of packets, signed a mail subsidy contract for a route operating between New York and Liverpool. The contract specified that he was to build five ships, each being at least two thousand tons and having one-thousand-horsepower engines. These ships were to provide a total of twenty round trips per year; for this service Collins would receive $385,000 annually for ten years. Collins's vision was not tempered by the realities of the balance sheet, however, for he was determined to make his new ships the finest in the world. They would be much larger than required by contract, and they would offer luxuries never before seen at sea; of his first four, *Arctic* and *Atlantic* were twenty-eight thousand tons and *Baltic* and *Pacific* were only slightly smaller.

The subsidy legislation specified that these ships were to be built "under the inspection of a naval constructor in the employ of the Navy Department . . . and [be] so constructed as to render them convertible at the least possible cost, into war steamers of the first class." This provision is a reminder that in the nineteenth century, as in the twentieth, a degree of public service and bureaucratic oversight was imposed when industries were provided with public funds. The engines installed in Collins's ships were in fact designed by U.S. naval engineers who had studied British propulsion units and were of a size and power never before constructed in the United States; they drove Collins's paddle-wheel ships across the Atlantic at record speeds. Yet the navy was not entirely satisfied. In November 1848 Commodore Matthew C. Perry, brother of the hero of the Battle of Lake Erie and himself later commander of the naval squadron that opened Japanese ports to American ships, was appointed as the first general superintending agent of mail steamers. He had not participated in developing the original Collins contracts and probably would not have approved them if he had. Perry thought Collins's ships were extravagantly showy and worried that converting them into warships would be unnecessarily expensive.[17] But Collins's ships excelled in civilian service and quickly upstaged the Cunard fleet.[18] Unfortunately, however, their performance was more than offset by the expense of operating them. The ships' wooden hulls were heavily strained by engine vibration and needed frequent repair. Congress repeatedly raised Collins's subsidy, but even after it had reached $850,000 in 1852 the company still failed to produce a profit.

Financial disappointment was soon compounded by tragic loss. In 1854 *Arctic* collided with a French vessel off Cape Race and most of the passengers and crew, including Collins's wife and children, were lost. The loss was instructive as well as tragic. At that time Americans did not consider transverse bulkheads important, although European shipbuilders had begun to install them. The French vessel was so equipped and managed to limp back to Halifax, while Collins's *Arctic* went to the bottom.

Two years later *Pacific* disappeared without a trace. Collins, undaunted, began construction of a fifth steamer, *Adriatic*, a ship of 4,100 tons that would surpass all previous liners in speed and luxury. But by this time Congress had had enough. A bill to abolish the entire subsidy system was introduced in 1855, and although it was defeated, it reflected a growing reluctance to promote American shipping at public expense. Unlike Great Britain, America had no overseas empire; its frontier lay to the west. Congress had been willing to provide start-up funds to support the development of the new technology, but it was reluctant to provide continuous subsidies for private enterprises incapable of operating at a profit.

Sectional rivalries were also a factor. Shipping companies were owned by northern investors and ships were built in the North. The only subsidized service sailing from the South connected Charleston, South Carolina, and the West Indies. Southern congressmen therefore saw little reason to expand or even continue the subsidy system. Their resistance and Collins's continuing losses finally overwhelmed the Dramatic Line, which ceased operations in 1858. His great ship *Adriatic* was subsequently sold to British owners. Southerners saw this as a natural denouement. The economic links between Great Britain and the American South were growing stronger as the cotton trade flourished, and the South was happy to have Britain provide the necessary carrying services. In June 1858, Jefferson Davis, the future leader of the Confederacy, addressed this issue in a speech he gave while still a senator from Mississippi. "I see no reasons why," Davis said, "if we can get our mails carried in British vessels across the Atlantic, we should establish a line of American vessels merely that we may compete with them in a race across the Atlantic."[19] It is an argument shippers have been making to carriers ever since.

Collins's costly fleet was never able to earn a profit and Congress was unwilling to continue meeting the rapidly increasing cost of sustaining it. Americans engaged in the protected coastal trade enjoyed greater success. The West Indies Line on the East Coast and the Pacific Mail Line on the West Coast had the advantage of being protected from foreign competition. Only American-flag ships could transport passengers, mail, and cargo between the Atlantic and Pacific via the Isthmus of Panama; this restriction became immensely valuable when gold was discovered in California in 1849. The shelter from competition that these companies and the coastal operators enjoyed enabled them to invest profitably in the latest steam propulsion technology, and they kept the U.S. shipping industry in touch with new developments in steam navigation. Yet Great Britain, with its larger and more highly developed industrial base, continued to lead the way in steam navigation for almost a century. In 1856, when Samuel Cunard decided to build all his future liners out of iron and the Inman Line equipped all its ships with screw propellers, the United States simply lacked the industrial resources needed to meet such challenges.

The Repeal of the British Navigation Acts, 1849

European merchants and political leaders have warily circled around one another for centuries. Merchants hoped to create wealth by buying cheaply and later selling at higher prices. Political authorities, not encumbered by the need to make a profit, tended to believe that men and materials should be applied to the purposes of the state. From the fifteenth to the eighteenth centuries the policy of mercantilism provided a comprehensive and roughly coordinated set of ideas that spoke to the interests of both merchants and monarchs. Mercantilism allowed merchants to operate under the protection of the state. In return for its protection, they agreed to conduct their trade according to rules laid down by the state, rules designed to ensure that in the course of enriching themselves, merchants would also increase their nation's wealth and serve its purposes. But toward the end of the eighteenth century certain aspects of the venerable practice of mercantilism were called into question, and a political case was made for easing the state's control of trade.

The classic argument for increased freedom in trade is contained in Adam Smith's *An Inquiry into the Nature and Causes of the Wealth of Nations* (1776). Smith was not attacking the need for a strong state, nor was he arguing that trade should be freed from all political constraints; quite the contrary, he considered Great Britain's Navigation Acts both necessary and effective. He simply urged that if the goal of economic policy is to increase the wealth of the nation, then the state should in principle minimize the restrictions it imposes on trade. Merchants, not legislators, know best how to organize and conduct trade and industry so as to create wealth. Smith's argument has, of course, been of enormous historical consequence, and his book is regarded as the fountainhead of modern economic science. As developed by David Ricardo and others, his argument gave rise to the modern doctrines of laissez-faire economics and the liberal state.

The nineteenth-century campaign to remove legal restrictions on the British grain trade was a classic instance of applying Smith's ideas to a concrete policy. The importation of grain into Great Britain had been controlled since the twelfth century, and the regulations governing this trade were known collectively as the Corn Laws. One goal of these laws was to ensure that Great Britain could feed itself; a perhaps unintended consequence of protecting the domestic grain market was to raise the price of grain in Great Britain and the value of the land on which it was grown. Serious problems arose in the late eighteenth century, however, when Britain began to experience annual food shortages caused by both a rapid increase in population and wartime disruptions in trade. During the last decade of the Napoleonic Wars there were several bad harvests and serious food riots. These circumstances drove up the price of food and created severe difficulties for industrialists employing thousands of workers who depended on their wages to buy enough food to survive. One proposed solution, stoutly resisted by the landowners, was to repeal the Corn Laws, so that grain supply could follow demand and prices would not fluctuate so widely; this was the action championed by Richard Cobden and the Anti-Corn Law League. This argument eventually prevailed in 1846, when Prime Minister Robert Peel successfully moved for the repeal of the Corn Laws.

Shortly thereafter it was the shipowners' and shipbuilders' turn to come under attack from the free traders. Parliament appointed a Select Committee in 1846 to look into the operation of the nation's Navigation Laws. For five months the committee met and heard some thirty witnesses. Although the committee remained deeply divided and never issued a final report, one of its members, J. Lewis Ricardo, a rising young member of parliament and a nephew of the noted economist David Ricardo, had impeccable free trade credentials. In 1847 Ricardo published a book based on the hearings titled *The Anatomy of the Navigation Laws*. In his preface to this summary of questions raised by the committee and answers provided by some of the witnesses, Ricardo declared that in his view the hearings had demonstrated the "impolicy and mischievous tendency of the maritime laws of Great Britain."[20]

Ricardo's book provides revealing illustrations of British and American shipowners' perceptions of their industry in the first half of the nineteenth century. Defensiveness and anxiety in the face of change pervaded British shipping. Like the landowners who had defended the Corn Laws as fundamental to the welfare of the nation and the existence of a strong yeomanry, the shipowners insisted that the Navigation Laws were essential to the preservation of British shipping and the defense of the realm. As early as 1815, when Great Britain signed its first reciprocity treaties, the shipowners saw disaster on the horizon. A British Shipowner's Society report published in 1833 gloomily concluded that "the long continued and still existing depression of the shipping interest, the partial production and great aggravation of distress caused by the continued changes in our Navigation system . . . the embarrassment, decay and ruin of the British shipowner may now be reviewed as incontrovertible propositions."[21]

Ricardo recounted these predictions of doom and decay so he could point out how far off the mark they were. Britain's ocean tonnage had in fact continued to grow every year since the 1815 treaties, and whereas British registered tonnage stood at 2,635,000 in 1833, in 1847 it had grown to 3,817,000 tons.[22] Having demonstrated that these beliefs had no basis in fact, Ricardo summarily assigned them to the realm of faith. He quoted a witness named Richmond as saying that the British shipowner would "scarcely ever move without the navigation Act." A well-worn copy of the act, Ricardo observed, was the shipowner's "constant companion, carried about with such affection as the saints of old carried their precious relics; and never a relic was believed to have worked more miracles than this same act."[23] In the twentieth century American shipowners engaged in a similar sanctification of legislation, the object of their veneration being the Jones Act of 1920, which reaffirmed their monopoly of U.S. coastal shipping.

Ricardo reported that several witnesses complimented American shipping practices. Captain Briggs, an American shipmaster, went out of his way to assure the Select Committee that British officers and crews were in no way inferior to their American counterparts: "[O]f the English captains who had entered the American service, some are the best. . . . They make voyages as quick, and sometimes quicker, than any Americans." If the Americans had the advantage, Briggs insisted, it was because they had a more effective system of incentives. American masters were still paid a percentage of the freight earned on the voyage, whereas British masters were paid by the month. This difference encouraged the American master to drive his vessel harder and to take greater care of his cargo. As Briggs noted, "[A] man who is out in bad weather, if paid by the month will be more likely to make for the first port he can, because his pay continues to run on notwithstanding the delay in the voyage." Briggs also observed that American ships were crewed by a higher class of men than those of British ships, and again the difference was one of incentives: "[Americans] have greater inducement to enter into it, the pay is so much greater, that it is a motive for them to follow the sea as a profession."[24] Yet even though American mariners were healthier, better educated, more productive, and more highly paid than their British counterparts, total American operating costs were no greater than British costs. American ships operated with smaller crews than those of comparable British ships; they also were turned around faster in port and hence had higher utilization rates. Briggs's testimony is clear evidence that American commercial practices commanded respect in Great Britain at midcentury.

According to Ricardo, the Select Committee also examined and found wanting a number of traditional arguments for protection. Witnesses demonstrated that American shipbuilding costs were not substantially lower than British costs and that American labor was more expensive than British labor. Timber in America was no longer plentiful near shipbuilding sites and most of the iron used for strapping and chain was imported from England. Copper for sheathing and hemp for cordage also had to be imported. Prices for American ships were competitive not because American materials and labor were cheap but because the yards they were built in were well managed and their designs and methods of construction were state of the art.[25] According to the testimony heard by the Select Committee, it truly was a golden age for American wooden ship construction.

Ricardo's book was written with political purpose, and it quickly became the bible of the free traders seeking to repeal the Navigation Laws. As Parliament began to debate the issue, attention focused on the likely consequences of allowing foreign ships access to British markets. If foreign-built ships were allowed to qualify for British registry, would the death knell have been sounded for British shipbuilding? And if foreign owned and registered ships were allowed to carry British trade without restriction, would it spell the end of British shipping? These were weighty questions and it required considerable self-confidence and courage to press on. Nonetheless, in May 1848, after many delays, the following resolution for repeal was placed before Commons: "That it is expedient to remove the restrictions which prevent the free carriage of goods by sea to and from the United Kingdom, and the British possessions abroad, subject nevertheless, to such control by Her Majesty in Council as may be necessary and to amend the laws for the Registration of Ships and Seamen."[26] When Parliament reconvened in January the free traders gained ground. In April, following a third reading of the bill, Commons voted by a comfortable margin to repeal the Navigation Laws.[27]

This repeal, it should be noted, was fundamentally different from earlier agreements to reduce trade restrictions reciprocally, for the repeal was unilateral and unconditional. Of course the bill's advocates hoped that other nations, especially the United States, would follow Britain's lead and repeal their own navigation laws. At the outset it seemed likely that the United States would do so. The American minister in Great Britain, George Bancroft, was an ardent free trader and had given assurances that America would respond in kind. But a new president, General Zachary Taylor, was being elected at the time Parliament was considering repeal. American policy was suddenly reversed and became strongly protectionist. American shipowners and shipbuilders firmly opposed opening the U.S. registry to foreign-built ships or the coastal trade to foreign competition.

American protectionism was motivated by a potent combination of gold fever and imperial competition, both of which directed attention to the Central American isthmus separating the Atlantic and Pacific Oceans. The discovery of gold in California set off a rush to the West, with some prospectors undertaking the arduous overland route and others braving the ocean passage. As the nation looked increasingly westward in the decades before there was a transcontinental railroad, the dream of building an interoceanic canal in Central America was reawakened. The favored route ran through Nicaragua, connecting Lake Nicaragua and the San Juan River. If America was to build and control such a canal, the East and West Coasts of its great continental empire would be linked by a protected and greatly shortened water route. As the California trade boomed, elected officials saw no reason to share the new-found wealth and prospects with other nations.

British interest in an interoceanic canal in Central America further stiffened American resistance to following its lead. Britain had little inclination to honor the Monroe Doctrine, and it had good reason to look for shorter sailing routes to the West. With Canada stretching to the Pacific, Britain, too, wanted an easier water route from the East Coast of America to the West. Building a canal across the Suez Isthmus was a dream that would not be realized until the 1860s. In the 1840s and '50s Britain therefore actively explored the possibility of opening a canal across Central America. Its ships bound for China and Japan could then be sent westward, a route that would spare them the need to sail around Africa. British initiatives in Central America made the United States very nervous and defensive, which is not a good state of mind for experimental reductions in trade barriers.

In March Bancroft received new instructions and was forced to tell the British government that the new administration was not prepared to reciprocate.[28] This reversal caused consternation in Parliament but did not stop the bill's progress. Although the protectionists made a last stand in the House of Lords, the bill passed and, in June 1849, became law. The protection that the Navigation Laws had provided for British shipping for centuries had been abandoned.

This historical shift in policy elicited a number of different responses. Although all British envoys were instructed to ask their host countries to lift their restrictions on shipping, only a few other nations were willing to face unfettered competition. The Scandinavians, having little to lose, were delighted to be offered an opportunity to carry Britain's trade, and they quickly eliminated their own protectionist regulations. The Americans, not being obliged to reciprocate and having the most to gain, looked forward with pleasure to competing for British cargoes and selling Great Britain ships while keeping their own coastal trade and registry firmly shut.[29] In London the insurance underwriters helped allay the fears of the British shipbuilders by changing ship classification rules so as to increase the cost of American ships.[30] The discovery of gold in California and the war in the Crimea also helped by creating sharply higher demand for ships and shipping at an opportune moment. In the longer run, British shipbuilders adapted to the new realities of international competition and, having the advantage of working in the first nation to industrialize, they soon established their dominance in the new technologies of steam and iron. Great Britain, already the world's dominant maritime power, had opted for free trade in shipping and shipbuilding. It was a bold and risky decision, but it certainly paid off. It demonstrated, among other things, that one did not have to be a mercantilist to be a successful imperialist.[31]

Would Adam Smith have approved of Britain's unilateral repeal of its Navigation Laws? Smith, knowing that military and commercial competition among nations is unrelenting and unforgiving, was a strong proponent of a powerful navy. And in accordance with British policy ever since the time of Queen Elizabeth, he considered the merchant marine the training ground for the sailors who would man naval vessels in time of war. This is what he had in mind when he wrote that "as defence, however, is of much more importance than opulence, the Act of Navigation is perhaps the wisest of all the commercial regulations of England."[32] But had Smith lived to the middle of the nineteenth century, he well might have recognized that the venerable Elizabethan policy was going out of date. The future of the navy lay in steam and iron, and Britain had no rivals in the new technology. Thus at the time the Navigation Laws were in fact repealed, Smith could still have avowed the primacy of defense while also welcoming the move to free shipping.

The New Nationalism and California Gold

The clipper ship and the California goldfields were made for one another. Clippers had reached their mature form shortly before gold was discovered in 1849 in the recently acquired California territory, and the speedy ocean-roaming ships provided the fastest and, for a confident young nation, the most fitting way to reach the new El Dorado. The thousands of men determined to make their fortunes in the western goldfields could get there by undertaking the arduous overland trip, by taking the steamer to the Isthmus of Panama if they could afford it, or by doing what most of their compatriots did, which was to board a clipper for the long swift run around Cape Horn. Speed was essential and the clippers were in a league of their own. Midcentury steamers chugged along at six or seven knots; when conditions were right the clippers roared past them at more than twice that speed. Donald McKay's *Sovereign of the Seas* was reported to have made an eastbound passage from Honolulu to New York in eighty-two days, at times averaging better than fifteen knots. His *Flying Cloud* covered 374 miles in a single day while westbound from New York to San Francisco.[33] Since speed was of the essence and demand was intense, the clippers commanded premium rates. Freight rates soared to $25 per ton, twice the previous rate, and some clippers earned as much as $125,000 in a single voyage, enough to fully pay off the cost of building such a ship.[34] Only American-flag vessels could ply this intercoastal trade, but the restriction was not of great importance so long as the clippers were setting the pace.

Of course it was a bubble; such rates and such a demand could not be sustained. But in 1854, just as the California rush was leveling off, war broke out in the Crimea. This was an overseas conflict for the European powers involved, chiefly France and Britain against Russia, and France and Britain were both soon chartering American ships to transport their troops and supplies. As war-related needs pulled other nations' ships off their normal trade routes, American shipowners moved in to fill the vacancies. American shipyards were soon producing ships at a record pace. In 1855 more than five hundred ships were launched in the United States. In the ten-year period 1847–57 windjammers totaling almost 4,400,000 tons were built in America; this was twice the tonnage built in Great Britain.[35] It was indeed a bubble, but it was a great decade nonetheless.

Not all ships that sailed around Cape Horn for San Francisco were clippers, and not all clippers stuck to that trade. After reaching California many clippers sailed in ballast to the Chinese treaty ports of Shanghai, Swatow, and Hong Kong, and then journeyed on to India, where they loaded tea for an around-the-world voyage to England. The stolid ships of the British East India Company were left sitting at their docks as the American clippers, which charged twice the Indiaman's freight rate but could get to London in half the time, loaded the high-value cargoes. It was reported that when the New York clipper *Oriental* arrived in London, crowds gathered on the dock to admire its graceful lines and towering rig. The London *Times* chided British shipbuilders, urging them to use their "long practised skill, steady industry and dogged determination" to compete with the "youth, ingenuity and ardor" of the United States.[36] Since British shipbuilders could no longer retreat behind a wall of protection and had no subsidies to offset the advantages enjoyed by their competitors, they either had to meet the challenge or quit the field. Choosing to compete, they soon found ways to match the achievements of their American cousins.

In the late 1850s British shipbuilders figured out how to use iron for certain heavily stressed hull components and began building ships that could match the American clippers. They followed the American hull lines and rigging plans but used iron in many of the structural members, such as the frames, beams, and keelson. They then planked these metal skeletons with teak and sheathed the ships' underbodies in copper to prevent fouling in the warm water of the tropics. The resulting ships, called composite clippers, could sail with the American clippers but weighed less and had more internal space for cargo. They were also cheaper to build and in time recaptured the Far East trade for the British. These British clippers, such as the legendary *Thermopylae* and *Cutty Sark*, live on in their nation's maritime history exactly like their American counterparts.[37]

During the California boom the Pacific Mail Line also flourished, its steam-driven ships being especially advantageous against the northwesters that blow down the California coast. Started in 1846 with the assistance of a mail subsidy, this company had the good fortune to get established in the West Coast trade just before the gold rush. As business and profits soared, new and larger vessels were ordered to replace the original one-thousand-ton side-wheelers, and in ten years the company owned twenty steam-powered ships. By 1856 they had carried 175,000 passengers from the Isthmus of Panama to California and had transported $200 million in gold as return cargo.[38] But by then the goldfields were mined out and the rush was over. Clippers were suddenly in vast oversupply, freight rates returned to normal, and Pacific Mail Line went back to providing routine communication service between the East Coast and California. After 1858 it was the only surviving American line with a mail contract. The boom was indeed over and the Civil War was only three years away. The surviving ships that could still be sailed were increasingly manned by hard-driven foreigners, and their owners had to settle for such unglamorous cargoes as grain, guano, and coal. The golden age had come to an end, a new iron age was beginning.

In the 1860s the advantages of Britain's commanding lead in iron production and machine tools made itself felt at sea. The ships built in Glasgow and other coastal cities were not glorious or memorable, but they were reliable and efficient; and as these innumerable steamers puffed along on their scheduled routes or crisscrossed the seas as tramps in search of cargoes, they drove the remaining sailing freighters further and further to the margins of commerce. The Americans, despite their great seagoing past, could not build iron steamers at British prices, and since American-flag ships had to be built in America, its foreign-going merchant marine rapidly declined. The protected coastal trade continued to grow as thousands of wooden schooners, many of them huge multimasted vessels, carried cargoes to and from locations that the railroads could not yet reach or at rates the railroads could not match. But for those like George Hoar, a former senator from Massachusetts who had known the glory days, nothing would ever be the same: "I can remember very well the time when the names of the great shipbuilders, Donald and Lauchlan McKay and their brothers, were famous all around the world. They were building and commanding the marvelous clipper ships for which the shipyards of New England were unrivaled. It was a contest which enlisted the feeling and the pride of the whole people of the country. There was no boy's play of yacht-racing in those days. The strife was between nations, and the prize was the commerce of the world."[39]

4

THE CIVIL WAR AND
THE TURN TO THE WEST,
1860–1880

The Civil War erupted at a fateful moment for the U.S. merchant marine. The shift from sail and wood to steam and iron was proceeding apace and had the United States not suddenly plunged into fratricidal war, a way might have been found to navigate the transition without putting the nation's maritime industry at risk. By the 1860s it had become clear that the American shipbuilding industry would have to be thoroughly transformed if it were to compete with its British counterpart in the construction of steam-powered iron ships. Given these circumstances, it would have been reasonable to let U.S. operators buy foreign-built steamships and register them under the U.S. flag while U.S. shipyards learned to be internationally competitive in the new technology. But the U.S. iron industry was protected from foreign competition by high tariffs and did not want to see its domestic market invaded, and U.S. shipbuilders were equally unwilling to have their protected market thrown open to international competition. Had the technological challenge been addressed over time in a commercial context, economic reason eventually might have prevailed. But the outbreak of war made political concerns paramount as shipbuilding was rapidly mobilized into the wartime economy. The brief moment in which U.S. shipbuilders might have begun adapting to the changing technology of oceangoing shipping passed unnoticed; the shipbuilders' determination to protect and serve a strictly national market was not seriously challenged.

The war battered the merchant marine on two sides. On the one hand it gave the shipyards a lot of naval work at government rates, thereby freeing them from pressure to remain internationally competitive while adopting the new technologies. On the other hand it decimated the North's existing commercial fleet. The Union government purchased the best of the North's steam vessels to enforce its blockade. Confederate commerce raiders wreaked havoc on all kinds of Union ships at sea and drove insurance rates so high that many owners were compelled to sell their ships to British operators or register them under the "Red Ensign," so they could sail as neutrals protected by the British flag. At the end of the war the U.S. foreign-going fleet had been reduced to a shadow of its former self. After the war the shipbuilders allied with the protected domestic industries, including the operators of coastal shipping lines, to ensure that foreign-built ships would continue to be excluded from U.S.-flag registry. American shipowners, barred from purchasing British-built steamships and unable to find comparable ships built in America, were simply unable to recapture their former share of the nation's foreign trade.

THE WAR BETWEEN THE STATES

Although the Civil War began with an assault on Fort Sumter in the harbor of Charleston, South Carolina, it was, like most wars, waged primarily on land rather than at sea. The Northern navy, lacking the ships needed to defend Yankee merchantmen from Southern commerce raiders and unwilling to use those it had in convoy operations, committed itself to three tasks: establishing a tight blockade along the southeastern Atlantic and Gulf coasts, providing supporting gunfire for the army, and transporting troops and equipment along the coasts and up rivers. Before the war the South, whose economy depended on trade with Great Britain, had relied on the North and the British for shipping services; once the war began it struggled mightily to break the Northern blockade and destroy Northern shipping. While the two belligerents managed to inflict considerable damage on each other, the naval war was at all times secondary to the bloody army campaigns conducted far from the oceans.

Although economic differences were a major cause of the war, the North and the South were not direct commercial competitors. The economy of the South was based on agriculture and the use of slave labor, while that of the North centered on commerce and manufacturing utilizing free labor. The South had an extensive cotton trade with Great Britain and advocated free trade and low tariffs, while the North favored tariff protection of its developing industries. Given these differences and the North's antipathy to slavery, the South sought refuge in the doctrine of states' rights. Whether newly created states in the West would be slave or free became one of the war's precipitating issues. When the election of Abraham Lincoln signaled to the Southern states that the extension of slavery would be thwarted, South Carolina decided to secede from the Union. Its decision to attack the federal fort in Charleston in April 1861 has rightly been called "the most disastrous miscalculation in American political history."[1]

The demons of war, once unleashed, devoured lives and wealth on both sides before finally being stilled by the destruction of the South and the North's military victory. Neither the North nor the South was prepared for war. Although the North had a few ships in its navy, most were deep-draft oceangoing vessels ill-suited to the needs of the coming war. The South essentially had no naval forces or ships it could requisition for war service. Aware of their vulnerabilities, both sides looked to history for strategic lessons. The North recalled that during the War of 1812 British control of coastal waters enabled its commanders to restrict commerce and launch amphibious raids; blockade therefore became the North's strategic focus. The South, counting on covert assistance from the British and realizing that the North had few seagoing naval vessels, focused on commerce raiding and set out to punish the North, as the Americans had punished the British in two previous wars, by attacking its merchant ships offshore.[2]

At the beginning of the war the U.S. Navy consisted of thirty steamships and about sixty obsolete sailing vessels. With the outbreak of hostilities the Union government requisitioned all the steamers it could find in the coastal merchant fleet and began a huge building program. Converted and armed merchant cruisers were formed into squadrons that were sent to blockade Southern ports. The best of them were just able to match the speed of the British-built blockade runners that the South soon deployed. By 1863 the North was able to begin to reinforce its blockading squadrons with fast, newly constructed steamships.[3]

The Civil War marked a turning point in the history of commerce raiding, as the capture of prizes gave way to the routine destruction on the high seas of ships and cargoes. After settling on a strategy of commerce raiding, the Confederacy followed traditional practice by issuing Letters of Marque and Reprisal. It was a venerable way to wage war at sea and had circumstances been different, the stratagem might have been effective. Little came of it, however. There were few Southern ships and crews capable of acting as privateers, and the tight Northern blockade made it virtually impossible to bring captured prizes into Southern ports to be sold.

Equally important was British determination to end the practice of privateering. As the world's dominant seapower and champion of free-trade imperialism, Great Britain exerted its considerable might toward this end during the middle of the nineteenth century. Under British guidance a Paris Declaration outlawing privateering had been issued in 1856; it was subsequently adopted by all industrialized nations except the United States.[4] If the South wished to pursue the strategy of raiding Northern commerce, it would have to find a new way to do so.

The new technology of steam propulsion helped the South devise an alternative to privateering. The offshore commerce of the North was carried in sailing ships; the South's legally neutral but covert ally was the foremost builder of oceangoing steam auxiliaries. The Confederacy therefore had Great Britain build it a number of fast, blue-water raiders that normally operated under sail but also had auxiliary steam power. These swift vessels could easily overhaul and destroy Northern merchantmen. The most famous of these ships, *Florida* and *Alabama*, were ostensibly built as merchant vessels, although their real purpose was widely known. They were armed offshore, *Florida* in the Bahamas and *Alabama* in the Azores. Industrialization, having increased the destructiveness of armies by supplying them with railroads and rapid-fire rifles, spawned yet another novel weapon, the steam-auxiliary commerce raider. Down to the end of World War II heavily armed, fast surface raiders, lineal descendants of these Southern raiders, continued to threaten and destroy merchant ships during wartime, even though early in World War I it had been discovered that the submarine, a more recent technological innovation, could do the job more efficiently.

The story of *Alabama*'s success as a commerce raider and its eventual destruction by the Union ship *Kearsarge* is one of the great sagas of the Civil War. *Alabama* was commanded by Raphael Semmes, a former U.S. naval officer, but its crew was composed primarily of British seamen—the Confederate government promised to pay them prize money at the end of the war. For twenty-two months the *Alabama* roamed the oceans, cruising seventy-five thousand miles without serious mishap while capturing sixty-four Northern merchantmen, all but ten of which were burned at sea.[5] On August 24, 1864, *Alabama* was forced to engage the Union ship *Kearsarge* off Cherbourg, France. The ships were nearly identical in size, speed, and armament; but Captain John A. Winslow of *Kearsarge* had the advantage of commanding a Yankee crew eager to put a stop to Southern depredation of their countrymen's ships. Winslow's crew was also more highly trained. Because Semmes had to conserve his ammunition, his gun crews had had little practice; *Alabama* had had one rather inconsequential encounter with an armed ship, the small USS *Hatteras*, which it sank. The engagement with *Kearsarge* was brief and decisive; in little more than an hour *Alabama* was shot to pieces and sunk. Ironically, Semmes and Winslow had served together as young lieutenants during the war with Mexico.

Year	Tons
1860	17,418
1861	26,649
1862	117,756
1863	222,199
1864	300,865
1865	133,832

Table 4.1: Tonnage of American Shipping Sold to Foreign Owners.
Source: Data from Winthrop L. Marvin, *The American Merchant Marine: Its History and Romance from 1620 to 1902* (New York: Scribner's, 1902), 338.

The sinking of *Alabama* did not bring Southern commerce raiding to an end, but no other raider matched its record of destruction. The greatest damage done to Northern shipping was not the loss of ships, however, even though a total of 110,000 tons were destroyed, but rather the loss of cargoes to foreign shipowners. Southern attacks drove cargo insurance rates so high that Northern ships were left sitting empty at their docks. At the beginning of hostilities the extra war-risk premium was 1 to 3 percent of the insured value of the cargo. By 1864 it had increased to 4 percent; the next year the New York Chamber of Commerce told the secretary of the navy, "[T]he war premium alone on American vessels carrying neutral cargoes exceeds the whole freight for cargoes carried in neutral ships."[6] This is a fact that deserves a moment's reflection. The United States was engaged in a civil war, not a war involving Great Powers, and neither side in the war had a fleet that could challenge Great Britain's command of the sea. Thus Great Britain's status as a neutral, and its ability to defend its ships' rights as neutrals, meant that the North had no way of keeping its commerce from draining away to ships not subject to attack by Southern raiders. From a commercial point of view the situation appeared dire and irremediable.

Of course the U.S. government might have assumed responsibility for paying the war-risk premium on vessel insurance, but the United States was at that time a second-rate power and this solution would not be used until World War I. Meanwhile Northern shipowners pleaded with the government to protect their ships on the high seas, but little could be done. Blockading ships were occasionally dispatched to hunt down Southern raiders, but no coordinated effort was ever mounted. Northern ships that were not destroyed left the field as foreigners took over the carrying trade. In a July 1863 letter to Gideon Wells, the secretary of the navy, the Boston financier R. B. Forbes noted that "of the 180 vessels in New York, 146 were under foreign flag." The consequences were not hard to predict: "Our commerce will soon be entirely in the hands of foreigners unless our trade is protected by every means within the power of the government."[7] The Southern strategy was certainly taking its toll.

A commercial ship is a capital investment; if it is not making money, it must be laid up or sold. Northern shipowners who could not attract cargoes could not simply wait for the war's end; many of them therefore took the only course of action open to them by operating under a foreign flag or selling their ships to foreigners, often at prices far below their former value. Over 800,000 tons of American shipping was reregistered in neutral countries during the Civil War. An offshore fleet that totaled 2,490,894 tons in 1861 had shrunk by war's end to half that size, 1,387,756 tons. Three quarters of the loss was attributed to the work of Confederate raiders. As Captain Semmes told the crew of *Alabama* just before engaging *Kearsarge*, "[I]t is not too much to say that you have destroyed, and driven for protection under neutral flags, over half of the enemy's commerce, which at the beginning of the war covered every sea."[8] What Semmes could not know, and what drove the final nail in the coffin of the Northern merchant marine, was that after the war had ended none of those ships that sought such protection were allowed to return to U.S. registry.

The Union blockade, together with the Northern army's success in isolating port cities from their hinterland, became increasingly effective as the war dragged on and naval support for the grinding land war helped break the South's resistance. When the war began Union general Winfield Scott had laid out a master plan that called for encircling the Confederacy; the fall of Vicksburg in July 1863 gave the Union control of the Mississippi from its headwaters to the Gulf of Mexico and completed the encirclement. The South was now doomed; but two more years of punishing campaigns led by Generals Sheridan, Sherman, and Grant would be required to bring the war to a close.

When it did end, the South's economy was in shambles. The cotton trade, which had been so important to American shipping, had almost ceased. It would be fifteen years before it again reached prewar levels. But postwar cotton was then carried almost entirely in foreign bottoms rather than U.S. ships. The U.S. overseas merchant marine was one of the many casualties of this terrible war.

THE SECOND AMERICAN REVOLUTION

The American colonies fought side by side to obtain their independence from Great Britain, and the leaders of the new nation came in equal proportion from the North and the South. But as the regions' economies diverged and as industrialization on the British model began to transform the North more rapidly than the South, tensions arose that eventually led to war. The war itself exacerbated the differences between the regions by destroying the planter aristocracy that had contributed so much to the nation's early history and by stoking the furnace of industrialization in the North. At the war's end the South faced a generation of agonizing reconstruction while the North was poised on the brink of a tumultuous industrial expansion that one historian has aptly called "the Second American Revolution."[9]

Many factors combined to turn the burgeoning nation's attention away from the sea and toward its inland frontier, yet it is still amazing how rapidly and completely this reorientation took place. The post–Civil War United States had a greatly diminished foreign-trading merchant marine, for reasons that have already been described. Since the nation had no foreign possessions or overseas imperial ambitions, it had no need for a navy either. The commercial coastal fleet continued to grow as internal traffic boomed, but this trade was protected from foreign competition by legislation and did not draw the nation's attention beyond its own shores. Dramatic internal developments commanded increased attention. The 1846–48 Mexican War, sparked by the 1845 annexation of Texas, ended with the addition of California and the territories that later became Arizona and New Mexico. Now stretching "from sea to shining sea," the United States had, at least in the eyes of its millions of settlers of European descent, vast stretches of vacant land waiting for eager farmers, ranchers, and miners to cultivate and develop. Plowing the soil rather than plowing the seas would now be the occupation of Americans living on the frontier. To get to that frontier and to send back the products of their labor, they would make use of that rapidly advancing mechanical marvel the iron horse. It is no exaggeration to say that the railroad built America, at least during its period of rapid industrialization following the Civil War. It was the lonely whistle of the freight train, rather than the rhythmic chuff of the steam engine deep in the bowels of a ship, that fired youthful imaginations in America in the latter half of the nineteenth century.

Northern cities grew rapidly in the war-driven economy of the 1860s. From 1862 to 1864, 180 new factories were established in Philadelphia alone. New industries were founded and consolidated by individuals whose family names have become part of the American lexicon: Carnegie in iron and steel, Remington in firearms, Rockefeller in oil, and Weyerhauser in lumber.[10] Population grew rapidly and became increasingly concentrated in urban areas, both in the older eastern ports and in such newer cities as Chicago, St. Louis, Pittsburgh, and San Francisco. Between 1865 and 1875, 3.25 million new immigrants settled on the farms and in the cities of the North and West.[11] Industry, internal trade, exploitation of natural resources, and farming were the new sources of wealth. After suffering the indignities of an Atlantic passage in steerage, newly arrived immigrants plunged into American life without a second glance back at the sea.

And what opportunities awaited them! The 1870 census indicated that the per capita wealth in the North had doubled in the previous decade; during the war years the value of the Pennsylvania oil industry increased from $150,000 to $38 million. Kerosene refined from petroleum, "rock oil," quickly displaced the widely used illuminant whale oil, and what was left of the New England whaling fleet, after it had been ravaged by Southern commerce raiders, suddenly lost its major market. After 1865 coal production tripled and iron ore production in Minnesota increased ten times.

Tying it all together was the railroad. The major lines in the eastern half of the nation had been laid down before the Civil War and played a crucial role in the movement of men and equipment during military campaigns. But following the war railroad building became a true mania as the newly expanded nation wrapped itself in an increasingly dense network of rails. Track mileage doubled in the decade following the war. The government promoted this growth in the West with extensive grants of land for every mile of rail laid down and provided government credits as well. Bond financing, with local and state governments guaranteeing payment, funded equal portions of new track and public scandal. The leaders of the consolidated railroads established themselves as captains of industry, or as some preferred to call them, robber barons, titles that attest to their enormous public influence. Their names still carry the aura of great wealth and power: Vanderbilt, Gould, Huntington, and Stanford. The level of investment in railroads was far greater than anything ever seen in the merchant marine. And as the nineteenth century closed, the newly consolidated railroad corporations set the pattern for hierarchical, centralized management in America. The reach and power of the railroads posed a challenge to America's political traditions, a challenge that still reverberates in late-twentieth-century debates over antitrust policy.[12]

The railroad first crossed the Mississippi in 1854, an event that put it in direct competition for freight with riverboats throughout the Mississippi

River basin. Rail cities, such as Chicago and Kansas City, could now compete with river cities, such as St. Louis and New Orleans, as major transportation hubs. Fifteen years later the golden spike was driven at Promontory, Utah, joining the Union Pacific Railroad to the rail line reaching east from California. But the rate of building did not slacken as other regions of the West rushed to be connected by rail. James J. Hill headed the Great Northern Railroad. Chartered in 1864, it crossed Wisconsin, Minnesota, and much of what was then Indian territory to connect Lake Superior to Puget Sound. Hill's vision marched on ahead. He not only encouraged settlement of the lands along the route, but also sent agents to China and Japan to develop new markets for American goods. Then, having reached the limits of the continental frontier, he established a steamship line to connect his West Coast terminal with the Orient.[13] It was leaps such as Hill's that brought the oceanic frontier back into the American industrial and political consciousness in the closing decades of the nineteenth century.

"THE CAUSES OF THE REDUCTION OF AMERICAN TONNAGE"

While the American economy in the North and West surged forward after the Civil War, the merchant marine languished. Its failure to recover was unmistakable. When the war ended in 1865 the United States was carrying only 32.2 percent of its foreign trade, which was roughly half the percentage it had carried before the war. Five years later the situation had not improved. The wartime naval construction boom was also over and shipbuilders were desperate for orders. Congress was aware that if it did not take positive action the shipbuilding and shipping industries would continue to deteriorate, but then, as ever since, Congress found itself hamstrung by competing interest groups.[14] A new maritime policy was definitely needed, but need alone was not enough, and in the end nothing was done.

Different sectors of the maritime industry looked to Washington for different forms of support, but there was nothing novel in that—much of the day-to-day work of the federal government has always been reconciling and compromising competing claims made by constituents. What was unsettling about the way the government addressed maritime policy problems in the post–Civil War years was its irresolution, its failure to put in place policies capable of bringing order and prosperity to what had so recently been a vital industry. The maritime industry itself was clearly incapable of fashioning a coherent program of its own, one that could command the political support needed to succeed as legislation. The industry's friends in Congress were no more effective; each of them was beholden to interests that represented only one or two sectors of the industry and hence they could not secure passage of suitable legislation. The difficulties both groups faced can be easily understood, but since the failure of leadership they exhibited doomed the industry to continued decline, their inaction cannot be easily excused.

By the time the war was over the merchant ships requisitioned by the navy were worn out and none of them returned to commercial service. Nonetheless, the merchant marine could have been given a shot in the arm by allowing ships that had shifted to foreign registry during the war to return to the U.S. flag, but that was not to be. The economic realities of international commerce are often at odds with national sentiments, and the North, having watched as British-built Southern raiders destroyed its men and ships, was not about to forget that many had fled the battle and sought protection under the British flag. Shipowners who had reflagged their vessels were widely condemned as traitors. For years certain members of Congress railed at Great Britain for being a treacherous ally of the South and called the Northern ships that sailed under the British flag "Anglo-Confederate pirates" and the ships that England built for the South "English piratical vessels sailing under the Confederate flag."[15] Sentiments such as these led Congress to pass a law in 1866 prohibiting the return of U.S. ships that had sought protection under neutral flags during the war: "Be it enacted by the Senate and House of Representatives of the United States of America in Congress assembled, that no ship of America which has been recorded or registered as an American vessel, pursuant to law, and which shall have been licensed or otherwise authorized to sail under a foreign flag, and to have the protection of any foreign government during the existence of the rebellion, shall be deemed or registered as an American vessel."[16]

Having denied the industry the use of ships that had been forced to flee the dangers of war, Congress felt compelled to examine other options for supporting the merchant marine. A Select Committee of the House of Representatives was created in 1868 and charged "to inquire into and report . . . the causes of the great reduction of American tonnage engaged in the foreign carrying trade . . . [and to] report what measures are necessary to increase our ocean tonnage, revive our navigation interests and regain for our country the position it once held among the nations as a great maritime power."[17]

Congressman John Lynch of Maine, the state that led the nation in the construction of wooden sailing vessels, chaired this Select Committee—he also served as chairman of the House Committee on Commerce. Like the

Select Committee of the House of Commons that had investigated British shipping two decades earlier, the Lynch committee knew it was looking at an industry at the crossroads; but unlike the Commons committee, the Lynch committee was controlled by the American shipbuilders rather than by those who carried goods to and from foreign ports. It is therefore not surprising that the committee decided the laws governing the maritime industry did not need to be fundamentally changed. The Lynch committee's report, published in 1870, thus stands as one in a long line of committee and commission reports that, after presenting ample and often conflicting evidence about the maritime industry, attempted little and achieved less. The committee's report contains revealing historical evidence, but it did not in the end alter the policies under which this industry operated.

The Lynch committee did consider the possibility of allowing shipowners to purchase foreign-built ships and register them in the United States; this was known as the "free ship" policy. The requirement that a ship be U.S. built to obtain U.S. registry had not been a serious handicap to American operators before the Civil War, for at that time U.S. shipyards built the best and cheapest merchant ships available. But steady improvement in steam and iron ships had eliminated this competitive advantage, and after the Civil War foreign-built ships looked very attractive to U.S. operators. A Mr. Snow, speaking as a representative of the New York Ship Owners' Association, told the committee that U.S. shipowners could not compete with foreign owners if their competitors could buy ships more cheaply. What U.S. shipowners wanted, Snow said, was "a free navigation law, similar to the law passed in England some twenty years ago."[18] A Mr. Hinchen, another member of the New York association, discouraged the committee from thinking that U.S. shipyards could quickly become competitive as builders of steam-driven iron ships: "The first building of them (iron vessels) would be an experiment, and very few shipowners were willing to try that experiment." Furthermore, Hinchen noted, it would "take from four to five years to acquire the facility in building iron ships as cheap and as good as they were now built on the Clyde, and that period of four or five years might be fatal to the interest of commerce in this country."[19] These men spoke from experience; they could not afford to put their businesses at risk while U.S. shipbuilders learned how to fashion iron hulls and steam engines.

Many witnesses pointed out that most of the difficulties the committee lingered over arose from a single unnecessary assumption: that ships in U.S. registry must be U.S. built. Why, they asked, was America the only major nation that imposed this requirement? Captain John Codman, the leading advocate for uncoupling shipbuilding and ship operation, told the committee why he favored "free ships." Eighteen years later he was still insisting that coupling these two issues together was "the great mistake of all Congressional legislation, or rather Congressional obstruction, from the report of Mr. Lynch's Committee in 1870 down to the present day."[20]

Congressman Lynch, however, was not about to see shipbuilding uncoupled from ship operation; he focused his attention instead on finding out what the shipbuilders needed to become competitive in the new technologies. The proposals they introduced and the claims they made were not always well informed. High American tariffs, designed to protect inefficient domestic producers, certainly did make the cost of essential materials, such as iron and steel, relatively more expensive for U.S. shipbuilders. To offset this cost differential the builders proposed that they be given drawback payments to offset the tariffs paid on imported materials and the higher cost of domestically produced materials. But aside from this not unreasonable request, the shipbuilders' claims lacked plausibility. When asked how they would overcome the disadvantage of wage scales that were twice what British yards paid, they said higher American productivity would make up the difference. When asked how long it would take for Maine shipbuilders to learn to build iron ships competitively, they replied no more than two years. One builder, a Mr. Stimers, said, "[T]he change is very simple . . . it requires only a different arrangement of materials, which is very easily learned."[21] He also breezily asserted that "American iron was much better than English iron." If the members of the committee accepted these claims, it was only because they were predisposed to believe these shipbuilders from Maine.

Indirectly, unwillingly, and ultimately unsuccessfully, the committee searched for ways to ease the transition from the past to the present and the future. The past was wood and wind, and for the Maine shipwrights, who were beginning to build ever larger wooden schooners for the coastal trade, wood and wind were the present and future as well. But for Congress and the New York shipowners involved in foreign trade, the present and the future belonged to iron and steam. The committee was prepared to propose that regular subsidies patterned on the earlier mail subsidy program be provided for ships in liner trade, the understanding being that only steam-driven iron ships would qualify. The shipwrights countered with a proposal that bounties be paid to all vessel operators, whatever the type of ship or service provided. Theirs was a proposal to routinely subsidize an entire industry, even those parts of it employing a declining technology in a protected national market. The shipwrights' justification was not international competitiveness, but rather that hoary communitarian claim, still so often heard

in defense of outdated industrial policies, that the government has an obligation to protect and preserve traditional ways of life.

Shipbuilding and ship operating had indeed been communal enterprises as well as profit-making ventures in earlier periods of American history, but it is still something of a shock to encounter statements from the
second half of the nineteenth century insisting the federal government has
an obligation to use public funds to sustain traditional forms of communal
life in the face of technological change and industrial transformation. Yet
Lynch made sure the committee stood by his Maine constituents. The
committee's final report contains a glowing account of an earlier moral
economy whose superiority to the selfish financial interests of the capital-
ist is taken as obvious: "The shipping of the United States has never been
built, and only to a limited extent has it been owned, by capitalists, but by
men of moderate means; the lumberman, the mechanic, the shipmaster,
and the merchant, each having an interest in the production, the sailing,
and the freight of the ship, independent of the profits to be directly derived
from its earnings." The final report also dismissed as irrelevant all tramp
shipping and freight services that did not directly serve America's producers and consumers, again reflecting the provincial populism of the local
communities that built wooden ships for the protected coastal trade: "It is
estimated that more than sixty percent of the tonnage of the United States
engaged in the foreign carrying trade is engaged in freighting between the
ports of foreign countries, and many of these vessels never return after sailing from our ports. What advantage would it be to our country to have the
real or nominal ownership of this class of vessels, built, fitted, manned, and
provisioned for their voyage in a foreign port, touching only at the ports of
the United States to obtain an American register, and then pursuing their
voyage, perhaps never to return."[22]

However politically motivated these views may have been, they certainly were authentic, in the sense of representing strongly held local
beliefs in the nineteenth century. Could one build a rational and effective
national maritime policy upon such foundations? It seems unlikely, and the
Lynch committee certainly did not do so. To put the point charitably,
Lynch demonstrated that his dedication to American democracy was
greater than his concern for international trade. His priorities were politically reasonable, but they made for bad maritime policy. None of the final
report's other proposals could overcome the burden imposed by the committee's insistence that ships that had sailed under foreign flags during the
Civil War not be allowed to return to U.S. registry and that the requirement that all U.S.-registry ships be U.S. built be preserved. To save a tra-

ditional way of life, the committee was prepared to sacrifice the nation's entire foreign-trading maritime industry.

But as always, the issues involved were more complex than such simplifications suggest. The memory of the nation in arms was still fresh in the 1860s, and the Lynch committee did not ignore the connections between the maritime industry and naval power. They attempted to turn this linkage into a further justification for making the preservation of shipbuilding the dominant concern of federal maritime policy. The committee concluded that if the ships that reflagged during the Civil War were allowed to return or if the "free ship" position was adopted, these policies "would deprive us of the mechanical skill requisite to build our navy in time of war, or oblige us to maintain it at great expense connected with the government works in time of peace." Protection always has a purpose, but someone must always pay the bill as well. In this case a protected market for ships was preserved to ensure that the nation would have the capacity to build ships in time of war; the bill for this policy would be paid by shipowners and their customers. In fact, although this clearly was not the intention, the policy further accelerated the virtual elimination of U.S. shipping from all trade routes open to foreign competition.

Having locked itself into a policy founded on protecting American shipbuilding, the committee did what little it could to prop up the rest of the industry. It proposed that drawbacks be provided to reduce the cost of materials used in shipbuilding. Stores for vessels engaged in foreign trade were to be drawn out of bond, as was allowed by most European nations. Ships engaged in foreign trade were to be exempted from tonnage taxes imposed by the states, although most of these taxes had already been declared unconstitutional. A mail subsidy system for ships engaged in foreign trade was to be provided, the justification being that the government was already giving lavish land grants and monies to builders of western railroads and funding expensive improvements of inland waterways. The rest of the report was largely wishful thinking of the following sort: "The testimony taken by the committee is nearly unanimous that by offering to our citizens the same encouragement and protection as is afforded by other commercial nations to their citizens, our shipping can be built and lines of ocean steamers established as fast as the requirements of business demand, and that there would be a present demand for ships if they could be cheaply supplied." The higher cost of American labor would be offset by the low cost of lumber, "a large amount of which is used even in the construction of iron vessels," and by "greater efficiency and skill." The committee, knowing full well what it would recommend before it began collecting evidence, proceeded unerringly to its predetermined target.

The legislative proposals the committee drew up never had a chance. Unlike the railroads and inland waterways industry, the oceangoing maritime industry was largely unorganized and politically ineffective in the nineteenth century. After World War II, when big shipping companies and big labor unions spoke for the industry, its interests were more powerfully represented in Washington, but before that the industry played a minor role in the formation of national policy. The centerpiece of the legislation Lynch proposed was a system of tariff drawbacks for shipbuilding materials that were imported and an equivalent bounty of fifteen dollars per ton for ships build with American iron. The prospect of protecting the iron industry with high tariff walls and then subsidizing it when its products were used to build ships was a bit too much for Lynch's congressional colleagues; their opposition was seconded by the disappointed advocates of "free ships." Lynch's bill died in the House after failing to secure enough votes to qualify for a third reading. His committee had thoroughly investigated and described "the causes of the reduction of American tonnage," but they had failed to provide a workable remedy.[23]

Following the Civil War the U.S. merchant marine settled into a long decline from its golden age. This decline was not inevitable, but it could only have been avoided by boldly seizing new ways of performing old tasks as technology changed. Yet while Great Britain and then Germany showed the world how to build modern freighters and organize large steamship companies, the United States clung to traditional modes of operation and its protected markets. And indeed it seemed that America could well afford to do so because it was specially blessed. Americans had a continent to settle and could afford to turn their backs on the strenuous and often lonely effort required to gain and hold a place in international shipping. Blissful images of "amber waves of grain" replaced the venturesome hardships of seafaring. Although America had been settled and had gained its independence as a maritime nation, the coming of the railroad turned its eyes westward and committed the nation to becoming a continental power. Given the magnitude of the opportunities that beckoned on the western frontier, it can hardly be surprising that few mourned the maritime opportunities that were being abandoned.

5

MARITIME DECLINE, NAVAL REVIVAL, 1880–1914

The gradual decline of the U.S. merchant marine engaged in foreign trade, a decline which had started before the Civil War, was dramatically accelerated by the war itself and the policies adopted at its conclusion. American shipowners, forced to compete for cargoes in a political climate that favored protection rather than competition, found themselves operating at a distinct disadvantage in the more profitable foreign trades. They did not simply walk off the field. As the United States was being transformed by rapid industrialization and the creation of a national economy, the shipowners searched for ways to overcome the handicaps that federal maritime policy imposed on them and to maintain an American presence in world trade. But none of the moves available to them could offset the advantages enjoyed by their British and European competitors, and the percentage of American trade carried in U.S.-flag vessels declined steadily.

In the final decades of the nineteenth century the United States began to realize that it had become a major power in the global economy and had to develop the diplomatic and military instruments needed to protect its interests from the other Great Powers. The U.S. Navy, but not the merchant marine, was therefore reinvigorated with new ships, new doctrines, and new manpower. While the navy had long been committed to protecting American overseas commerce wherever it was conducted, in the early years of the twentieth century it responded to A. T. Mahan's new analysis of seapower by concentrating its major ships into a single Atlantic fleet. This was an appropriate strategy for a nation reaching for naval preeminence; but it provided little protection for the commerce of a country like the United States, which traded worldwide yet had a second-rate navy.

The inadequacies of America's naval and commercial situation were thrown into sharp relief when Europe went to war in 1914. The hidden costs of consigning the nation's foreign trade to foreign shipping suddenly had to be paid, and as the war dragged on, it became obvious that even if the United States were able to avoid direct involvement, it could not continue to act as a Great Power unless it had increased power at its disposal. Thus in 1916 three major "readiness" bills were passed, one each for the army, the navy, and the merchant marine. Under the press of war the federal government stepped in to reverse the long decline of the American merchant marine. The basic policies it adopted would continue to sustain the U.S. merchant marine engaged in foreign trade, to the extent that it was sustained, until the end of the twentieth century.

SUBSIDIES AND PROTECTION

The Lynch committee's failure to reform and invigorate federal maritime policy cast a pall over the remainder of the nineteenth century. The "free ship" campaign was now politically dead, and whenever Congress took up maritime issues, it assumed its primary task was reviving American shipbuilding. This state of affairs left ship operators engaged in foreign commerce with two choices, other than simply abandoning the business. They could either use the institutional options provided by international capitalism and register their ships under foreign flags or they could look to the government for subsidies that would offset the cost of purchasing and operating American-built ships under the U.S. flag. Both options were pursued.

In the nineteenth century there were no national registries designed and maintained primarily to service foreign-owned ships; that kind of registry, now known as a flag of convenience, first appeared in the twentieth century. Nonetheless, ship operators engaged in foreign trade could avoid economically disadvantageous registry requirements by establishing a company in a country that did not have punitive registry provisions and then operating their ships under that country's flag. Several American corporations, including the Chesapeake and Ohio Railroad, W. R. Grace and Company, the United Fruit Company, and Anglo-American Oil Company, did this in the nineteenth century. By 1901 there were 670,000 tons of such U.S.-owned, foreign-flagged ships engaged in commerce. Taken together, they constituted by far the largest part of the American-owned tonnage engaged in foreign trade.[1]

Because Congress was determined to defend the requirement that U.S. ships be U.S. built, it had little maneuvering room on issues of maritime policy. It could, of course, have acted directly by appropriating funds to build a national merchant fleet to be operated either by a federal agency or under contract by commercial shipping companies. This was the solution Woodrow Wilson embraced in 1916, but he did so only when forced to by a national crisis more severe than any since the Civil War. In the latter half of the nineteenth century the federal government had no reason, and little inclination, to engage directly in industrial enterprise. A less overt approach would have to be found.

An indirect way of getting at the problem was to create a demand for American-built ships; the easiest way to do this was to subsidize operators, so they could afford to order and operate domestically built ships. But this proposal raised a host of administrative problems and called for carefully crafted legislation. A subsidy program designed to promote the building and operation of U.S.-flag ships would have to meet at least three distinct challenges: (1) it would have to provide enough money to ensure that operators would in fact order and operate U.S.-built ships under existing registry requirements but not provide so much support that the shipowners would no longer feel compelled to operate their ships competitively; (2) it would have to be designed to promote technological progress, industrial development, and cost consciousness among shipbuilders; and (3) administrative responsibility for the program would have to be assigned to a government agency capable of ensuring that the funds allocated to support the maritime industry were properly and effectively utilized. Failure to meet any of these challenges would inevitably result in the subsidy program being characterized as a waste of the taxpayers' money and would assure its eventual termination. These daunting challenges called for considerable political and administrative sophistication; unfortunately, the historical record reveals that these qualities were in very short supply.

Congress, never eager to invent when it can copy, modeled its post–Civil War maritime subsidy program on the earlier system of mail subsidies (see chapter 3). The mail subsidy system offered two distinct legislative advantages. On the one hand it utilized an existing governmental agency, the Office of the Postmaster General, as the conduit for getting appropriated funds to the targeted industry. This convenience enabled Congress to avoid creating a new office or commission, which would have enlarged the federal bureaucracy. The other advantage was that the primary criteria for efficient mail service, speed and reliability, were precisely the advantages of steam-powered over sail-powered shipping. This fortuitous alignment of interests provided Congress with two forms of cover. First, it could plausibly claim that it was providing shipping subsidies to ensure there would be speedy and reliable mail service on the subsidized routes, a claim that would convince taxpayers that they were receiving something of value for their money. One may wonder, however, if this claim could have survived a rigorous cost-benefit analysis. Second, linking payments to improved mail service enabled Congress to create a "technology forcing" subsidy system that linked the rate at which funds were allocated to the size and speed of the ships deployed on the subsidized routes. By insisting on this feature of the program, Congress was able to avoid more comprehensive demands for a straightforward subsidy-per-ton bounty system for all American-flagged vessels, whether sail or steam and whether engaged in foreign or coastal trade. It should also be noted that ships built to operate with mail subsidies were supposed to be for rapid conversion into naval auxiliaries in time of emergency. At the outset, therefore, the mail subsidy system looked quite appealing.

But the mail subsidy system had serious shortcomings as well. The most threatening were (1) that it essentially had to rely on the operators to determine how great a subsidy was needed to make operation on a given route possible and (2) that the federal office charged with administering the program lacked the resources and expertise needed to ensure that it was properly run. In the absence of informed and diligent administrative oversight, procedural failure and abuse were bound to occur, and they most certainly did. Although the primary fault lay in the design of the legislation, the blame, unfortunately but predictably, was most easily (and in some cases justifiably) assigned to the operators who received the subsidies. In the populist democracy of nineteenth-century America, the greedy businessman was a very useful villain.

Even before the Civil War had ended Congress had agreed to pay up to $150,000 per year to subsidize the cost of a monthly sailing to Brazil; this, together with a contribution from the Brazilian government, was enough to make the service viable. A year later, in 1865, Congress authorized a subsidy of $500,000 to the Pacific Mail Company to provide a monthly service to the Sandwich (Hawaiian) Islands and the Far East. Before the service was inaugurated, however, the company had its contract amended, but without any reduction in subsidy, to eliminate its obligation to have its ships call at Hawaii.[2] Since Congress was not in a position to challenge the cost estimates of companies that bid for mail subsidy contracts, the pattern of low-bid entry and subsequent cost escalation became commonplace. Once it had its initial contract, the Pacific Mail Company lobbied intensely in Washington and obtained the backing of President Grant and his secretary of the navy. A few years later the company argued that to meet growing British and Japanese competition, its subsidy should be doubled, so that its ships could sail twice rather than once a month. In June 1872, after a bitter struggle, Congress voted the funds needed for the expanded service.

The victory proved to be Pyrrhic, however, for the political vulnerability of the mail subsidy system turned out to be too great to defend. In 1872, an election year, the Republican Party's platform declared that "it is the duty of the general government to adopt such measures as may tend to encourage and restore American commerce and shipbuilding."[3] While the party, which championed high tariffs, went on record as wanting to aid the merchant marine, it had little success in securing additional funds for subsidy.

Two years later scandal erupted. A congressional Investigating Committee showed that the Pacific Mail Company had spent some $890,000 when lobbying for its increased subsidy. Most of this had been used to inflate the price of its stock, so that certain stockholders, including some influential members of Congress, could make a killing. The remaining funds, some $335,000, could not be traced. No evidence of direct bribes was ever presented, but the political damage was devastating.[4] Congress refused to fund the Pacific Mail Company's increased subsidy and would not renew the contract when it expired. The subsidy for service to Brazil was also terminated when its contract expired, although no wrongdoing had been alleged. Congress may have been eager to help the foreign-going merchant marine, but they were not about to pay the political costs associated with the mail subsidy system.

In the absence of purposeful and effective federal assistance, the merchant marine engaged in foreign trade continued to decline. British-built steamers had captured most of the higher-value trade, where speed and reliability were important; but much of the world's low-value bulk trade still moved in sailing vessels, which had the advantage of not having to buy or carry coal. Since most of the American-flag ships engaged in foreign trade were still made of wood and powered by wind, they were reduced to competing for marginal commodity cargoes. In the late nineteenth century the U.S. flag could still be seen on ships in California ports, loading grain for Europe; on the River Plate in South America, discharging cargoes of lumber from the American Northwest; and in the islands off the coast of Peru, loading guano for use as fertilizer on America's farms. Hauling such low-profit materials was a miserable finale to a once proud national tradition of seafaring.

In 1890 Congress again addressed the decline of the merchant marine, and again it set out to subsidize the high cost of building and operating U.S.-flag ships. A proposal was put forward to provide bounty payments, based solely on the size of the vessel, for ships engaged in foreign trade, a scheme favored by the builders and operators of wooden sailing ships. A bill to this effect was passed by the Senate, but it died in the House. The other approach, which eventually passed both houses of Congress, called for reviving the system of providing specific subsidies for specified services. William P. Frye, a senator from Maine, was the champion of this approach and the sponsor of the Ocean Mail Act of 1891. This bill defined four levels of payment according to the size and speed of the vessel. The highest level of subsidy was reserved for iron or steel screw-propellered ships of eight thousand tons or more capable of a speed of twenty knots. Declining levels of subsidy were offered down to the fourth class, which required that the ship be steam-driven, although it could be built of iron or wood, and that it be able to maintain a speed of twelve knots.[5]

The new legislation did result in significant new operating contracts and shipbuilding. The Pacific Mail Company regained its subsidy for the Far East and obtained new subsidies for services from New York to the port of Colon on the east coast of the Panamanian Isthmus and for a connecting service from Panama's west coast to San Francisco. Three newcomers also received contracts: the Red D Line, which many years later was purchased by the Grace Line, for service between New York and Venezuela; the New York and Cuba Mail Line, for service between New York, Havana, and Vera Cruz, Mexico; and the Oceanic Steamship Line, which in 1900 provided sailings to Hawaii and Australia from San Francisco and later became part of Matson Navigation. These initiatives succeeded in getting a handful of modern American ships on a few of the world's trade routes. They could only continue to sail, however, so long as Congress continued to subsidize them. The combination of restrictive registry requirements and federal subsidies thus kept the most modern part of the American foreign-going merchant marine hostage to political will rather than economic performance.

Intense competition on the North Atlantic generated unusual problems that required some adjustments of the mail subsidy system. In 1886 the American-owned International Navigation Company bought the British Inman Line, which operated mail steamers between Liverpool and New York. The British government had been providing the Inman Line with a mail subsidy, and the new owners expected this subsidy to continue. Immediately after buying the Inman Line they ordered two fast new liners from British builders, but before the new vessels were delivered, the Inman Line's subsidy contract expired and the British government, which was also subsidizing two other British lines on the North Atlantic run, declined to renew it. This turn of events left the American owners of the two new ships nearing completion on the Clyde in a highly exposed position. They appealed to Congress and proposed a trade-off. If they were allowed to register the two British-built ships under the American flag, they would order two sister ships to be built in a U.S. yard and then operate all four ships under a U.S. subsidy contract. Congress agreed and passed a special act in 1892 providing the necessary funds.[6] Nothing could have more dramatically demonstrated the extent to which the fate of the U.S. merchant marine depended on congressional action or how readily registry requirements could be altered by the stroke of a pen when authorized by Congress.

The two new American-built ships, *St. Louis* and *St. Paul*, were constructed in the William Cramp shipyard in Philadelphia; they were the largest and finest steel-hulled steamships built in the country up to the time. President and Mrs. Cleveland attended the launching, with Mrs. Cleveland serving as sponsor. Their presence highlighted the national significance of the event. International Navigation's four-ship fleet operated successfully for a number of years and provided further evidence that the Postal Aid Law of 1891 did provide a much needed shot in the arm for the American merchant marine.

Subsidies helped, but they had the distinct political disadvantage of being highly visible and hence vulnerable, primarily because the funds for subsidies had to be appropriated annually. It was far better, from the shipbuilders' and operators' point of view, to operate under legislation that protected U.S.-flag shipping from competition rather than to use tax revenues to offset high U.S. costs in open markets. The restriction of coastal trade to U.S.-owned, U.S.-built, and U.S.-crewed vessels had distinct political advantages. Although it raised the cost of transportation over what it would have been had the trade not been protected, it did not require the government to appropriate funds to reimburse private companies. The higher costs were paid instead by those who used the transportation services and ultimately, of course, by the producers and consumers of the goods moved, while the government remained uninvolved on the sidelines. Thus, so long as U.S. registry rules required U.S. construction of U.S.-flagged ships, calls for protectionist laws that would reserve cargoes for U.S. vessels continued to be heard.

The traditional and least offensive method of providing trade protection was to impose a discriminating tariff, a tariff that U.S. ships did not have to pay. At the very end of the nineteenth century Senator Stephen B. Elkins resuscitated this idea when he proposed that a new 10 percent ad valorem tariff on all "goods, wares and merchandise imported into the United States" be imposed unless, of course, those goods were carried on U.S.-flag ships.[7] When arguing in support of his bill, Elkins lamented the condition of the merchant marine in an America that was otherwise booming. The fault, he said, lay in the policy of reciprocity that began in 1828, when the United States passed a law authorizing treaties that prohibit discriminating tariffs. Ignoring the registry restrictions that made U.S. shipping uncompetitive, Elkins called for greater protection of American shipping from free-market competition. When told his proposed law would abrogate existing maritime treaties, Elkins declared that his tariff was preferable "to continuing a policy of maritime reciprocity."[8] He had no patience with the free-ship argument; many industries received tariff protection against foreign competitors, he noted, why not the shipbuilders? He also pointed out that the system of mail subsidies had not halted the industry's long decline. Elkins's tariff proposal was designed to capture increased cargo tonnage for U.S. ships. His bill was so badly flawed, however, that it was never reported out of committee. The cargo protectionist principle he articulated reappeared repeatedly in subsequent legislative proposals, however, and in the twentieth century became a mainstay in the web of legislation that kept a reduced U.S. maritime industry afloat.

Elkins was correct when he noted that the 1891 Postal Aid Law did not halt the slide of the U.S. merchant marine. By 1905, when Great Britain had 17 million tons of shipping engaged in foreign trade and Germany had 3.5 million tons, America had less than 1 million foreign-going tons. France and Italy had more; the United States was in fifth place worldwide. The composition of America's fleet was even more troubling than its relative size. Great Britain had a fleet of 11,365 ships engaged in foreign trade, with less than 20 percent being under sail. Of America's 1,333 ships in foreign trade, 965, or 73 percent, were sail powered. Furthermore, these declines in America's relative standing occurred at a time when its foreign trade was booming—foreign trade increased fivefold in the four decades following the Civil War.[9] By the turn of the century U.S.-flag ships were carrying only 8 percent of the nation's exports and imports. It appeared that, so long as America's registry laws remained unchanged, there would be no end to the long decline of its merchant marine.

COASTAL SHIPPING

Although America's merchant fleet engaged in foreign trade had practically disappeared by 1914, its coastal fleet, which was protected from foreign competition by some of the nation's oldest laws, steadily increased in size. The United States had the good fortune to be endowed with some five thousand miles of ocean coastline along its Atlantic, Pacific, and Gulf of Mexico shores, an extensive freshwater coast along the Great Lakes, numerous and widely distributed natural harbors near major trade routes, and mighty navigable rivers that penetrated deep into the nation's heartland. These natural avenues of communication were especially valuable before the completion of the national web of railroads, and they continued to provide important commercial links well into the twentieth century. As the nation grew and its economy became more interconnected, coastal traffic boomed.

There were only 68,000 tons of shipping in the coastal fleet in 1789, when the first discriminating duties were enacted to favor domestic shipping. Five years later the fleet tonnage had doubled, and by the beginning of the nineteenth century it had grown to 245,000 tons. By 1818 it exceeded 500,000 tons and was roughly as large as the offshore fleet.[10] By 1880 there were approximately 2,638,000 tons of shipping employed in the various coastal trades.[11] After the United States had acquired Alaska, Hawaii, and several former Spanish colonies, its practice of protecting its coastal trade was extended offshore and began to look quite similar to the older policy of mercantilism, which the traditional colonial powers had used to keep interlopers from trading with their overseas colonies. By the beginning of the twentieth century America's protective shipping legislation was more restrictive than that of any other industrialized nation; by 1910 its coastal fleet totaled nearly 6,700,000 tons. At that point its international fleet was barely one-tenth as large.

From the sailor's point of view, working a coast is a very different undertaking than crossing an ocean. Coastal vessels must be handy, capable of sailing close to the wind and maneuvering rapidly and frequently. They anchor and dock more often and must always be alert to tides, currents, and threatening shorelines. Unlike the great clippers and other full-rigged ships, coastal vessels cannot make their way to the trade winds and then run for hundreds or even thousands of miles before steady breezes. American shipbuilders responded to the special needs of the coastal trade by constructing a distinct type of vessel, the two-masted schooner. Its fore-and-aft sail plan, as opposed to the square-rigged sails of the packets and clippers, enabled it to sail closer to the wind, tack more easily, and operate with smaller crews (schooner sailors did not normally have to be sent aloft to make or reduce sail). For decades these inexpensive and sturdy ships carried cargoes of cotton, sugar, and yellow-pine lumber northward along the Gulf of Mexico and Atlantic coasts; on their return voyages they carried manufactured goods to the South and, in the latter half of the nineteenth century, ice from Maine and anthracite coal from Pennsylvania. Similar schooners worked the Northwest coast, carrying lumber from Washington and Oregon down to the booming cities on the California coast.

During the second half of the nineteenth century schooner design evolved in two distinctive ways. American coastal schooners, often with sharply raked masts, became famous for their speed and were widely used in situations where speed mattered most, such as yacht racing and smuggling. More significant for the merchant marine was the enormous expansion in the size of schooners as operators in the coastal trades struggled to hold their share of the market in the age of steam, iron, and railroads. In the 1870s the length and tonnage of wooden schooners was increased considerably, and many were equipped with donkey boilers and steam winches, so that their small crews could handle the additional sails and heavier anchors that were installed. As the ships got longer, the number of masts was increased: in 1880 the first four-masted schooner was built in Bath, Maine; ten years later five-masted schooners were being built. Between 1900 and 1910 huge six-masted schooners of more than three thousand tons were produced, but by then this line of development was pressing up against its structural and economical limits. A few large steel-hulled schooners and one-seven masted monster were built, but the end was not far off. As railroads reached more and more coastal cities and as coastal steamers took over the passenger and high-value trades, the coastal schooners were left working low-value bulk cargoes, such as lumber, brick, stone, and coal. By 1914 they also had to compete with coastal barges towed by steam-powered tugs, and many of the schooners with sound hulls were themselves unceremoniously dismasted and converted into barges. A number of them remained in this service well into the 1930s.

The opening of the Panama Canal in 1914 stimulated a burst of modernization in coastal shipping, including new construction and expanded liner service. The American-Hawaiian Steamship Company, which provided service between the U.S. East Coast and San Francisco and Honolulu, flourished as travel through the canal increased, and another important intercoastal line was started by Lewis Luckenbach. Both lines left the intercoastal trade during World War I, when there was more money to be made in the Atlantic trade, but both returned in 1920. They were able to hold their own against increasing rail competition in the coast-to-coast trade until the beginning of World War II, when the government chartered all their ships.[12]

A New Century and a Nation on the Move

During the final decades of the nineteenth century there were in fact two maritime scenarios being played out in America. One, discussed above, was essentially industrial and civilian, and despite numerous attempts to recapture the nation's former glory as an international maritime power, decline continued. The other scenario was military and diplomatic, with the United States being increasingly drawn into the global imperial competition that was then being carried on by the Great Powers. This involvement in Great Power confrontations was inevitable given America's determination to defend what it considered its special sphere of interest in the Western Hemisphere and its access to world markets. While many longed for the simpler days when Americans could prosper in isolation behind the nation's oceanic boundaries and tariff walls, few were prepared to accept indefinitely the economic limitations such withdrawal from world markets would impose. And so, while its merchant marine continued to languish, America's navy began to grow.

In the nineteenth century, ideas about the nature of war and its use as an instrument of state policy were still largely traditional. Although the American Civil War had given the world a glimpse of how wars would be waged in the age of industry and what their human costs would be, the horrors that accompanied submarine, gas, and air warfare, not to mention the use of missiles and nuclear weapons, still lay far in the future. In the nineteenth century both diplomatic negotiation and military intervention were considered legitimate means on a continuum of power, instruments that could be used as necessary to achieve national aims. One of the aims of nations, so long as there have been nation-states, has been to increase their aggregate wealth. For nineteenth-century strategists, therefore, defining the connections between the wellsprings of national wealth on the one hand and the uses of military force on the other was a problem of central importance. And among those who puzzled over the relation between the wealth that flows from maritime trade and the possession and use of naval power, no one was more penetrating or influential that the American naval officer, historian, and strategist Alfred T. Mahan.

Mahan's best known work, *The Influence of Seapower on History (1660–1783)*, was published in 1890 after being delivered as a series of lectures at the U.S. Naval War College in Newport, Rhode Island. His argument, while far from simple, is fairly straightforward. Maritime trade arises naturally from the human inclination to engage in exchange, and trade creates wealth. The seas are a "great common" that lies beyond the reach of the police powers established to protect life and property in civilized states; nations therefore create armed navies to protect their merchant shipping from attack and seizure at sea. In a world of competing maritime empires, the nation with the most powerful battle fleet will be the greatest power. In short, according to Mahan, seapower has an enormous and seldom acknowledged influence on history. Mahan developed this argument in a way that contemporary strategists found utterly persuasive. He reviewed the maritime history of the British Empire and presented his conclusions as the inescapable lessons of the historical development of the world's leading seapower. The linkage he "discovered" between the growth of imperial wealth and the establishment of naval supremacy was widely seen as a strategic conclusion that compelled assent and pointed the way forward for Great Britain's would-be rivals. After Mahan, anyone who thought about geopolitics had to come to terms with his account of the influence of seapower.

The significance of Mahan's doctrine for the United States was far from clear, however, for America was a non-imperial continental power. In practice, however, Mahan provided the central text for the burgeoning campaign to build a stronger navy. At the time the book appeared William C. Whitney, the secretary of the navy, complained of the deplorable state of the U.S. Navy, saying it did not have "the strength to fight nor the speed to run away."[13] Mahan was a prolific and persuasive author and his writings were used to convince those in power that they should strengthen the navy. The United States was becoming a Great Power and it had to develop the means needed to protect its interests and project its power. Exercising maritime power required overseas bases to serve as coaling stations and provide other essential logistic support. Mahan pointed this out in 1893, in an article commenting on U.S. interests in the Hawaiian Islands, when he said it was "imperative to take possession, when it can righteously be done, of such maritime positions as contribute to secure command."[14] Righteousness evidently lay in the eye of the taker, but since Americans were beginning to discover their "manifest destiny," he did not agonize unduly over justifications for action. Sailing under the Mahanian banner, and motivated by a complex amalgam of interests and beliefs, America soon made a late but unmistakable entry into the global imperial sweepstakes that occupied the Great Powers at the end of the nineteenth century.

America cut its imperial teeth in the Caribbean and Latin America, which it considered within its "natural" sphere of influence. In 1895 a dispute arose between Venezuela and Great Britain over the boundary of British Guiana, and the United States offered to arbitrate. When Great Britain indicated reluctance to accept the offer, America threatened war. The incident was defused by Great Britain's refusal to accept the challenge, but the larger message was clear. The United States was prepared to act belligerently to exclude foreign nations from a hemisphere over which it felt it had the right to exercise a degree of hegemony. The Monroe Doctrine, which had been in place for seventy years, was being reasserted with new force and urgency. The United States, having chosen its ground, was evidently eager to demonstrate its resolve.

An opportunity soon arose in Cuba, where the Spanish government was vigorously suppressing a long-simmering revolt. The American press was publishing increasingly sensationalized accounts of Spanish atrocities, sugar plantations were being destroyed, and American investments were being put at risk. Sentiment for intervention was already on the rise when, in February 1898, the battleship *Maine* exploded and sank in Havana Harbor. A Naval Board of Inquiry concluded the cause was an underwater mine, which they presumed but did not assert that the Spaniards had planted, and "Remember the Maine" became the battle cry of those seeking retribution. President McKinley soon demanded that the Spanish government agree to an immediate armistice, American mediation, and independence for Cuba. Only the last demand was resisted; yet on April 1, 1898, the president sent Congress a message that was an invitation to war with Spain. Twenty-three days later Congress passed a joint resolution accepting the invitation.

Although the American army was unorganized, lacked equipment, and was hopelessly unprepared for war, the navy appeared to be ready for the coming engagement. A fleet of battleships, cruisers, and gunboats had been built and was standing by, yet the navy had almost none of the auxiliary ships needed to support operations at a distance from U.S. shores. Thus the navy entered the first American war to be fought overseas since 1812 without the ships needed to sustain a modern battle fleet in foreign waters.

The first significant encounter occurred in the Philippines at the end of April. Under cover of darkness Commodore George Dewey's small Asiatic squadron, accompanied by two recently purchased British transports loaded with coal and supplies, entered Manila Bay, where an ill-prepared Spanish squadron lay at anchor under the protection of the shore batteries. In the one-sided battle that took place the next morning, Dewey issued his famous order to the commanding officer of his flagship—"You may fire when ready, Gridley"—and the harbor fortifications quickly surrendered after the Spanish ships had been destroyed. It was only several months later that American troops arrived and forced the surrender of Manila itself. They had been immobilized until emergency legislation was passed authorizing the transfer of recently purchased foreign-built transports and supply ships to American registry. When these support vessels were available, the army moved in to occupy the archipelago.

While the Philippine Islands were being subdued, the navy was busily chartering foreign colliers to fuel its fleet in the Caribbean. The four liners of the American Navigation Line were also obtained and armed, and three of them were sent out to scout for the small Spanish force of cruisers and destroyers that was hurrying from Spain to the Caribbean. These ships were eventually found at Santiago, on Cuba's southern coast, and at the end of May the U.S. Navy began a blockade designed to keep them out of action. By the end of this brief war the navy had mobilized twelve ships from the merchant marine, nearly all that were suitable, for service as naval auxiliaries. The army meanwhile was chartering roughly thirty coastal vessels to transport troops from Tampa, Florida, to a landing beach near Santiago.[15] Rear Admiral William T. Sampson meanwhile did his best to keep his squadron at sea off Cuba. His first step, on June 7, was to seize and hold Guantanamo Bay, a spacious harbor forty miles east of Santiago, in which he could refuel his coal-hungry combatant ships. Blockaded by sea and besieged by land, the Spanish squadron in Santiago attempted to escape on July 3. The American forces intercepted them and all six Spanish ships perished. The cost to the U.S. forces was one fatality. This essentially ended the war, and a preliminary peace treaty was signed in Washington on August 12, 1898. Short as the war had been, it had demonstrated that the navy's lack of suitable auxiliary ships could seriously restrict the use of its battle fleet.

Other offshore islands that became U.S. territories and part of its cabotage trade area were also added at this time. While the Spanish-American War was still in progress, Congress ratified the annexation of Hawaii after the native monarch had been deposed by an opportunistic American-led coup. Both Great Britain and Japan had coveted this mid-Pacific chain of islands, and the American annexation of Hawaii demonstrated that the United States, like the other Great Powers, was prepared to seize territories it considered important to its trade and defense. The United States also occupied the formerly Spanish islands of Puerto Rico and Guam shortly before the peace talks began. In time all these territories, as well as Alaska, were included in the protected "coastal" trade of the United States. An attempt in 1908 to include the Philippines failed, however, because the sugar interests were against it. While the sugar barons had worked hard promoting American intervention in Cuba to protect their investments there, they did not wish to see the cost of bringing Philippine sugar to the United States increased by requiring that it be carried in U.S.-flag ships. Once again the shape of empire was determined more by the play of commercial interests than by the logic of geography or the law.

The Spanish-American War demonstrated how ill-prepared the navy was to sustain military action offshore; a year later the Boer War between Britain and the Dutch settlers in South Africa demonstrated how vulnerable the United States was to disruption of the shipping services provided by other nations and how much it depended upon those services. As Britain prepared to dispatch and sustain an army to a front that was six thousand miles away, it pressed into national service many of the ships that normally carried American commerce. The sudden reduction in available tonnage drove freight rates through the roof and seriously disrupted the North Atlantic trade, forcing Americans to acknowledge that their economy was not nearly as self-sufficient as they imagined. A congressional committee later complained that the inflated freight rates that American exporters had been forced to pay to those British ships that remained in the Atlantic trade had offset Britain's cost of waging this war.[16]

The plight of the merchant marine was well publicized during the early years of the twentieth century, but nothing of consequence was done to alleviate it. Winthrop L. Marvin, in his 1902 book, *The American Merchant Marine, Its History and Romance,* gave a well-informed account of its problems, and in that same year President Theodore Roosevelt called for an expert commission to investigate the industry and suggest what might be done. The commission that resulted was chaired by Senator Jacob Gallinger of New Hampshire and took testimony from three hundred witnesses during the latter half of 1904. Once again extended hearings produced a wealth of evidence, some of which was revealing, some of which was uninformed special pleading. Although recommendations were forthcoming, no action was taken.

The Gallinger Commission determined what was already widely known, that American shipbuilding costs and crew wages were higher than those of their foreign competitors. Nonetheless, the members of the commission were not about to accept the mandate of international economics and suggest that U.S. costs be reduced to world rates. They therefore searched diligently for a noneconomic explanation of low foreign costs. This search led them to foreign subsidies, and although they produced no evidence to support their claims, they insisted that such subsidies were a major reason for the decline of the U.S. merchant marine. It was true, and prominently noted, that Great Britain did subsidize some of its passenger ships, but the fact that Britain's few subsidized liners constituted no more than 5 percent of its fleet was conveniently ignored. The bulk of Britain's fleet, like the fleets of the Scandinavian countries and other competitors, was made up of tramps and cargo ships that received no government support. France was an exception, but the bounty system it adopted, which provided large payments to its merchant fleet, was widely acknowledged to be a miserable failure.[17] Not surprisingly, this analysis of the problem led the commission to propose again that U.S. subsidies be increased. In 1905 it drafted a bill designed to "promote the national defense" by creating naval reserve forces and to expand the merchant marine by establishing new shipping lines supported by mail contracts. Although President Theodore Roosevelt endorsed the bill, it aroused all the old passions and ultimately failed in Congress. The similar bills that Senator Gallinger continued to sponsor up until 1911 all failed as well.

Stalemate and drift ensued. Congress was gridlocked on maritime policy, with the advocates of free ships effectively blocked by defenders of high tariffs and protection. We can see in retrospect that the one enduring piece of legislation that did emerge from congress was a harbinger of the merchant marine's increasing dependence on national security arguments. The Military Transportation Act, passed in 1904, stipulated that all U.S. military cargoes must be carried in American-flag ships. Interestingly, however, the act did not require that these ships be American built, thereby significantly reducing the cost of shipping to the military services. It is both sobering and amusing to think of the disproportion between the events that led to the stipulation that military cargoes be carried in U.S.-flag ships, chiefly the experience of the Spanish-American War, and the huge volume of cargo that was directed to U.S. ships by this act during the remainder of the twentieth century.

The absence of an adequate naval auxiliary was again made plain when President Roosevelt sent "the Great White Fleet" around the world between 1907 and 1909. Intended to demonstrate America's naval might and its ability to deploy its Atlantic fleet to the western Pacific, the venture almost turned into a fiasco when it was realized that the navy was unable to support its fleet so far from its home base. The day was saved by hastily purchasing and chartering a ragtag collection of foreign tramps to either trail after the fleet or meet it along the way. It seemed that no matter how often the need for adequate logistic support was demonstrated, the navy simply did not want to learn the lesson.

In 1912, as construction of the Panama Canal neared completion, Congress seized the opportunity to benefit the merchant marine by passing the Panama Canal Act. The canal opened in 1914, as World War I was beginning, and although it was heavily used by many naval and commercial vessels during the war, construction work was not finally completed until 1920. One of the two great interoceanic canals begun in the nineteenth century, the Suez Canal being the other, it had immense strategic and commercial importance. The canal reduced the distance from New York to San Francisco by more than half and opened trade between East

Coast ports and the west coast of South America and the Orient. Not surprisingly, the railroads had strenuously opposed the construction of the canal, for it introduced competition in the coast-to-coast freight business. Thus one purpose of the Panama Canal Act of 1912 was to amend the Interstate Commerce Act of 1887 so as to specifically prohibit collusion by the railroads, which might engage in selective rate-cutting to limit competition by ocean carriers using the new canal. Congress also tried to help U.S. ocean carriers by stipulating that American intercoastal vessels could use the canal toll free. This provision clearly violated a 1901 treaty with Great Britain prohibiting any discrimination in canal tolls, but President Taft signed the bill into law nonetheless. His successor, Woodrow Wilson, was more responsive to foreign protests, however, and persuaded Congress to repeal this part of the act.

ATTEMPTS AT CORPORATE CONSOLIDATION IN THE MERCHANT MARINE

In the merchant marine, as in other industries, the use of new materials in the latter half of the nineteenth century, especially iron and steel, and new sources of energy, most notably the steam engine, brought about dramatic increases in productive capacity and industrial competition. As the costs and risks of business increased, owners and managers naturally looked for ways to limit the introduction of excess carrying capacity and the ruinous price competition that followed. One way those engaged in the carrying trades could control capacity and competition was to establish conferences that included all the shipowners engaged in a particular trade. The conference acted as a cartel by limiting entry, setting rates, and allocating shares of the trade among its members. European nations did not treat cartels as illegal restraints on trade, as they were viewed in the United States.

A more commanding way to control competition was to create a trust that effectively monopolized trade on a given route by bringing ownership of all the ships engaged in that trade under a single set of managers. In America, however, the formation of powerful industrial trusts excited considerable political opposition. As John D. Rockefeller and Andrew Carnegie demonstrated, centralizing ownership and control could lead to enormous cost reductions as well as immense personal fortunes. But for reasons that are still only partially understood, many attempts to consolidate industries into vertically integrated corporations failed. As the experience of industrialization revealed, only some industries could be concentrated in ways that would effectively exclude new entrants and provide the levels of profit needed to pay the huge sums of money that had to be expended to create a trust. As it turned out, shipbuilding and the maritime carrying trade were not among those industries that were successfully consolidated into dominant trusts.

J. P. Morgan's consolidation of the American steel industry provides a case study of why industrialists sought to form trusts. Before consolidation the U.S. steel industry experienced extreme fluctuations in price. Since the American market was protected, prices for American steel sold in the states were generally higher than prices in Britain, where American producers had to compete in a more open market.[18] Prices changed rapidly and dramatically before consolidation: in 1900 the price of American steel in Pittsburgh dropped from a high of $58.24 per ton in January to a low of $24.08 in August. When steel production outstripped its market, American producers dumped their excess on the foreign market, but since a high tariff protected their domestic market from British competition, they could set a domestic price well above what they charged their British customers. American shipbuilders were outraged at being obliged to pay more for American steel than their British competitors. J. P. Morgan finally imposed order on the industry in 1902 when he merged a number of smaller steel companies he controlled with Carnegie Steel, which he had recently acquired, to form the U.S. Steel Corporation. American steel quickly rose to a price slightly above the British price and stabilized at about $40 per ton.

Between 1900 and 1902 the shipbuilder Lewis Nixon attempted to do something similar in the shipbuilding industry. He formed the United States Shipbuilding Corporation and with his associates acquired a number of major shipyards, the most prominent being Newport News Shipbuilding, Union Iron Works, Bath Iron Works, Harlan and Hollingsworth, and the machinery manufacturer Hyde Windlass. As is usually the case in such acquisitions, Nixon borrowed heavily and had to pay inflated prices for the companies he sought, but he expected to be able to meet his obligations once he controlled the industry. The attempt failed in 1905, however, and swiftly collapsed into bankruptcy.[19]

Between 1901 and 1907 a similar attempt was made to consolidate all the major coastal lines operating between New England and the Gulf of Mexico. Charles Morse, who headed the Eastern Steamship Company, quickly acquired the Clyde Steamship Company, the New York Texas Steamship Company (the Mallory Line), the New York and Cuba Mail Line, and the Puerto Rico Steamship Company. Unlike the lines engaged in foreign trade, however, the coastal lines faced competition from another mode of transportation, and this proved fatal to Morse's bid for control of the market. The New Haven Railroad reacted to his acquisition campaign by purchasing the Maine Steamship Company, which operated between Portland and Boston, and the Merchant and Miners Company, which operated to U.S. South Atlantic ports out of New York. Morse thus found himself unable to raise freight rates to pay off his indebtedness and was soon driven into receivership. The various lines he had purchased were reestablished as separate companies and continued to operate in the coastal trades until their ships were acquired by the government during World War I.[20]

In 1902, the year he formed U.S. Steel, J. P. Morgan also set out to gain control of liner shipping in the North Atlantic. He created a new shipping company, the International Mercantile Marine (IMM), and began buying existing lines. At its largest IMM owned 136 ships, or about one-third of all vessels engaged in the Atlantic trade. Morgan's first purchases were the British-owned Leyland and White Star lines. He also acquired the American-owned International Navigation Company, which operated the only American-flag ships in this vast fleet: *St. Louis, St. Paul, New York,* and *Philadelphia.* These purchases gave Morgan the largest privately owned commercial fleet in the world, one which was equal in tonnage to the entire French merchant marine. He did not yet have control of the trade, however, so he continued expanding his empire. He negotiated a profit pooling agreement with the two largest German lines, Hamburg-American and North German Lloyd, and rationalized their share of the trade. This consortium then purchased the Holland-American Line. At that stage only Cunard, the dominant Atlantic carrier, lay outside Morgan's control, and the British government was not about to let him capture this line, too.

Bad luck and the economics of the shipping industry then began to undermine Morgan's campaign. Not having gained the degree of control over the North Atlantic trade that he had sought, he could not force up shipping rates so as to generate the revenue stream he needed to service the vast debts he had incurred while acquiring his fleet. The slump in shipping following the financial panic of 1907 further depressed earnings. In 1912 the loss of *Titanic*, the crown jewel of the White Star Line, was another serious setback. Cunard, with the backing of the British government, continued to resist Morgan's efforts at takeover, and in 1914 the IMM was forced to default on its bonds. It survived, thanks largely to the shipping boom that accompanied the beginning of World War I, but was finally liquidated in 1937. Morgan had failed, not because he lacked adequate resources or managerial talent—nor was he simply defeated by bad fortune. As he could not have known then, but as we can see now, shipping is an industry in which it is effectively impossible to exclude entry by newcomers. This ease of entry meant that if shipping rates were stabilized at levels that adequately rewarded investment, new competitors would enter the trade and trigger a new round of price cutting and eventual bankruptcy. Morgan's failure to replicate in shipping the kind of consolidation he achieved in steel can, like the related phenomena of shipping conferences and European cartels, be fully explained in terms of the economics of international shipping.[21]

Even before the collapse of the IMM, Morgan's maneuvers were attracting political attention. In 1912, the year Woodrow Wilson led the Democrats to victory in the presidential election, Congressman Joshua A. W. Alexander of Missouri called for an investigation of the Atlantic Shipping Trust. Alexander, reflecting the widespread antitrust sentiment that Wilson shared, questioned the legitimacy of a number of practices that many Americans found offensive. Morgan's pooling arrangements and rationalization of German shipping services clearly constituted attempts to restrict trade. Steamship conferences, which set shipping rates on specific trade routes, were also criticized. Conference use of "fighting ships" to defend against new entrants in the trade was thought especially objectionable. A fighting ship would arrange to arrive in a port served by conference members just before a ship operated by someone trying to break into the trade. The fighting ship would then capture all the available cargo by offering uneconomically low freight rates that would shut out the challenger. Although the use of such a defensive technique in an industry that could not be consolidated into a tight monopoly made perfectly good economic sense, it was clearly a public-relations catastrophe in a nation that insists small businessmen have a right to fair treatment in the marketplace.

In 1913 Alexander became chairman of the House Merchant Marine and Fisheries Committee and began a formal investigation of existing shipping practices. At the same time the Justice Department brought action under the Sherman Anti-Trust Law against members of the North Atlantic Steamship Conference for pooling passenger traffic. Once he had been inaugurated, Wilson's first priority was to reduce the high levels of tariff protection that his Republican predecessors had maintained. When the Underwood Tariff had been passed, however, he joined those eager to legislate tighter controls over trusts and monopolies. The Clayton Act, which prohibited various types of unfair trade practices not covered by the Sherman Act, was soon passed; it was followed by the establishment of the Federal Trade Commission, charged with investigating unfair competition and violations of antitrust laws. American business, which had grown luxuriantly during the age of industrialization, was now to be forced to conform more closely to the cultural expectations of American democracy. While the beliefs that informed this campaign are indisputably worthy on their own terms, the economic consequences of applying them to industry were unpredictable and in many cases proved to be severely negative. The age of freewheeling business was over, but it was far from clear how or how well the new order would function. Thomas McCraw has nicely summarized the central dilemma of antitrust policy in the following comment on the beliefs of Louis Brandeis, Woodrow Wilson's chief antitrust adviser: "[Brandeis] symbolizes one of the characteristic shortcomings of the American regulatory tradition: a disinclination to persist in hard economic analysis that may lead away from strong political preference. This shortcoming appears vividly in the subsequent tangles of twentieth-century antitrust policy, in the continuing institutional schizophrenia of the Federal Trade Commission, and in the frequent unwillingness of legislators to act on the unpopular principle that protection is usually anticompetitive and anticonsumerist, even when it is small business that is being protected."[22]

The steady decline of the merchant marine transformed America's relationship with the sea. The experience of seafaring, to which Richard Henry Dana had reacted with legal and benevolent concern, was portrayed in increasingly abstract and primal terms, with Herman Melville and Jack London providing romantic myths of moral confrontation and physical challenge. And in fact the social world of the sailor, never one of ease or equality in the best of times, was becoming increasingly coarse and brutal as American ships were forced to scrounge for the slim profits that could occasionally be realized on the margins of foreign trade.

It can hardly be surprising, therefore, that many concluded that Americans should go to sea primarily as sailors in the U.S. Navy, whose mission was to defend U.S. interests worldwide. American shipyards would build ships for the navy; the carrying trade for American exports and imports could safely be left to the ships of other nations. Such a strategy seemed reasonable and advantageous, but the outbreak of war in Europe quickly revealed it to be a dangerous delusion. In the first half of the twentieth century America discovered, while fighting two world wars, that it could not survive without merchant shipping over which it had some control. If the commercial world failed to meet this need, the government would have to do so. And so, as the memory of the golden age was displaced by the reality of world war, a new merchant marine began to take shape in America.

PART II

WAR-IMPELLED INDUSTRIES

6

THE VORTEX OF WAR
AND ITS AFTERMATH,
1914–1930

The balancing act that the European Great Powers had long maintained collapsed like a house of cards in 1914. At the outset it was thought that the war that followed would be brief and that America would remain largely uninvolved, but this illusion was shattered long before German submarines claimed American lives and the United States began preparing for war. As soon as the belligerents began putting their economies and their merchant fleets on a war footing, Americans discovered that their economy was vulnerable in ways few had anticipated.

Much of the country's agricultural and industrial output was produced for foreign markets; foreign shipping provided its primary links with these customers. Decades of peace and prosperity had lulled the nation into the comfortable assumption that there would always be plenty of ships to carry its goods. The warning that Thomas Jefferson had conveyed to Congress in 1793, when serving as the first secretary of state, had either been forgotten or dismissed as an archaic concern from a distant and very different past: "When those nations who may be our principal carriers shall be at war with each other, if we have not within ourselves the means of transportation, our products must be exported in belligerent vessels at increased expense, or perish on our hands." In 1914, when the conditions that Jefferson warned against appeared, the government was forced to act.

The severity of the problem posed by "our products perishing on our hands" became evident in 1914; a comprehensive legislative response was proposed in 1915, it was passed in 1916. In this brief period the U.S. merchant marine left one historical era and entered another. Laissez-faire, legislative inaction, intermittent subsidization, and complaints about carriers who colluded to limit competition were suddenly seen as not constituting an adequate national maritime policy. The nation needed shipping services that it had assumed the commercial merchant marine would provide, even when constrained by mercantilist laws. That assumption was now shown to be false. The government was therefore obliged to step in, first to remove some of the constraints that had prevented foreign-built ships from registering under the U.S. flag, then to provide new ships and shipping services at public expense. As the country mobilized its resources, initially to defend its status as a neutral and later to join the war as a belligerent, having an adequate merchant marine engaged in foreign trade was recognized as an essential part of the nation's preparedness. Those who cried for protection and those who put their faith in unfettered market forces suddenly found themselves marching to the beat of a command economy. Under the press of war Woodrow Wilson built a new national merchant marine that in a few years made the United States once again a leading seafaring nation. At the end of the twentieth century American maritime policy still bears the stamp of his forceful and innovative response to the crisis of World War I.

WAR IN EUROPE AND THE GREAT NEUTRAL

War in Europe threw international trade into turmoil. The United States, with few ships trading offshore, could only watch with helpless anxiety as normal services were disrupted. Great Britain, France, and Italy immediately diverted much of their shipping to supporting their war efforts. The British navy set up a blockade that kept German merchant vessels idle in their home ports or in neutral harbors of refuge. Many Allied ships also remained in port to avoid being attacked by German surface raiders. The sudden scarcity of carriers drove up freight rates to dizzying heights; the threat posed by German U-boats sent insurance premiums soaring. Transportation costs on routes that included war zones increased as much as 700 percent.[1] Seagoing commerce was thoroughly disrupted, and U.S. exports to Europe virtually ceased.

This interruption of ocean shipping immediately caused havoc in major sectors of U.S. farming and industry. Goods piled up on docks; railroad cars could not be unloaded; trains sat idle on the tracks; cars were not available to carry products from farms, mines, and factories—in short, much of the nation's transportation system and the economy it served came to a halt. In August 1913, America had exported 257,000 bales of cotton; a year later, less than one-tenth of this amount had been sent overseas. Not surprisingly, between July and December 1914 the price of a bale of cotton declined from $62.50 to $36.25.[2] Although America was not directly involved in the European war, its economy was most definitely under siege.

President Wilson was quick to acknowledge the emergency and realized that the government would have to intervene forcefully. Indeed, a generation of political Progressivism had persuaded most Americans that market forces could not always be relied on to serve the nation well and that governments must occasionally take the initiative in curbing excessive business practices and ensuring that pressing social needs are met. The Mahanian doctrine, with its emphasis on the tight links between foreign trade, seapower, and national prosperity, was still dominant; it was widely understood that national power rests primarily on economic strength rather than military might. Thus Wilson did not need to be convinced that the United States must be able to get its products to market, even if that meant altering existing laws and deploying government resources. Indeed, the crisis that diverted European shipping from world trade was seen as offering Americans a heaven-sent opportunity. Once the immediate problem had been dealt with, the United States could look forward to taking over much of the trade abandoned by the European powers, especially in Latin America. A New York newspaper heralded, "Europe's tragic extremity becomes . . . America's golden opportunity."[3] So far as America was concerned, at least before the war bogged down in the trenches on the western front, the game of global economic competition was still on.

But first the immediate crisis had to be addressed, and Wilson did so with vigor. He had Treasury Secretary William McAdoo, who played a leading role in implementing Wilson's maritime policy, introduce legislation establishing a Bureau of War Risk Insurance to underwrite potential losses for American shipowners. Wilson also liberalized the ship registry provisions of the 1912 Panama Canal Act. This act permitted the transfer to American registry of foreign-built ships five years old or less. At the time the act was passed advocates of "free ships" considered this a significant victory, but few ships had in fact been reregistered under the U.S. flag. When World War I began Wilson had the act changed to eliminate the age restriction and give the president the power to waive certain stipulations requiring that the officers and crews of American-flag ships be U.S. citizens. Since U.S. neutrality in a world at war suddenly made sailing under the U.S. flag highly attractive, shipowners responded swiftly to these changes in the law. More than half of the five hundred thousand tons of American-owned shipping that had been operating under foreign flags switched to the U.S. flag.[4] Many foreign-owned ships were also purchased and reflagged. The rapid increase in freight rates for foreign trade also drew many coastal vessels into offshore operation, a switch that was accelerated by the availability of government war-risk insurance. To encourage acquisition of ships to sustain the reduced coastal fleet, the government suspended, for the duration of the war, the laws that prohibited foreign tonnage from entering the coastal trades. As has often happened in American history, a war-related crisis brought about a legislative revolution that had remained beyond reach during the preceding decades of peace.

But Wilson and McAdoo went much further than merely liberalizing existing protectionist legislation. While commercial ship operators welcomed government war-risk insurance and foreign ship transfers, they were dismayed by the additional proposal that a government corporation be established to acquire and operate as many as fifty ships. This initiative was defeated in 1915 when it was first considered by Congress, but when resubmitted as part of a more comprehensive proposal to build a national merchant marine that would also serve as a naval auxiliary, it passed in 1916. McAdoo went even further, arguing that the government also needed to regulate the business of shipping. The rapid run-up in ocean freight rates and the shifting of ships to routes that were more profitable, as opposed to the routes that most needed to be served, were in his view characteristic failures of a market economy to serve the public interest. The antitrust tradition of Progressive politics, reinforced by the national emergency created by world war, strongly inclined Wilson and McAdoo to bring the operation of the shipping industry under close federal control. War had revealed that the business of shipping was too important to be left to the operators.

Rather surprisingly, however, the international shipping industry showed considerable resilience in its response to the challenge of war. As Wilson issued pleas and threats in defense of America's right to trade as a neutral, the disruption of commerce across the Atlantic began to work itself out. New ships were acquired, the order books of American shipyards filled up, and freight began to move more regularly. American shipowners' prospects were exceedingly bright. Ships that could have been chartered for one dollar per ton per month before the war were being chartered for fourteen dollars per ton if operated outside the war zone and as much as twenty-one dollars within the zone. With charter rates this high, it was possible to pay for the entire cost of a ship in a single voyage.[5] During the first year and a half of the war it was thought that the conflict would end soon with one side or the other making a massive breakthrough. It therefore appeared unlikely that America would become directly involved or that the nation's economy would have to be put on a war footing. Shipowners, like American munitions makers, seized the day and set out to make their fortunes as neutrals whose services were much in demand.

But the war did not end quickly, and the belligerents were forced to modify their strategies accordingly. The western front bogged down into a brutal new form of trench warfare based on siege and attrition. The war became as senseless as it was hopeless, a battle in which armies were commanded by "men without imagination, to execute a policy devoid of imagination, devised by men without imagination."[6] The ability to obtain supplies by sea became crucial. Germany's economy was being crippled by the British blockade, a blockade the German surface fleet was incapable of breaking. The only response available to Germany was to use submarines to impose their own blockade on Great Britain and its Allies. But if Germany was to avoid bringing the United States into the war, it had to respect America's claims as a neutral. Neutrals traditionally had the right to engage in trade during wartime so long as they did not carry contraband cargoes to belligerents. Of course it was exceedingly difficult to police neutral trade without offending those not at war, but if the nonbelligerents were relatively powerless, there was little they could do except protest, as Jefferson and his successors had discovered long ago. The shocking fact discovered in World War I, however, was that when the blockading vessels were submarines, it was impossible to employ the standard procedures for searching neutral vessels.

It was also impossible to abide by the normal rules of war when using submarines to raid enemy commerce. Traditionally, a ship of war intending to capture or sink a commercial ship would first notify it and allow the crew and passengers to disembark. But submarines only had a decisive advantage so long as their presence was undetected, an advantage they surrendered when they surfaced and made themselves known. If submarines were to be effective as blockading vessels and commerce raiders, they had to shoot their torpedoes first, and if they did that, they would inevitably take the lives of citizens of both belligerent and neutral countries and occasionally sink neutral ships. This was the dilemma Germany found itself in as the war ground on: if it waged all-out submarine warfare, it might be able to win the land war before America could send enough men and supplies to Europe to tip the balance in favor of the Allies; if it restricted its submarine warfare so as to avoid bringing America into the war, it could not impose a blockade of sufficient effectiveness to win the war of attrition. It was a harsh choice, a devilishly appropriate dilemma for a hellish war.

In May 1915, a German U-boat sank the British passenger liner *Lusitania* without warning; 1,198 lives were lost, including 128 Americans. The public was outraged by this unannounced attack on civilians, and Wilson vigorously protested the taking of American lives. In August another British passenger ship was torpedoed, and three more Americans were killed. Germany, sensing that additional provocations might push the United States into war, promised there would be no further attacks on passenger ships without suitable warning. Germany also shifted most of its submarine operations to the Mediterranean, where few Americans would be encountered. But in March 1916, a U-boat sank the British Channel steamer *Sussex*, again with the loss of several American lives. By this time the United States had extended so many loans to the Allies and was selling them so much war materiel that its claims of neutrality were becoming threadbare; yet Wilson again threatened Germany if such attacks were not halted. The Germans, eager to keep the United States on the margins of the war, again promised to avoid unforewarned sinkings, and the rest of the year passed without further serious incidents. At the end of 1916, as Wilson completed his first term as president and began campaigning for reelection, he could still present himself to the electorate as the man who "kept us out of war."

In retrospect we can see that the events of 1916 were steps on a path that led to war, but at the time Wilson believed he still might be able to persuade the European nations to lay down their arms and make peace. Throughout the year he tried again and again to get the belligerents to accept his offers of mediation and pledge themselves to postwar disarmament and the creation of a League of Nations. Over time he found that the experience of war hardened attitudes and that when dealing with nations in arms, only those who commanded power could speak with authority. During the course of the year he therefore insisted that the United States increase its "preparedness," not in anticipation of going to war, but so that it could stand on its own as one of the Great Powers in the culminating diplomatic struggle Wilson hoped to orchestrate.

Three major preparedness acts were passed in 1916. The Army Act doubled the regular army and vastly expanded the National Guard, yet this increase did not exactly transform the United States into a nation in arms. Previously, the army consisted of only 110,000 men, little artillery, no tanks, and fifty-five airplanes. Indeed, Wilson had been so opposed to any possible U.S. involvement in the war that when he learned that the army's general staff had drawn up contingency plans for possible operations against the Germans, a task that seemed to be their obvious responsibility, he ordered them to stop doing so immediately.[7] The Navy Act of 1916 committed the United States to a major one-year shipbuilding program, chiefly of cruisers and battleships. Admirals at the time measured the strength of their navies in terms of their capital ships and battle fleets, and

Wilson was determined that the United States be respected. Assistant Secretary of the Navy Franklin D. Roosevelt realized, however, that the investment in capital ships had to be justified in terms of their ability to protect commerce. He carried his case to the nation, arguing in speeches and articles, "[W]e must create a Navy not only to protect our shores and our possessions, but our merchant ships in time of war, no matter where they go."[8] One of Roosevelt's biographers has said that Roosevelt's seven and a half years as assistant secretary of the navy provided him with "the most thorough training conceivable for his later service as War President."[9]

The third major piece of preparedness legislation passed that year was the Shipping Act of 1916. The concern here was more economic than military and had become evident as soon as the outbreak of hostilities had disrupted shipping services in the North Atlantic. Wilson and McAdoo had tried to get such an act passed in 1915 but had been blocked by opposition to creating a maritime fleet built and owned by the government. In 1916 they added the need for a naval auxiliary to their previous arguments and carried the day. Although the operators still opposed the bill, American farmers acknowledged the need for sufficient tonnage under national control to move their exports, and the bill passed. This single piece of legislation opened an era which is only now drawing to a close, an era in which the federal government funded, regulated, and dominated the U.S. merchant marine.

The Shipping Act was drafted by Congressman Joshua Alexander of Missouri, chairman of the House Merchant Marine and Fisheries Committee. In 1913 and 1914 Alexander had conducted hearings on regulating the maritime industry, one of his particular concerns being the operators' trade-route conferences that set rates and excluded nonmembers. Alexander, with Wilson's concurrence, therefore made sure that regulation of the conferences was included in the proposed legislation. The act that emerged created a five-man Shipping Board with authority to acquire and operate a government fleet. A subsidiary organization, the Emergency Fleet Corporation (EFC), was assigned the task of procuring ships. The Shipping Board was given broad regulatory powers to correct the abuses uncovered by Alexander's committee. All liner operators were required to file their freight rates with the Shipping Board, the use of "fighting ships" to exclude carriers who were not conference members was prohibited, and deferred rebates for shippers who used only vessels owned by conference members were outlawed. Alexander did not try to eliminate the conferences, however; he knew that foreign governments could not be compelled to adhere to U.S. antitrust legislation. At the time, all foreign governments permitted closed conference systems, which is to say conferences closed to outsiders. The new law nonetheless stipulated that conferences in which Americans participated must be open to all and that in the future any shipowner could operate in the U.S. trade, whether or not he was a party to a conference agreement. The government thus treated the shipping industry as a public utility by exempting it from certain features of antitrust law, while at the same time bringing many of its operations under federal regulation for the first time. After 1916 the industry was thus doubly hedged in by federal intervention, by antitrust regulation on one side and by the creation of a national fleet on the other. The only significant concession the operators got written into the new law was a requirement that the government dispose of its fleet within five years of the end of hostilities in Europe.

THE UNITED STATES ENTERS THE WAR

German calculations of the costs of unlimited submarine warfare changed as the war progressed. During the few months in 1915 when unrestricted submarine attacks had been permitted they had inflicted enormous losses on Allied shipping, sinking fifteen merchant ships for every submarine lost.[10] While the balance appeared to favor the Germans, it was not clear at that point that there were enough submarines to sustain the campaign. But by the end of 1916 German submarine production had risen to a point where the submarines being lost could be replaced. As the conflict dragged on, the German high command was increasingly eager to wage all-out war and bring Great Britain to its knees. They accepted the risk that America might join the Allies if their submarines were unleashed, but the United States was so weak militarily that it was thought it could have little effect on events in Europe. Thus, on the last day of January 1917, Germany notified the United States that unlimited submarine warfare would begin the next day.

Wilson was appalled and genuinely surprised. Although his efforts to move both sides toward peace had been unavailing, he could not believe that a civilized nation would engage in such a barbaric form of warfare. Almost immediately he severed diplomatic relations with Germany. In mid-March four American ships were sunk with the loss of thirty-six lives. The nation was ready to commit. On April 2 Wilson asked Congress for a formal declaration of war; four days later it was forthcoming.

At first it appeared that the German gamble would pay off. They were now able to keep two-thirds of their fleet of 120 submarines at sea. Whereas during the 1915 campaign they had sunk 120,000 tons of Allied shipping per month, during April 1917 alone they sank 875,000 tons. One out of every four ships that left Great Britain during this period did not return.[11] The Allies could not survive if they continued to lose ships at such a rate.

Just before America's entry into the war Rear Admiral William S. Sims had been sent to London to serve as the U.S. Navy's liaison with Great Britain. At his first meeting with Admiral John Jellicoe, First Sea Lord of the Admiralty, Sims was given a copy of a memorandum that described the damage being inflicted by German submarines. In the first quarter of 1917, 1,300,000 tons of British and neutral shipping had been sunk. When Sims asked what the consequences of the sinkings were, Jellicoe said, "[T]hey will win, unless we can stop these losses." When Sims asked if there was any solution to the problem, Jellicoe replied, "[A]bsolutely none that we can see now."[12]

While Jellicoe expressed the Royal Navy's official view, which is to say the view of its leaders who thought first of the battle fleet of capital ships, an alternative strategy was being considered further down the ranks. A group of British captains and flag officers argued for reviving the system of convoy protection that had proven so effective during the Napoleonic wars, but the senior officers would not hear of it. Since a convoy can only proceed at the speed of its slowest ship, they said, restricting cargo ships to convoys would delay the arrival of vital supplies. Furthermore, they asserted, without bothering to provide any supporting evidence, merchant ship captains were incapable of maintaining close formation, which is essential for effective convoying. Finally, the admirals argued, there were far too few escort vessels available to provide convoy protection for the vast number of ships entering and leaving British harbors. These and other arguments against sailing in convoy were accepted by many merchant captains as well.[13]

Sims was sufficiently concerned about the loss of Allied shipping to give the convoy idea careful consideration. He was also appalled by the cavalier way the admirals dismissed it. He had the advantage of having direct access to the prime minister, David Lloyd George, and found that he had long since concluded that convoys offered the best hope for reducing the huge shipping losses. The prime minister soon ordered the Admiralty to make a serious study of the convoy question; the task was assigned to a Commander Reginald Henderson, one of the foremost advocates of the convoy system. Henderson reexamined the claim that there were too few escort vessels to protect convoys, and his findings left the admirals' case against convoying in tatters. The claim turned out to be an artifact of the Admiralty's desire to minimize the public's awareness of the losses being sustained. To do this, they included in their weekly report of sink-ings a figure for all the arrivals and departures of ships in British ports. Since the number of sinkings seemed small in comparison to the total number of sailings and arrivals, which the Admiralty reported as about 2,500 per week, the losses appeared acceptable. What Henderson found, however, was that of the 2,500 movements, only 120 to 145 per week involved oceangoing ships forced to transit the U-boat danger zone. These were the only ships needing convoy protection; the others ships being reported were mostly engaged in coastal traffic. The Admiralty had argued that it did not have enough escort vessels to protect 2,500 ships; Hender-son demonstrated that it had enough escorts to provide convoy protection for the ships threatened by submarine warfare. In May 1917, Sims was able to inform Washington that use of the convoy system would soon com-mence.[14]

The United States entered the war in April; in May six of its destroy-ers arrived in Queenstown harbor to begin convoy duty. Many more destroyers were to follow. Sims's headquarters while serving as comman-der of the U.S. Naval Forces Operating in European Waters were in London, near the Admiralty. By fall the convoy system had dramatically reduced ship losses and the first battle of the Atlantic was being won. Sims was widely praised by Lloyd George and others for his role in persuading the cabinet to require use of the convoy system. The U.S. Navy responded as well. Production of capital ships was halted so that all available facilities could be used to construct wooden submarine-chasers and fast, steel destroyers. The "four piper" destroyers turned out to be highly effective in convoy duty and were to prove their worth again in this service twenty years later.

The German general staff brought the United States into the war in a carefully calculated manner and at a time when its expectations seemed reasonable. Events soon demonstrated, however, that the generals had misjudged the will and capacity of the American people, probably because they were misled by America's fervent isolationism during the war's early years. Once mobilized, the United States proved to be a formidable foe. By autumn of 1918 it had an army of 2 million men in France and another 2 million being trained to join them. The intervention of the "arsenal of democracy" eventually proved decisive.

THE EMERGENCY SHIPBUILDING PROGRAM

Implementation of the Shipping Act of 1916 got off to a slow start. The five commissioners who were to make up the Shipping Board were not appointed until March 1917, and it was August before the board itself began to function effectively. Relations between the chairman of the Shipping Board and the general manager of the Emergency Fleet Corporation (EFC), which was to acquire or build ships for the board, then became so strained that both had to be replaced. Wilson appointed Edward N. Hurley, the chairman of the Federal Trade Commission, as the new chairman of the board. A succession of admirals were appointed as general managers of the EFC until Charles Piez, a prominent mid-western industrialist, proved equal to the task; he then held the position until the end of the war.[15]

The extent of the commitment the United States would have to make to the Allied cause only became evident as 1917 wore on. The European nations had already mustered up nearly all their available men and the front was in danger of collapsing as the Allied troops became exhausted, with some even becoming mutinous. The United States soon realized it would have to deploy a huge field army and sustain it three thousand miles from its home base. In June 1917, Congress gave the president some of the tools he needed to do the job when it appropriated $750 million to purchase existing ships and build new ones. It also granted the president broad powers to requisition ships already in operation and added to the government fleet the four hundred ships under construction or on order for private owners. German ships interned in American ports were seized; several of these were large passenger liners that were used as troop transports. In later months the initial appropriation of three-quarters of a billion dollars was increased to $2.9 billion, a figure so large there was no chance of its being spent promptly or efficiently. To put this sum into perspective, it was twice the value of the entire world fleet engaged in international trade prior to 1914.[16] Under Wilson's leadership, and with the enormous resources made available by the nation's burgeoning industrial economy, the United States set out to build the cargo fleet it needed for victory in war and peace.

By the time this enormous flood of money was being applied to ship acquisition, the ways and order books of existing shipyards were already filled with navy and private contracts; considerable innovation was therefore in order. The wooden-ship builders sprang to the call, and a number of large schooners were constructed. Composite ships with wooden planking on metal frames were also built, as were several ferro-cement hulls. Command economies and the exigencies of war sometimes bring forth unlikely experiments and surprisingly atavistic responses.

The EFC's first major task, however, was to set up new shipyards in which the steam-powered, steel ships needed for the war effort could be built. The most notable of these yards was on Hog Island, located in the Delaware River across from Philadelphia, where fifty building ways were constructed. Because these shipyards were to be managed and manned by people with little or no shipbuilding experience, the designs of the new ships were simplified as much as possible, with machinery and equipment being subcontracted to manufacturers throughout the country. This crash program continued to operate well past the armistice of 1918, and a huge fleet of sufficiently well-built ships was turned out in a very few years. It was an outstanding achievement, but at what a cost! The EFC awarded contracts paying $145 per ton for a type of ship being built in Great Britain for as little as $75 per ton. By the time the program was completed in 1921, the Shipping Board and the EFC had spent $3 billion for ship construction. The average cost for the ships delivered was $200 per deadweight ton. As McAdoo later said, "[A]ppalling prices were paid for everything that had to do with a ship. Engines and other equipment were purchased at such a staggering cost that I fancied more than once that the machinery we were buying must be made of silver instead of iron and steel."[17]

This extravagant but necessary building program, largely carried out by men unburdened by the routines of earlier practice, gave the United States a huge new cargo fleet capable of dominating international trade once the war ended. It was also a fleet that was largely freed from dependency on remote coaling stations. Since their inception steamships had been fueled with coal. By 1914 Great Britain, having great coal resources at home and a global empire, had established a worldwide network of coal-bunkering stations that serviced much of the world's shipping. Early in the war Britain had instituted a licensing system that made possible tighter control of this coal supply system. Britain resented the fact that the U.S. government was building a huge modern cargo fleet at public expense while all the resources of the European powers were being devoured in a war of attrition, and it had no intention of surrendering its dominance of world trade once the war was over. Rationing bunkers was one way it could exercise some control over the trade of its rivals. But the United States, which was the world's leading oil producer at this time, did not wish to be dependent on its foremost commercial rival for fuel. The EFC therefore decided that the boilers of its war-built fleet would be oil fired. It turned out to be a shrewd decision.

Coal and oil each have distinctive advantages and disadvantages as boiler fuel. European naval officers normally preferred coal as a fuel because it was stored in bunkers located between the outer hull and the vulnerable boilers and would therefore help smother explosions due to gunfire. Coal was also less flammable than oil and posed less of a risk of fire when the ship was under attack. Since Great Britain, Germany, and France had extensive coal reserves but no domestic sources of petroleum, they naturally built coal-fired ships. But using oil as fuel had offsetting advantages. Ships could carry enough oil for relatively long voyages without refueling; oil could be carried in the ship's double bottom, thereby freeing up more space for cargo; oil-fired ships did not need large gangs of men to shovel fuel into the boilers; and the tedious and labor intensive process of refueling with coal could also be avoided. Because of its advantages, and because the United States had ample domestic oil supplies, the U.S. Navy had switched to oil for all its new construction before World War I.[18]

When the war was over the EFC's shipbuilding program was subjected to the usual charges of waste and inefficiency. If measured against the prices normally paid for comparable ships, the war-built fleet was outlandishly expensive, but war is by its very nature wasteful. Indeed, if the shipbuilding program deserved to be indicted, it was for responding too slowly to its task, not for spending too much. Not a single ship was delivered under government contract in 1917; the government yard at Hog Island did not deliver a ship until after the armistice was signed in 1918. The program was also continued long after hostilities had ended; keels were laid for one-third of the entire fleet after the war was over. But wars are fought, and peace agreements are hammered out, without benefit of the knowledge granted by hindsight. In 1917 many thought that the war might extend into the next decade, and the armistice was agreed to long before a lasting peace agreement was reached. When it was over, the United States had available a vast fleet of ships that would form the basis of its new merchant marine in the years ahead. Some of these ships would continue to operate through the next war as well.

THE MERCHANT MARINE ACT OF 1920

As American intervention began to tip the balance of war to the advantage of the Allies, the problems involved in negotiating a lasting peace had to be faced. The United States had become a deeply involved Great Power, but how would it use its new might? Wilson believed he had identified the fundamental obstacle to establishing world order. Internationally, the war had been caused by selfish and ultimately disastrous commercial and imperial competition. A lasting peace could therefore only be arrived at if all nations agreed to international accords that would regulate and mitigate the insupportable consequences of competition among Great Powers. Wilson determined that the United States would use its new strength and its moral authority to this end. The pathologies of industrial capitalism were also evident on the domestic front, where industrialists too often pursued profit for their stockholders while ignoring public needs. The U.S. government had to ensure that the nation had the resources it required and that they were deployed in ways that served the legitimate interests of the nation. These were the concerns that shaped Wilson's domestic and international maritime policy as the war wound down and the struggle for peace began.

The British were extremely concerned about America's determination to become a major participant in world shipping after the war was over. Wilson believed such a move was crucial to preventing a repetition of the vicious imperial competition that had brought on the Great War. He therefore did not terminate the EFC shipbuilding program when the armistice was declared and directed that all the ships built at government expense be restricted to national use only. His shipping offensive was directed by Edward Hurley, the chairman of the Shipping Board, a man who shared Wilson's desire to extend to world trade the kinds of industrial regulation that the Progressives were attempting to impose on American industry. Their goal was to constrain commercial rivalry through international harmonization of maritime costs and freight rates, the dream being that trade would then flourish without the accompanying wastage of capital and human resources that led, ultimately, to war. But the notion of regulating international trade was one the British, who had done very well by aggressively pursuing free trade, did not welcome.

Wilson and Hurley proposed that all nations adopt the standards for maritime labor that had been established for U.S.-flag ships just before the war. The standards they had in mind had become law in the United States after a long campaign to improve the lot of the U.S. seaman through unionization and congressional action. The "emancipation of seamen," one of the great labor crusades of the Progressive era, reached its climax with the passage and signing of the Seaman's Bill in 1915. The Great Emancipator was Andrew Furuseth, the head of the International Seamen's Union (ISU) and a man of such implacable determination and virtue that he was frequently compared to Lincoln and Jesus. A Norwegian immigrant who began his career working on the lumber schooners in the Pacific Northwest, Furuseth dedicated his life to protecting U.S. jobs for U.S. sailors, to organizing U.S. sailors into a union of disciplined workers who were proud of their skills and their labor, and to securing the passage of federal legislation that would free sailors from the "bondage" of contracts that prevented them from collecting their wages and leaving their employment when they wished. His goals and achievements were generally in accord with the growing American response to industrialization, but they were not the sort of thing British shipowners wished to see applied to international commerce.

Furuseth's chief legislative ally was the prominent Progressive senator Robert La Follette of Wisconsin. They first introduced a Seaman's Bill in 1910; it was passed in 1912 but was vetoed by President Taft. The bill was passed again in 1915, and President Wilson signed it after meeting privately with La Follette and Furuseth. This law established minimum levels for crew sizes and skills on U.S.-flag vessels and required that at least three-quarters of the crew be able to understand English. This language requirement was an indirect way of outlawing the hiring of Chinese crews, such as those that had served very effectively on many American ships sailing in the Pacific.[19] The law also abolished imprisonment for desertion and gave sailors the right to demand half their wages at any port of call. It was argued that these latter provisions merely guaranteed to sailors the rights that other free laborers had long enjoyed. What they meant in practice, if applied internationally, was that sailors' wages would be forced up to the level required to hire labor in the most expensive ports. For example, a sailor who signed on in Liverpool could collect half his wages and leave the ship in New York; he could then offer to sign on again at New York wages. Wilson, Hurley, and La Follette had no problem with that, but the British and other foreign shippers certainly did.

Although Furuseth's International Seaman's Union (ISU) enrolled many merchant sailors before the war, it lacked bargaining power and did not dramatically alter labor relations in the shipping industry. The vast expansion of the U.S. fleet during the war did have a profound effect, however, as the government, which was more receptive than private operators to union demands, became the major employer. By the war's end union membership had grown to 117,000, seamen's wages had risen to $85 per month, and seamen serving on deck had for the first time achieved an eight-hour workday. These conditions were the product of a command economy, however, and could not be sustained in international commerce unless heavily subsidized. With the sudden decline in government cargoes at the war's end, competition for trade became intense and the shipping industry slipped into deep depression. In 1921 the owners cut wages by 15 percent and eliminated overtime payments. Seamen, supported by the ships' engineers, struck in protest, but when the owners managed to continue operations with nonunion crews, the strike collapsed. As the depression continued, the Shipowners' Association drove seamen's wages down to $47.50 per month, the prevailing rate before the war; the Shipping Board was soon forced to accept the new wage rate as well.[20] Federal regulations and the union strategy of collective bargaining proved to be impotent when faced with the disruptions created by throwing a huge war-built fleet onto an intensely competitive international market.

Wilson's attempt to use his new fleet of cargo ships as a bargaining lever in the peace negotiations was equally unsuccessful. His intention had been to put these ships on all the major trade routes so as to force the British and the other Allies to agree to his peace plans. But the aftermath of the war proved to be even more dire than he had anticipated, and Wilson soon found he could not deploy his assets as he had intended. Economic collapse and social chaos stalked Germany and most Eastern European nations, where millions were facing starvation. Further east, the specter of Bolshevism, now firmly established in Russia, threatened to reap the whirlwind and spread westward. Wilson, while trying to prevent the reestablishment of prewar forms of commercial and imperial competition, was forced to address more immediate postwar emergencies. Europe had to be fed, and Herbert Hoover was appointed director of the relief programs. Most of the aid had to be transported in U.S. vessels, and Wilson ordered Hurley to pull government-owned ships off commercial trade routes and make them available for the relief effort. He also sought to have the Germans use their prewar commercial fleet for this purpose, but the Allies took these ships as war reparations and opposed Wilson's plan for their use. As the U.S. commitment to providing war relief grew, the Allies found more and more opportunities to place their ships in international trade without accepting the conditions Wilson sought to impose. In the end Wilson was unable to dictate a new international order in ocean shipping, despite the massive fleet under his control.

Even domestically Wilson and Hurley found they had less control than they expected over shipping. Hurley had warned Wilson that without government direction shipowners would choose to operate primarily in those trades that provided the best return, a pattern of behavior that only a Progressive could find surprising. Like his predecessor McAdoo, Hurley did not trust the free market to allocate resources in ways that would best serve the public. He therefore continued to construct a comprehensive plan for all international trade routes, his intention being to operate government ships at government expense on those routes that did not attract private operators. He never completed the plan or its implementation, but the program lived on in subsequent federal subsidy programs for the merchant marine. The struggle between market responsiveness and federal planning was far from over.

Government ownership and operation of a large fleet of merchant ships was coming to an end, however. By 1919 there was growing pressure to get the government out of the shipping business. The Shipping Act of 1916 allowed the government five years after the end of hostilities to divest itself of the war-built fleet, but that was now seen as too long. In 1920 new legislation was passed defining the terms and conditions to be followed in transferring the government fleet to private ownership. These provisions were, however, just one part of a bill called the Jones Act, named after its congressional author, Washington senator Wesley Jones, chairman of the Senate Committee on Commerce. The legislation addressed many facets of maritime policy and sought to define a new basis for assuring the continued existence and vitality of the U.S. merchant marine. The Merchant Marine Act of 1920 was itself a landmark.

The act's preamble defined for the first time the maritime goals of the United States, and in essence these goals continue to define federal maritime policy today: "It is hereby declared the policy of the United States to do whatever may be necessary to develop and encourage the maintenance of a merchant marine . . . sufficient to carry the greater portion of its commerce and serve as a naval or military auxiliary in time of war or national emergency, ultimately to be owned and operated privately by citizens of the United States." In addition to stipulating how government ships were to be offered to American owners, the act established a greater degree of federal oversight than had existed before. It carried forward McAdoo's and Hurley's program of identifying a set of essential trade routes by authorizing the Shipping Board to operate vessels on those routes not adequately served by U.S. operators. It also strengthened the antitrust penalties provided in the 1916 Shipping Act, saying, for instance, that any conference employing "fighting ships or offering deferred rebates would be penalized by having the ships of its members barred from entry to U.S. ports."[21]

The act also sought to help the U.S. industry by removing an antitrust restriction affecting insurance. American operators had been forced to depend on the London market for insurance, and they had complained for years than the London underwriters manipulated their rates to favor British owners. In an effort to break this monopoly the 1920 act allowed U.S. insurance underwriters to form combinations to insure American vessels. Resentment against British refusal to accept Wilson's and Hurley's proposals took many forms.

No one found British maritime attitudes more irritating than Admiral William S. Benson, Hurley's successor as chairman of the Shipping Board. Hurley was so opposed to selling the best government ships to private interests at rock-bottom prices that he felt compelled to resign when the 1920 act was passed. Benson, who had served as chief of naval operations, had made something of a career of stating his animosity toward the British. When Rear Admiral Sims was ordered to London in 1917 to coordinate naval operations, Benson had warned him, "[D]on't let the British pull the wool over your eyes, it is none of our business pulling their chestnuts out of the fire. We would as soon fight the British as the Germans."[22] Had Wilson let Benson set policy, the United States and Great Britain would probably have begun a commercial war as soon as the guns on the western front fell silent.

The 1920 act sought to strengthen the merchant marine in other ways as well. In 1913 the authors of the Underwood Tariff included in their bill a system of direct discrimination, a 5 percent reduction in all duties on goods imported in American ships. In 1915 the Supreme Court negated this provision because it violated existing commercial treaties. The 1920 act therefore proposed an indirect nontariff discrimination to achieve the same ends; it granted preferential inland railroad rates to cargoes carried in American ships. The act acknowledged that this provision violated the "most favored nation" clause contained in many U.S. commercial treaties and simply directed the president to abrogate any treaty that prohibited such preferential rates. Although stung by his defeats at Versailles, Wilson found this tactic more than he bargained for. The battle to regulate international trade had been lost and there was no reason to create more friction among the Allies. He signed the act, but four days later the Shipping Board was forced to declare a ninety-day postponement in its implementation of preferential rail rates. The moratorium was extended the next year, which gave Wilson enough time to get Congress to reverse this part of the legislation.

The best-known feature of the 1920 act, and the part that is usually referred to as the Jones Act, strengthened the existing coastal shipping law. Once again Congress included the Philippine Islands in the territories defined as "coastal" shipping, but when foreign shipping companies threatened to boycott Manila if the Philippines were brought under cabotage protection, the Shipping Board decided not to implement this provision of the act. Congress certainly had picked up the tone of injured self-righteousness that dominated American foreign affairs following World War I, but many of the policies it proposed were more embarrassing than effective. International trade proved to be less susceptible to regulation and more resistant to mercantilist protection than the politicians had ever imagined.

As the 1920s began, the Shipping Board put its almost 10 million ton war-built fleet on the block. Most of the ships were steel cargo vessels that could be employed profitably in world trade. The Dollar Steamship Line purchased seven of the twenty-three combination cargo-passenger liners in

1923 for its around-the-world service. These ships, referred to as 502s or 535s, according to their length, were the most commercially attractive vessels in the war-built fleet. When prices were reduced two years later, the Dollar Line also purchased the five ships that Pacific Mail had been operating for the Shipping Board and continued to run them in the Far East service. American Mail Line, a Dollar Line subsidiary, bought five of the remaining 502s and 535s to operate from ports in the Pacific Northwest to the Orient. The Munson Line purchased five of the combination ships for its service from New York to the east coast of South America.[23]

The indirect subsidy provided by making good ships available to American operators at extremely low prices greatly increased U.S. presence on the world's ocean trade routes. The retreat from public ownership and operation continued; by the mid-1920s the only routes which the government continued to serve were in the North Atlantic, where the government-owned United States Lines operated. During the years right after the war cheap ships and plentiful cargoes made for profitable operations, but overseas rivals soon reappeared on ocean trade routes with newer and more efficient ships of their own. At that point the higher cost of operating American ships again became a handicap, and the advantage of having a low-cost, war-built fleet began to fade.

THE MERCHANT MARINE ACT OF 1928

Citizens rally around the flag with considerable enthusiasm when mobilizing for war, but demobilizing after a war, that is to say moving from a command economy to a market economy, can be a bewildering and dispiriting experience. In most cases demobilization involves drastically reducing the number of men and women in uniform, shutting down war production, and disposing of most of the surviving war materiel. But because the war-built merchant fleet contained many ships having commercial utility, it created an indigestible lump that blocked the flow of orders once the demand for shipbuilding was again being set by market considerations. The authors of the 1920 act realized this and authorized low-cost ship construction loans as a way to keep the shipbuilding industry afloat, but no one could afford to use them. Between 1922 and 1928 no ships were built in the United States for world commerce. The effect of this prolonged period of shutdown was predictable. The Bethlehem Steel Corporation's many shipyards found just enough repair work to survive, as did the Newport News Shipbuilding Company. In Maine the Bath Iron Works temporarily suspended operations in 1924, while William Cramp and Sons in Philadelphia built their last ship, for the coastal trade, in 1927. New York

Shipbuilding of Camden, New Jersey, scraped through, as did the Sun Shipyard in Chester, Pennsylvania, which built several tankers for its parent company, the Sun Oil Company. By the mid 1920s all the government-built yards had disappeared.[24]

As the war-built fleet aged and faced increased competition from foreign carriers, it appeared that the huge wartime investment in reviving the merchant marine might suffer the same fate as the wartime investment in shipbuilding. This was deemed politically unacceptable, however, and the Shipping Board became desperate in its search for some way to rescue the industry. It finally decided to resurrect an earlier subsidy program. Since the 1920 Merchant Marine Act had left undisturbed the Ocean Mail Act of 1891, the Shipping Board decided to rewrite this mail subsidy act in a way that would keep afloat the operating companies involved in international trade. The patriotic duty to provide rapid and reliable mail service would provide much needed political cover for directing still more funds to an industry that seemed to have an insatiable appetite for public underwriting.

The Merchant Marine Act of 1928 created an expanded mail subsidy system. As in the earlier law, contracts were to be awarded for specific trade routes, with the level of funding varying with the size and speed of the vessels employed. Again borrowing from the earlier act, four classes of ships were established, with a maximum rate of twelve dollars per ton per mile for ships of twenty thousand tons or over capable of making twenty-four knots. No criteria for determining the effect of granting subsidies on mail service itself were included in the act. Also missing was a specific requirement that those receiving subsidies must contract for new construction. The board did, however, eventually make new construction a requirement for receiving a subsidy, and a number of outstanding ships were built under the act. Several of them established legendary reputations in the great age of passenger liners. *Manhattan* and *Washington*, passenger ships of twenty-four thousand tons, plied the North Atlantic for United States Lines. *Montery* and *Mariposa* served the Matson Line on the run to Australia, while *Lurline* sailed on Matson's Hawaii run. Stanley Dollar added two more passenger-cargo vessels to his fleet, *President Coolidge* and *President Hoover*. American Export Line ordered four combination vessels for the Mediterranean service, the "Four Aces": *Excaliber*, *Excambion*, *Exeter*, and *Exochorda*. Grace Lines built four ships that set new standards for luxury cruising: *Santa Elena*, *Santa Rosa*, *Santa Paula*, and *Santa Lucia*. In terms of output the 1928 act was a great success; the funds it provided led to the building of some sixty-four new ships and the modernization of a number of others.

The cost of success was exorbitant, however, and the gross mismanagement of the mail subsidy program, when brought to light, became an enormous political embarrassment. As the Great Depression, which began in October 1929, further soured American attitudes toward big business and as the Democrats pilloried the policies and practices of their Republican predecessors, the costs and collusion of the mail subsidy program became the stuff of scandal. By 1935 Senator Hugo Black, chairman of the Senate Investigating Committee, was characterizing it as a "saturnalia of waste, inefficiency . . . [and] exploitation of the public."[25] The process by which subsidies were awarded seemed reasonably well designed. The postmaster general would certify the specific trade routes to be established; the Shipping Board would certify that appropriate vessels existed to perform the service and would specify the frequency of service to be provided; the postmaster would then award contracts based on the lowest bid. But according to Senator Black, the outcome was cooked up in advance by officials who had been thoroughly captured by the industry:

> The agents, lobbyists, and representatives of steamship lines descended upon the Post Office Department, which acting in startling harmony with these alert and affable spokesmen of privilege, speedily reached decisions as to the "mail route" to be established, the character of service to be demanded (always a service then being operated by the prospective contractor, or which it was fully prepared to operate). . . . Having accomplished this preliminary spade work, the result of these cooperative labors was transmitted to the Shipping Board in the form of a "certification of the Postmaster General." This certification was accompanied by the representatives of the prospective contractors, who were equally at home in both places. Instantly and pliantly the Shipping Board "determined and certified" to the Postmaster General those things which the interdepartmental sub-committee and the prospective contractor had agreed upon in advance that the Shipping Board should certify. The route was then advertised and the contract let to the company, which had arranged in advance the service it was to render and the compensation it should receive.[26]

This was the system used to certify and negotiate subsidies for twenty-seven ocean-mail routes at the outset and, eventually, for forty-six separate contracts with thirty-one companies. A final report on the mail subsidy system prepared by the Maritime Commission in 1937 determined that the post office had provided steamship companies with some $200 million under subsidy contracts. This amounted to almost ten times the normal rate for the mail actually carried, but it should be remembered that providing cost-effective carriage of the mails was never a motivating concern for the mail subsidy system.[27]

7

MARITIME POLICY IN THE NEW DEAL, 1930–1939

The 1920s should have been happy years for most Americans, and for many they were. Those who had given their lives for their country in the Great War were appropriately mourned and memorialized, yet the United States had not suffered anything like the physical destruction or spiritual devastation visited on victors and vanquished alike in Europe. The war had, on balance, strengthened America's standing among the Great Powers, and as demobilization proceeded, the American economy boomed. More precisely, it boomed in those sectors, such as automobiles and entertainment, that expanded to meet growing consumer demand; but in those areas of the economy having excess capacity as a result of rapid wartime expansion, severe competition drove prices down and brought on industrial depression long before 1929. Foreign shipping was one of these overbuilt industries. In the absence of a war-induced demand to occupy the war-built fleet and the war-trained pool of manpower, utilization rates in commercial shipping fell to distressing levels. The glittering public image of first-class travel on subsidized passenger liners distracted attention from serious underemployment, falling wages, impotent unions, vanishing profits, absence of reinvestment, and the impending collapse of the U.S. merchant marine. The Great Depression that brought the Roaring Twenties to a close simply added to the misery of a shipping industry that was already sick.

As the depression deepened and President Roosevelt and his advisers wrestled with the Hydra-headed problem of reviving the American economy, it became evident that the merchant marine posed a number of peculiar challenges. Unlike the railroads or protected coastal shipping, overseas shipping had to contend directly with foreign competition. Ever since the end of the Civil War it had been clear that the United States was a high-cost carrier in international trade. The competitive disadvantages under which the industry labored were exacerbated by legislation that further increased American costs and reduced management options by regulating working conditions and limiting business practices that could have alleviated competitive pressures. In 1916, when it was realized that for both commercial and security reasons the United States had to have a merchant marine, the federal government resolved to fund it. Doing so in a wartime economy proved to be a fairly straightforward, although expensive, undertaking; the political challenge of designing and sustaining a commercially-operated, peacetime merchant marine proved to be far more daunting. The revived mail subsidy system of 1928 was so badly managed that it was correctly judged a failure. When in a time of general economic depression the government addressed this problem again, it acted with unprecedented boldness, setting aside the equivocations of the past and grasping the central difficulties with a directness never before seen in American maritime policy. The result was the landmark Merchant Marine Act of 1936, the most comprehensive maritime legislation ever enacted by the federal government. Although the world has changed greatly since this act was passed and the act itself has been repeatedly amended, it is still the basic legislation that defines American maritime policy today. At the end of the twentieth century it is easy to identify the inadequacies of the 1936 Merchant Marine Act, but its novelty, vigor, utility, and promise at the time it was passed should also be noted.

THE BLACK COMMITTEE REPORT

The authors of the 1936 act were able to take a comprehensive view of maritime policy because they were addressing a problem that had attracted attention yet defied solution for fifteen years. When President Warren Harding entered office in 1921, he appointed Albert D. Lasker as chairman of the Shipping Board. Lasker inherited a problem. Many congressmen wanted the government to retain ownership of the war-built fleet, but the contracts that the government had agreed to with the private operators who ran its ships made this an expensive proposition. The operators were paid a percentage of gross revenues plus certain commissions, an arrangement that provided no incentives to control costs. Furthermore, the operators increased their profits by establishing ancillary companies that provided stevedoring and ship chandler services for the government ships under contract. This highly profitable arrangement, known as "transfer pricing," was entirely legal but was clearly suspect. Lasker had great faith in private enterprise and considered the existing system wasteful. He insisted that the government did not know how to operate ships and estimated that it was losing $16 million per month by doing so. He recommended that the vessels be sold to private interests as quickly as possible and that the government provide large bounties to the operators who bought them. The Democratic minority in Congress considered this proposed raid on the Treasury outrageous. President Harding backed the bill Lasker introduced, and it cleared the House. Several Republican senators shared their Democratic colleagues distaste for it, however; and it failed in the Senate.

In 1929 the government set out to sell the United States Line, the last major operating company it owned, but the Shipping Board attempted to rig the choice of buyers in ways that were both obvious and embarrassing. Its buyer of choice was the International Mercantile Marine (IMM), but in the first round of bidding a Chicago banker, P. W. Chapman, outbid the IMM group. Chapman won the company, but when he encountered financial problems and asked the board to revise the payment schedule, it refused and foreclosed, bringing the company back under its control. When it was put up for sale again, Stanley Dollar, owner of the Dollar Line, won the bidding. The IMM group then insisted Dollar's bid was "not fully responsive," and the board quickly agreed. The board finally managed to sell the company to IMM, but only on terms that would have enabled Chapman to succeed in his original attempt to buy it and after the board had insisted that IMM pay Dollar enough to prevent litigation.[1] It was not a stellar example of democracy in action, as the press gleefully pointed out, and in 1931 an embarrassed President Hoover felt compelled to appoint a special investigating commission to review the Shipping Board's operations.

Scrutiny of the mail subsidy system that had been revived in 1928 proved even more damaging, as was indicated at the end of the previous chapter. When Franklin D. Roosevelt won the election in 1932 the Democrats had been out of the White House for thirty-eight years, except for the period when Wilson was president. Having also won only four congressional elections during that period, they thirsted for power and political legitimation after their years in the wilderness. With the economy in ruin and the Republicans in disgrace, they were not unwilling to find partisan scapegoats; happily for the Democrats, the long record of Republican administration offered rich pickings. In the maritime field the harvest was brought in by Senator Hugo L. Black of Alabama. He conducted a detailed examination of the mail subsidy system, held extensive public hearings, and in 1935 produced one of the most scathing reports ever produced by a congressional committee investigating a specific industry and a governmental agency.

Black's report began simply: in recent years the United States had tried to "create and maintain an adequate privately-owned American

Merchant Marine." In fact, however, "it was . . . neither adequate nor in any true sense privately owned." In a "plague on all your houses" judgment, the report identified the sources of this policy failure as bad law, corrupt administration, and greedy industrial exploitation: "First, this burden of costly failure rests upon the enactment of an ill-advised compromise law. Second, upon certain public officials who flagrantly betrayed their trust and mal-administered those laws. Third, upon those individuals who publicly posing as patriots, prostituted those laws for their private profit."[2]

The Black committee buttressed this sweeping condemnation with a wealth of detail, much of which indeed seemed outrageous. Even after mail subsidies were available, for instance, the Shipping Board continued to operate a number of ships through private companies by providing lump-sum payments in place of the previous cost-plus arrangements. The Lykes Brothers–Ripley Steamship Company, for instance, received $7,000 compensation per voyage yet claimed they were unable to make a profit. The company therefore asked that the payment be more than doubled, to $14,500, which was done without further investigation. But the Black committee did investigate, and it established that the operation had been profitable at the lower level of support. Between August 1930 and June 1933, Lykes had not only made profits from subsidiary companies that serviced government ships, but realized profits of $1,702,770 from the lump payments as well.[3]

The management of mail subsidy contracts was even worse. The committee report noted that only nine of the forty-three mail contracts that had been awarded were for less than the maximum allowable rate. When every one of those forty-three contracts was put out to bid, there was only one bidder. Only three of the contracts were awarded to a "person other than the persons for whom it had been planned in advance to award the contract." Even these three exceptions were soon eliminated, for this small group of contracts "was acquired in less than a year by the Line for which it was originally intended."[4] Mail-contract subsidy rates were set according to specified minimum vessel sizes and speeds. The contract for the *Margaret Lykes* required a speed of thirteen knots, but expert witnesses assured the Black committee that the ship could not go that fast. The committee estimated that through false claims Lykes had received $437,194 in over-payments and that other "unjustified and inexcusable classifications cost the government prior to January 1935 not less than $15,489,658."[5] The Grace Line had exploited the system by obtaining a subsidy for a coastal trade route. A Grace subsidiary, the Panama Mail Line, added a call at Havana to its voyage and applied for a mail contract. The Research Bureau of the Shipping Board notified the commissioners that the mail subsidy law did not provide support for coastal operators, but the contract was awarded anyway. This maneuver gave Grace Line a distinct advantage over the American Hawaiian Line and the Luckenbach Line, both of which operated in the same coastal service. And Grace Line was not the only coastal operator to receive a contract.[6]

The litany of misadministration continued. As the report plaintively noted, "it would seem elementary that aid in the form of postal contracts should not be extended to operators of foreign flag tonnage," yet it was. Many of the companies that received mail subsidy contracts were acting as agents for foreign companies that were competing with American companies. The United Fruit Company, which ran its fleet under foreign flag, was awarded contracts. The American Scantic Line, whose subsidiary company Moore-McCormack Lines operated Norwegian ships on the New York to the east coast of South America run, was also awarded contracts. The board was also supporting the Munson Line, one of Moore-McCormack's American competitors on that run, with $48,000 per voyage. Although none of these arrangements was illegal, the committee noted that they circumvented the intent of the mail subsidy law, which was to promote an American-flag merchant marine.[7]

Other legal but clearly improper arrangements were also uncovered. Companies were shifting costs to milk the system for all it was worth. The New York and Cuba Mail Line had a mail contract but was operating at a loss; its parent, the Gulf and West Indies Steamship Line, was at the same time showing a large profit thanks to the high charter rates it charged its subsidiary for the use of its ships. In one year the owners, having invested $3 million, paid themselves a dividend of $900,000.[8] The Grace Line, a subsidiary of W. R. Grace and Company, held several mail contracts. The parent company had diverse business interests, including banking, trading, and commercial aviation, and provided its officers with large salaries and benefits—its president, J. P. Grace, was paid $971,660 annually. The Black committee found such salaries inappropriate for a company receiving government support, but it realized that under current law it could do little more than complain.

Other executives were charged with having behaved with even more blatant disregard for the common good. The committee noted that John J. Farrell and his brother were "the Chief Officers of the complicated corporate structure of shipping companies and ship service companies which surround and almost concealed the subsidized American-South African Line, Inc.," later renamed the Farrell Line. Between 1926 and 1933 John

Farrell, although not an officer in any of these companies, had drawn $1,827,212 in salaries, dividends, and appreciated stock. As even Mr. Farrell admitted, when testifying before the committee, "these corporate returns and devices should be prohibited by law."[9] Stanley Dollar was also criticized for profiting through similar arrangements, as were the presidents of Export Steamship Corporation (Export Line), Waterman, IMM, Mississippi Shipping Company (later Delta Line), Scantic, Munson, and Lykes.

The committee's scorn was not restricted to the executives who ran the shipping companies, questionable as some of their behavior may have been. The government officials who should have made sure that public funds were being well and properly used were also pilloried. The Shipping Board commissioners and the responsible officials in the post office had failed in their duties. In 1930 the comptroller general had notified the Shipping Board that payment of lump-sum contracts was no longer legal, but the board had continued them nonetheless. The comptroller had also warned the board against awarding ocean-mail contracts without competitive bidding. The Black committee was astonished to discover that mail contracts were awarded on routes "where no mail moved." President Roosevelt's new postmaster general, James A. Farley, was eager to clean house and advised the president that "out of forty-three such active mail routes only twelve are of substantial value as mail carriers, eight are of slight postal value, twenty-three have no postal value whatever, and a number of them are actually detrimental to the speedy transmission of the mails."[10]

There were yet other failures to be catalogued. The 1928 act had in fact not resulted in any significant increase in tonnage, despite the vast funds expended. Only twenty of the forty-three contracts required that new ships be built. Of the fifty-one ships built to meet the requirements of subsidy contracts, twenty-eight had been completed by 1935. The report noted, "it appears that if all new ships required by the contracts are constructed and the old vessels now being operated remain in service, there will be in service when the contracts expire only sixty-five ships under seventeen years of age."[11] This is hardly a prediction that those who thought government funds were being used to underwrite a modern and competitive international merchant marine could find encouraging.

By the time the Black committee had completed its investigation, it had encountered most of the difficulties that typically arise when public funds are used to underwrite private industry. And since the committee was meeting during a period of industrial depression, when capitalism's failure to serve the public good seemed indisputable and the nation was looking to Washington for leadership, it found it convenient and gratifying to emphasize the moral failings of the shipping executives and their bureaucratic counterparts rather than the practical difficulties of yoking public and private resources to a common task. The committee evidently had no new ideas on how to grapple with the problem of turning individual self-regard into benevolent social action; it therefore concentrated on condemning the greed of the operators. Given the evidence provided, one can hardly say its condemnations were misdirected. The committee's final report was especially critical of the operators' reluctance to put their own money at risk: "This [reluctance] indicates that the ideal of a privately owned merchant marine set up by the Merchant Marine Act of 1920 and 1928 has in no sense been attained, nor is such an ideal likely to be attained (if attainable at all) under a system under which those who are in theory to become the private owners calculate substantially as follows: '2 and 2 are 4; the Government should give us 2, and inasmuch as we do not have the other 2, the government should loan us that 2 also.'"[12] Since private interests seemed determined to minimize their risks as individuals while maximizing the benefits they reaped from public funding, the Black committee declared the government's attempt to fund a privately owned merchant marine a complete failure: "In both the managing operator plan and the mail-contract plan, the losers were the taxpayers and the government, the operations and the profits were by and for private interests."[13]

After thoroughly discrediting the existing maritime programs, the Black committee concluded its withering report with a number of cautions and recommendations. It hoped that anyone bold enough to try yet again to design legislation that would use public funds to promote and sustain a privately owned merchant marine would consider these recommendations, as the authors of the 1936 act most definitely did. Indeed, the complexity and thoroughness of the 1936 act can only be understood when seen as emerging from the findings and recommendations of the Black committee.

The committee identified three alternatives for a public policy supporting the merchant marine: (1) government ownership and operation, (2) government ownership and private operation, and (3) private ownership and private operation. While the committee favored the first option, it acknowledged that it was not politically viable. It believed that the second option had been tried and discredited. This left option three as the only choice. Previous approaches to implementing the third option had failed; indeed, the committee strongly recommended that the Merchant Marine Act of 1928 be repealed and that the existing mail contracts be canceled. Yet it also recognized that the third option required some form of government subsidy to offset the higher cost of operating U.S.-flag ships in foreign trade. The committee's final report did not contain a design for a new subsidy system, but it did offer a number of safeguards that it believed should be built into any new system:

1. The subsidy should be administered by fearless, uncompromising men.
2. Subsidy, rather than being seen as a temporary subterfuge, should be established as an integral part of a permanent policy.
3. Subsidy should be available to all ships engaged in foreign commerce.
4. No subsidy should be paid to ships engaged in coastal or intercoastal shipping.
5. Construction subsidies should be paid directly to the shipbuilder, based on the differential of foreign shipbuilding costs. No special government loans should be provided.
6. Operating subsidies should also be based on foreign cost differentials.
7. No subsidy should be paid to an operator who fails to comply with government-required manning, wage scales, and labor conditions. The report continued by stating its belief that it is a primary duty to see that this portion of the subsidy (the operating differential) reaches its intended beneficiaries.
8. Both shipbuilding and ship operating subsidies should be subject to recapture of 75 percent of all profits over 6 percent.
9. No subsidy should be paid to an operator or shipbuilder who pays wages, salaries, or other compensation exceeding $17,500 in any one year to any officer, agent, or employee.
10. No operating subsidy should be paid to any line operating in competition with an unsubsidized American-flag line rendering adequate service.
11. The Subsidy Board should be disbanded and a new organization created to administer the revised subsidy program.
12. The regulatory functions of the Shipping Board should be transferred to the Interstate Commerce Commission.
13. No subsidies should be paid to a company where income resulting from a subsidy could be diverted to activities unrelated to American-flag, foreign-trade shipping.[14]

The Black committee report, written during a period in which the government's responsibility to structure and direct industry was widely accepted, was signed by the committee composed of three senators: Hugo R. Black, William H. King, and Pat McCarran. But Senator King, a Democrat from Utah, had serious reservations about the recommendation that a new subsidy system be constructed; he thought the emphasis should be on removing restrictions that made U.S.-flag ships uncompetitive in world trade. King noted that "if American citizens were not prevented from purchasing ships built in foreign countries and operating them under American registry . . . one of the obstacles to the realization of an effective merchant marine would be removed."[15] But as before, this approach to the problem found few supporters; in the 1930s providing public subsidies, rather than reducing regulatory burdens, remained the favored strategy in maritime policy.

THE MERCHANT MARINE ACT OF 1936

While the Black committee was doing its work, the maritime industry was also being examined by an Interdepartmental Committee on Shipping Policy chaired by the secretary of commerce. This second committee generally supported the merchant marine, but it did call for stricter government supervision of subsidy programs and recommended that a new Office of Maritime Affairs headed by an assistant secretary be created. Although this latter recommendation was not implemented until 1970, when an assistant secretary of commerce for Maritime Affairs was first appointed, the Interdepartmental Committee's recommendation that future subsidies be based on foreign cost differentials so that U.S. operators could compete on a par with foreign carriers was made part of the 1936 act.

In 1935 President Roosevelt endorsed and forwarded to Congress the reports of the Black committee and the Interdepartmental Committee. His accompanying memorandum showed that the president was well aware of the merchant marine's importance and of its need for sustained federal support. Roosevelt listed three reasons why the United States needed an adequate merchant marine. The first addressed American shipping needs in peacetime: "in time of peace, subsidies granted by other nations, shipping combines, and other restrictive or rebating methods may well be used to the detriment of American shippers." The second reason addressed American shipping needs in times of war when the United States is a neutral: "in the event of a major war in which the United States was not involved, U.S. commerce in the absence of an adequate American merchant marine, might find itself seriously crippled." The third reason addressed U.S. shipping and naval needs in a major war in which the United States is a belligerent: "in the event of war in which the United States itself might be engaged, American-flag ships are obviously needed not only for naval auxiliaries, but also for the maintenance of reasonable and necessary commercial intercourse with other nations." This list of traditional reasons for supporting the merchant marine led Roosevelt to propose forthright subsidies as the way to revitalize America's maritime heritage:

> In many instances in our history, the Congress has provided for various kinds of disguised subsidies to American shipping. . . . I propose that we end this subterfuge. If the Congress decides that it will maintain a reasonably adequate merchant marine, I believe that we can well afford honestly to call a subsidy by its right name. . . .
>
> An American merchant marine is one of our most firmly established traditions. It was, during the first half of our nation's existence, a great and growing asset. Since then, it has declined in importance and value. The time has come to square this traditional ideal with effective performance.[16]

The depression radically altered the political alignment on maritime policy. In the nineteenth century the central issue had been "free ships," with the Republicans stoutly protecting the American shipbuilding industry and the Democrats favoring less restriction on trade. But by the 1930s both sides of the aisle were supporting a thoroughly national merchant marine. The Democrats, newly allied with the labor movement, were deeply concerned with getting America back to work, a commitment that obliged them to protect American jobs. Defending the interests of the workingman was no longer a task that could be left to labor organizations alone. The maritime trade unions had been devastated by the shipping depression of the 1920s and played only a minor role in securing passage of the 1936 Act. The carriers were better organized, but since they were looking for government assistance, they were hardly in a position to press their "natural" opposition to oversight and regulation. The operators, as represented by their trade associations, therefore focused primarily on limiting government regulation of freight rates. The shipbuilders, meanwhile, used the opportunity to eliminate the opening that had been introduced for free ships in 1915; they successfully insisted that all future subsidies be restricted to ships built in the United States.

Roosevelt, rightly concerned with the task of administration, asked Congress to consider seriously many of the safeguards the Black committee had proposed to prevent abuse of whatever subsidy system was devised. He left the hard work of designing suitable legislation to Congress, however, a task that the Senate addressed sporadically during the next fifteen months. From the outset there was general consensus that America needed a merchant marine. And although the military services provided no supporting testimony, there was also general agreement on the need to sustain the merchant marine so that it would be available to serve as a naval auxiliary in time of war. In 1935 the armed forces were interested in having a national merchant marine at hand and on call when needed, but they carefully avoided saying anything that might make it appear that supporting the merchant marine was a proper use of funds provided for military readiness. This left as the central issue under debate how to design legislation that would provide effective federal support for the merchant marine while also including administrative provisions capable of preventing abuse of the new subsidy system.

The debate over how best to use public funds covered familiar ground. Senator Black strongly advocated outright government ownership. He insisted, with considerable reason, that if the Shipping Board had operated all the government-built ships itself, it could hardly have done a worse job, from the taxpayers' point of view, than the private operators. The shipowners were vehemently opposed to government ownership and operation, but in light of the record of their own stewardship, their arguments had to be somewhat circumspect. Ideology was on their side, however, and that proved to be enough. Government ownership and operation was portrayed as tantamount to socialism; private ownership was the American way. But even those who championed private ownership and operation acknowledged that if public funds were used to subsidize the building and operation of U.S.-flag ships, then there had to be adequate government oversight of the use of those funds. Striking the right balance between subsidizing private enterprise and controlling the use of public funds was indeed a thorny problem.

The story of how the 1936 Merchant Marine Act was drafted, revised, and ultimately passed is replete with the kinds of compromises, hidden agendas, and perilous moments that inevitably accompany the writing and enacting of complex laws. At first it appeared that Congress would never even draft a new maritime bill. In February of 1936, after several months of inactivity, President Roosevelt called the opposing congressional factions together in an attempt to get them to reconcile their differences, but nothing came of it. A month later the Senate Commerce Committee voted out a compromise Merchant Marine Bill but did not recommend its adoption. Several more months passed without its being considered by the full Senate. The Senate Appropriations Committee then forced the issue by threatening to withhold funding for all future mail subsidies unless the president's bill was brought to a vote. On June 19, 1936, one day before adjournment, the bill was ratified by the Senate; it was then rushed to the House where it passed without modification.

The bill that emerged strongly reflected both the larger concerns that occupied Congress during the depression years and the strategies the Democrats were using to get America back to work. It was in many ways a bill designed to stimulate employment. The government, through its Maritime Administration, would work with the seafarers' unions, the operators, and the Coast Guard to set manning levels, working conditions, and minimum wage levels. Rather than attempting to close the gap between the expectations of American seafarers, as spelled out in existing laws and labor agreements, and the very different practices on ships sailing under the flags of other nations, the bill directed the government to maintain high standards for American workers and subsidize the difference. The important thing was to get men back to work at decent jobs with decent pay; the economic and industrial consequences of the way it was done could be dealt with later.

The bill also reflected New Deal determination to restructure industries that had failed. From March to mid-June 1933, the memorable first hundred days of Roosevelt's administration, industries of fundamental importance, such as farming and banking, were reorganized and regulated under comprehensive legislation that remained in effect until the last years of the twentieth century. Agricultural policy, for instance, was guided by a commitment to restore economic "parity" between farm products and industrial products; if this was not done, farmers would have had no incentive to grow crops and bring them to market. To achieve parity and reinvigorate the vitally important agricultural sector of the American economy, the government decided to use its tax power to transfer funds from consumers to agricultural producers.[17] The subsequent program for subsidizing shipbuilding and foreign liner service obviously came from the same legislative mold. And just as New Deal farm policy "frankly envisaged the farmer as part of the national economy and assigned the federal government the decisive role in protecting farm income,"[18] so too did the Merchant Marine Act of 1936 make the federal government the guarantor of "reasonable" incomes for those working in maritime industries. Opponents of these bills also reacted in similar ways: a representative opposed to the Agricultural Adjustment Bill characterized it as "more bolshevistic than any law or regulation existing in Soviet Russia."[19] The maritime industry's reaction to the 1936 act was ideologically similar but tempered by a long history of dependence on various kinds of federal support.

Congress knew in a general sense what it wanted to do for the maritime industry and how it wanted to do it, but the bill that emerged was necessarily imprecise on topics that were still in dispute. Those assigned responsibility for implementing the new law therefore had considerable maneuvering room. Indeed, the law did not set targets for the number of ships to be built or for the percentage of U.S. foreign trade to be carried in U.S.-flag vessels. The preamble to the 1936 act repeated that of the 1920 act in calling for a fleet "sufficient to carry a greater portion of its [America's] commerce," except that the 1936 act spoke of a "substantial portion" and added the requirement that the fleet be U.S. built as well as U.S. flagged. But the meaning of "substantial portion" was left vague, and no requirement for reaching this elusive target was included. Maritime officials subsequently took the phrase to mean approximately half of all U.S. foreign trade, but achieving this goal never became a realistic objective.

The act that finally emerged in June 1936 was organized into eight sections or titles. Title 1, a general statement of policy, closely followed the preamble of the 1920 act. Title 2 established a new Maritime Commission to replace the Shipping Board, which in 1933 had been reduced to three members and made a part of the Department of Commerce. The five-member Maritime Commission assumed most of the Shipping Board's duties. These were broadly defined under three headings, the last of which reflected the recent loss of 134 lives in the 1934 *Morro Castle* disaster:

> First, the creation of an adequate and well-balanced merchant fleet, including vessels of all types, to provide shipping services on all routes essential for maintaining the flow of the foreign commerce of the United States, the vessels in such fleet to be so designed as to be readily and quickly convertible into transport and supply vessels in a time of national emergency. In planning the development of such a fleet, the Commission is directed to cooperate closely with the Navy Department as to national-defense needs and the possible speedy adaptation of the merchant fleet to national-defense requirements.
>
> Second, the ownership and the operation of such a merchant fleet by citizens of the United States insofar as may be practicable.

> Third, the planning of vessels designed to afford the best and most complete protection for passengers and crew against fire and all marine perils.[20]

Although charged with creating a merchant fleet "including vessels of all types," the Maritime Commission was authorized to provide subsidies only to ships operating in liner service as common carriers; no provisions were made for supporting U.S.-flag bulk carriers or ships restricted to carrying proprietary cargoes.

Shortly after its creation the Maritime Commission was obliged to focus all its resources on providing ships for a nation at war, yet it should be noted that the 1936 act also charged the commission with promoting the economic and commercial interests of the industry as well. The commission was directed to undertake studies that would promote the American merchant marine and "to cooperate with vessel owners in deciding means by which . . . the importers and exporters of the United States [could] be induced to give preference to vessels under United States registry."[21] This desire to bring into harmonious conjunction market-oriented commercial activities and a heavily subsidized and regulated industry appeared commendable in principle, even though it seems unlikely that such a marriage could have succeeded in the long run. This mandate was quickly forgotten, however, when the war-driven command economy took charge.

Title 3 of the new law brought working conditions and minimum wages directly under government supervision for the first time. All officers and crew members on subsidized ships were to be American citizens, except that 10 percent of the crew on passenger ships could be foreign seamen. The Maritime Commission was directed to define "minimum manning scales and minimum wage scales, and minimum working conditions for all officers and crews . . . on ships receiving operating differential subsidy."[22] Since crew costs on subsidized ships would obviously be higher than on foreign-flag ships, the government agreed to pay the differential. One of the unanticipated consequences of this commitment was that the wage levels established for subsidized operations soon set the standard for all U.S. shipping companies, whether subsidized or not. Title 4 terminated the existing ocean-mail contracts as of June 30, 1937, and transferred all the duties vested in the postmaster by the 1928 act to the new Maritime Commission.[23]

Titles 5 and 6 described the heart of the act's new subsidy system, the linked Construction Differential Subsidy (CDS) program and Operating Differential Subsidy (ODS) program. Both were only available for ships engaged in foreign-trade liner service. Title 5 laid out the process by which CDS payments were to be determined and paid. An operator who wanted to receive a construction subsidy to help pay for building a new ship in a U.S. shipyard first had to convince the Maritime Commission that "a new vessel of economical design was required to meet foreign competition and to promote the foreign commerce of the United States."[24] The navy was then given an opportunity to review the plans of the proposed ship to determine if it wanted any additional features, such as stronger decks or greater speed, that would make the vessel more useful to the military in time of emergency. The cost of any national defense features demanded by the navy would be paid for by the government. Once the final plans were approved, the commission would solicit competitive bids from U.S. shipyards and award the construction contract to the "lowest responsible bidder." At the same time, the commission and the future owners of the ship would enter into a contract that stipulated how much the operator would pay the commission for the ship. The law said that the cost to the owner was to correspond to "the estimated cost as determined by the Commission . . . to [build] such a vessel in a foreign yard."[25] The act stated that the construction differential paid by the commission should normally not exceed one-third of the ship's total cost. When there was evidence that the actual differential was higher, the subsidy could rise to 50 percent, which quickly became the norm as well as the maximum.[26]

The government imposed a number of constraints on owners who made use of the subsidy programs. Ships purchased through the CDS system had to remain documented under the laws of the United States for at least twenty years. The government also was authorized to buy the vessels that the new ships replaced; if it did so, the older ships could not again be placed in foreign trade.[27] Shipowners receiving operating differential subsidies were also required to establish a capital reserve fund (CRF). Each year the owner was to deposit into this fund a sum equal to the depreciation of vessels covered by the subsidy agreement. The fund was also to receive "deposits of proceeds from sale of vessels, indemnities on account of losses of vessels, earnings from the operation of vessels and receipts in the form of interest." Funds paid into the CRF were tax sheltered and could only be spent according to rules and regulations jointly prescribed by the commission and the secretary of the Treasury.[28] The purpose of the CRF program was to encourage shipowners to accumulate capital funds for the construction or acquisition of up-to-date vessels.

Title 6 described the Operating Differential Subsidy (ODS) program. Operators wishing to participate in this program had to commit to build or purchase the ships needed "to meet competitive conditions" on a particular trade route and to make a specified number of voyages on that route each year. They also had to agree to replace their vessels on a regular basis, which in practice meant every twenty years, or forfeit future subsidies. Meeting these conditions became part of a contract between the Maritime Commission and the operator. The commission's contractual obligation was to subsidize "the fair and reasonable cost of insurance, maintenance, repairs not compensated for by insurance, wages and subsistence of officers and crews, and any item of expense . . . [where] the applicant is at a substantial disadvantage in competition with vessels of a foreign country."[29] The government's subsidy was to be limited to no more than 75 percent of such costs. In return for receiving an operating subsidy, shipowners had to agree that half of any earnings in excess of 10 percent, when averaged over five years, would be returned to the government. The Black committee had recommended a more stringent rule for recapturing excess profits, but the operators nonetheless considered the terms set in the 1936 act a severe denial of any incentive to improve company performance.[30]

Title 7 authorized the commission to initiate construction of ships for the government's account should it find the shipbuilding plans of private operators insufficient. Such ships could then be sold or chartered to private operators.[31] This provision was utilized in 1952, when the commission built thirty-five Mariner-class ships, but it also created considerable disagreement within the maritime industry. Shipbuilders repeatedly called for "build and charter" programs that would keep their yards busy while not making comparisons between their costs and those of foreign yards. Such programs would also enable individuals who did not have enough funds to purchase ships to enter the shipping industry by chartering government-built ships. But the established shipping companies viewed such programs as disruptive, for they flooded the market with excess tonnage. Since most operators did not need additional ships, it was unlikely they would bid on government ships. But in as much as the law obliged the government to award charter contracts to the highest bidders, such programs created an avenue by which weakly backed new operators could enter into competition for available cargoes. Most established shipping companies therefore opposed government build-and-charter programs.

Title 8 imposed many of the safeguards against exploitation of public funding that had been recommended by the Black committee. The government was given the right to requisition vessels under construction after giving the owners proper compensation, a provision that was used on a number of occasions in wartime. Any contractor operating under an ODS agreement or operating a government-owned ship under title 7 was prohibited from employing anyone in anyway related to the company to provide "stevedoring, ship repair, ship chandler, tow boat or kindred services . . . to subsidized or chartered vessels."[32] Shipping companies were allowed to establish and employ companies to provide such services so long as they were wholly owned, in which case their profits would appear on the parent company's balance sheet and be subject to recapture. Subsidized companies were also prohibited from entering into any phase of the coastal or intercoastal trade. They also could not pay any of their officers or personnel more than twenty-five thousand dollars per year. This was a highly unusual provision to write into law, one that was not imposed on any other government contractor, not even the shipbuilders. If shipbuilders or operators wished to hire lobbyists, they had to obtain prior permission from the commission and file reports of their activities. The prohibitions against predatory shipping practices, such as the use of fighting ships, contained in the 1916 Shipping Act were strengthened; any company using them would lose its contract. In sum, the many specific provisions and prohibitions contained in the 1936 Merchant Marine Act reveal that the Black committee report powerfully shaped New Deal maritime policy.

The Merchant Marine Act of 1936 is a legislative landmark of unrivaled importance in the history of U.S. maritime policy. Although it contains the kinds of compromises and imprecisions that are inevitable in all contested and complex legislation, it was unprecedented in its comprehensiveness and has proven both flexible and long-lived in application. And yet it is well to remember that the design of legislation is only one of the factors that determines the success or failure of public policies. Skillful administration is also required, and while the Maritime Commission was often competently managed during its early days, it has also frequently exhibited the kind of uninspired institutional behavior that is rightly condemned as bureaucratic. As James Landis, one of the most astute authors of New Deal legislation, frequently remarked, the importance of vigorous and farsighted administration is too often ignored:

> Broadly speaking, the problems of legislation and of administration divide themselves into two phases. The first relates to the determination of what policies to pursue; the second, to the discovery of how to make the chosen policies effective. Political discussion commonly centers about the first. The choice of policy is attended by all the excitement of conflict among varying philosophies and among divers group pressures. Yet this phase is

transitory. . . . The administrative phase, on the other hand, is enduring in character. It requires a continuing effort of indefinite duration. . . . Yet, by a kind of perverse irony, this second phase all too often receives only casual attention.[33]

THE MARITIME COMMISSION TAKES CHARGE, 1937

The first task that the new Maritime Commission faced was to distinguish itself from the discredited Shipping Board that had mismanaged federal maritime policy in the 1920s. President Roosevelt, knowing that it would take vigorous leadership to bring the maritime industries into line with New Deal thinking, appointed Joseph P. Kennedy as the commission's chairman. Kennedy, remembered today primarily as President John F. Kennedy's father, had no prior experience in the maritime field, but he was a successful businessman and had proven himself a formidable administrator as head of the recently created Securities and Exchange Commission. The other commissioners who joined Kennedy when the Maritime Commission first met in April 1937 were also able men. Two of them brought navy backgrounds: Vice Admiral H. A. Wiley, who served as vice chairman, and Rear Admiral Emory S. Land, a naval engineer who had worked with Roosevelt during World War I. Thomas Woodward, a lawyer who had managed to retain a reputation for integrity while serving on the Shipping Board, also joined the commission, as did Edward Moran, a former congressman from Maine who maintained a healthy skepticism about the commission's mandate. When Joseph Kennedy resigned the chairmanship after ten months to become ambassador to Great Britain, he was replaced by Admiral Land, who remained as chairman throughout World War II.

The new commission was given no grace period to recruit and organize staff and obtain the funds needed to operate; it had to concern itself immediately with terminating the existing mail contracts and arranging new contracts that would ensure the nation's essential trade routes were served. The mail contract situation could not have been worse; the entire matter was entangled in complex litigation, the nightmare of every regulator and administrator. The companies holding mail contracts had filed claims against the Shipping Board for $166 million, while the Black committee had referred some of its findings to the Justice Department for possible prosecution. Kennedy, in a virtuoso display of negotiating skill, persuaded all parties, including the Justice Department, to drop all legal action a few days before the June deadline for ending the contracts. The claims that remained were settled for less than $1 million.[34]

Reorganizing foreign liner service so as to ensure that all essential trade routes were covered was equally daunting. Here the commission did what economic forces and prior subsidy systems had not done for the shipping industry, it reduced the number of companies and consolidated operations in the hands of a few financially strong corporations. The commission quickly realized that many of the thirty-two different companies that had held mail contracts with the Shipping Board were underfunded and would never be able to buy new ships. Kennedy set as a target reducing the number of companies to twelve. Only seventeen of the companies that had previously held mail contracts were invited to join the new subsidy program, and by 1939 only twelve shipping lines remained.[35] Several companies were driven off by the more stringent financial requirements imposed by the 1936 act; others withdrew because intercoastal shipping was no longer subsidized.

While consolidating service on two essential trade routes, the commission acquired several shipping lines by default. The Dollar Steamship Line was having trouble with debt service on two of its newer ships, *President Coolidge* and *President Hoover*, and had hoped to receive enough funds from mail contracts to meet its obligations. Instead of assuming responsibility for servicing these debts, however, the commission accepted most of the company's stock in return for paying off both mortgages. A year later the Dollar Line, now owned by the government, was renamed American President Lines (APL). Something similar occurred to the Panama-Pacific Line, an intercoastal service owned by the International Mercantile Marine. When it ceased to receive mail subsidies, the line sold its three ships to the government and went out of business. After the Munson Line went bankrupt the commission had to take over the service it had operated between ports on the U.S. East Coast and the east coast of South America. President Roosevelt had a special interest in this service, for he had visited South America during his first term and was determined to improve hemispheric relations. The commission designated its three reconditioned Panama-Pacific ships the "Good Neighbor Fleet"—they were operated by Moore-McCormack Lines and were assigned to the South America run. Both fleets were run for the commission by operating companies through charter arrangements.

The commission also addressed the problem of providing the American merchant marine with adequate ships. It estimated that up to fifty new ships would have to be built each year and developed designs of its own to standardize production. It designated its cargo ships C-1, C-2, C-3, and C-4, according to their size. During World War II two versions of the C-2 were mass produced: the EC-2, better known as the Liberty ship, and the VC-2, the bigger and faster Victory ship. Passenger ships were designated P-2 and P-3; they served as troop carriers or hospital ships during the war. The C-4s built during the war were also converted to troop carriers and hospital ships. The standard tanker was the T-2; it had a remarkably advanced design of sixteen thousand deadweight tons (DWT) with turboelectric propulsion and was the workhorse of the tanker fleet in the war.

In its first year the commission also began a study of the industrial economics of the maritime industry. This initiative clearly reflected the ambition and determination of the New Deal planners: if the government was going to take command of the industry, then the commission to which that responsibility was assigned had an obligation to understand the industry's structure and dynamics. The report that summarized the study's findings, *An Economic Survey of the American Merchant Marine*, was published in November 1937. It is a well-informed, clearly written, and unflinching description of its subject, and unlike most historical documents pertaining to the merchant marine, it can still be read with profit.

While *An Economic Survey* concentrates on economic analysis, it is no mere academic exercise. The focus throughout is on questions the commission had to answer: Does America need to support a maritime industry? Can that industry be made competitive? What will be the cost of doing so? The introduction opens with a remarkably direct and well-informed description of the business of shipping and a warning:

> Shipping is our oldest industry; it is also one of our most complex. Shipping, in the first place, is not a business in the usual sense of the word. It is, so far as the United States is concerned, an instrument of national policy, maintained at large cost to service the needs of commerce and defense. It is operated, meanwhile, as a private enterprise. We like to think of shipping as an example of individual initiative, sustained by investments and capable of being operated at a profit. In practice, however, the industry requires substantial Government support to survive. That entails some measure of Government control, which in turn means inflexibility, curtailment of investment, and perhaps in the end an increased need for subsidies.[36]

The introduction's account of the reasons for providing government support for the shipping industry is equally concise: "Careful examination of the arguments advanced on behalf of an American Merchant Marine indicates that there are only two sound considerations that justify the expenditure of public funds to maintain a foreign-going fleet by the United States: One of these considerations is the importance of shipping as a factor in the preservation and development of foreign commerce; the other is the relationship that exists between merchant vessels and national defense. Upon these two considerations must rest the case for the maintenance of a subsidized shipping establishment in the international carrying trades."[37]

A closer examination of commercial shipping enabled the commission to distinguish between valid and tendentious claims for supporting an American merchant marine. The commission noted that in 1937 the United States exported more than any other nation, with exports amounting to approximately 10 percent of all goods produced, and that only Great Britain imported more. But rather than assuming that a great trading nation had to have its own merchant marine, the commission asked if the existence of an American merchant fleet was of any particular importance to U.S. trade. It identified two advantages that appeared to be significant, continuity and quality of service. The commission considered continuity of service the more important and concluded that the availability of U.S.-flag ships provided "a measure of insurance against possible interruption of service." On the question of quality of service, the commission noted the great improvement that had occurred after World War I when newly established American lines forced foreign lines to improve service to stay competitive. Were Americans to withdraw from the major trade routes, the commission concluded, the services provided by foreign carriers would probably decline: "[S]uch a situation would again place our traders at a disadvantage. Dependence on foreign lines usually means dependence on lines that are primarily interested in the trade of their own countries. Their services out of American ports are likely to be less certain and permanent, and for the most part inferior to those offered their shippers at home."[38]

The commission examined the merchant marine's role in defense with equal candor. It was, they acknowledged, the foremost reason for providing government support for this industry, but they also noted that the United States "has renounced war as an instrument of national policy" and that there were no plans to send an expeditionary force abroad. Nonetheless, they concluded that "it would be well to prepare for any contingency." The commission had no interest, however, in engaging in Mahanian speculations about the nature of seapower or in trying to determine just how many ships and of what type might be needed for military purposes: "Although various claims have been made concerning the percentage of naval strength attributable to commercial vessels, the determination of such percentage is believed to be an impossible and unnecessary task. The question of relative importance, it might be said, is about on a par with endeavoring to determine the comparative value of the lungs and the heart to a human being. For the purposes of this study it is only necessary to state that a large volume of merchant tonnage is necessary to the effective functioning of the armed forces of the Nation in time of trouble."[39] The commission did acknowledge, however, that the intended dual use, both commercial and military, of U.S. merchant ships created a serious problem. The navy would require ships to have features that would render them commercially unprofitable. The navy's needs would also skew the allocation of capital funds for shipbuilding away from commercial considerations. The commission was especially worried about the navy's request for three hundred additional tankers, with at least twenty being capable of operating at naval fleet speeds. This requirement had to be given a high priority when the commission was putting together its building program.

The commission proceeded methodically when trying to determine "What are the shipping requirements of the United States?" It noted that American foreign trade was engaged in twenty trading areas served by some twenty-three trade routes emanating from the Atlantic, Gulf, and West coasts of the United States. It also looked carefully at unscheduled "tramp" steamer service and observed that the share of trade carried by such ships was declining. The United States did not participate in this trade and could only have done so if very heavy subsidies were provided; the commission therefore recommended, as the Lynch committee had many years earlier, that the United States leave the tramp trade to others. In 1970, when a later Maritime Administration attempted to support U.S. entry into the tramp trades, the wisdom of the earlier recommendation was confirmed. In 1937 the commission concluded that only liner service should be subsidized. It discouraged the building of large luxury passenger liners, however, saying, "[O]ur participation in international trade should be limited to the operation of vessels deemed necessary for the development of commerce and the preservation of our facilities for defense."[40]

The commission then took a closer look at the subsidized merchant marine as it existed in 1936. What it found was a very small industry with a sense of its own importance that vastly exceeded its position in the American economy. If one did not count tankers, which were largely owned and operated by the major oil companies, the U.S. merchant marine engaged in foreign trade consisted of 374 ships, half of which were subsidized. In 1936 the seventeen existing subsidized lines, together with two that could become eligible for subsidies in the future, had a net worth of $78 million, gross operating revenues of $122 million, and a net profit of only $4.1 million. Clearly, shipping was indisputably a small industry when measured by such usual standards as capitalization, total revenue, and employment. As the commission observed, on strictly business terms "these figures indicate a bleak outlook for vessels in foreign trade and a definite need for substantial government assistance."[41] How, the commission wondered, could such small companies compete successfully against the huge foreign-owned lines, many of which had fleets larger that the entire subsidized U.S. merchant marine? The British Peninsular and Oriental Steamship Line (P&0), for instance, had 339 ships, while the German Hamburg-American Line owned 247 and North German Lloyd had 229.[42] Such major companies had greater operating flexibility, more diversified trade routes, more stable earnings, and greater access to capital. What could not be foreseen in 1937 was the enormous destruction that the approaching war would soon inflict on these dominant foreign fleets.

The commission was especially dismayed by the shipping industry's investment situation. Why would anyone invest in an industry that had so many companies with poor earnings records, that lived in fear of disastrous strikes, that was burdened with high labor costs, that received subsidies from a government that did not stick to a consistent policy, and that was restricted in its operations by the laws that made subsidies available? The normal practice in business was to set aside a portion of the profits earned each year for future capital investment. Companies that earned high profits could also borrow money at reasonable terms in financial markets. But the shipping industry neither earned acceptable profits nor accumulated significant capital reserves. These handicaps were compounded by the restriction on profits imposed as a condition for receiving subsidies. It was, as the commission acknowledged, a vicious circle that would probably prevent the merchant marine from ever achieving commercial viability: "Subsidization, in the popular mind, is a device for the preservation of industries faced with extinction; it is not regarded as a proper instrumentality for guaranteeing profits. The moment a subsidized ship line creates substantial cash reserves and, perhaps, begins to pay dividends, there arises a demand for a reduction in the amount of aid."[43]

This depressing investment picture had a direct bearing on the question of vessel replacement. The commission concluded that among the unsubsidized carriers the oil companies were best able to build the new ships they would need, an estimated forty tankers in the following five years. The companies engaged in the unsubsidized but protected coastal trade had poor earnings records and probably would not be able to replace their fleets unless given further assistance or allowed to buy ships abroad. The subsidized lines would probably need thirteen new ships in each of the next five years, but there was little likelihood that private financing could be found for such a building program: "The brutal truth is that the American merchant marine has been living off its fat for the past 15 years; that is, we have been subsisting upon the war-built fleet. That fleet is now nearing the end of its useful life. Many of our operators built their business on vessels which they secured from the Government at prices as low as $5 a deadweight ton. Who is going to replace these vessels at $200 a ton? The Commission is forced to conclude that, from all present indications, it will have to be the Government."[44]

When the commission turned its attention to labor relations in the shipping industry, it had more cause for dismay. The attitudes of employers and employees were deemed equally "unenlightened"; seamen lived and worked in "deplorable" conditions, and labor relations were chaotic. While the commission was prepared to set new standards for crew quarters on all new ships, it was not empowered to take direct control of labor-management relations. It could therefore do little more than remind the two parties of their responsibilities, which it did in language that still rings true today:

> When a man puts foot on the deck of a ship he becomes part of a disciplined organism subject to the navigation laws of the United States. Seamen must recognize that the nature of their calling, which gives them a unique status under the law, also imposes upon them obligations not common to shore occupations. . . .
>
> Seamen and operators alike must not lose sight of the fact that shipping is primarily an international business. Losses incurred by domestic enterprise as a result of labor difficulties may be made up later. The situation with respect to shipping is vastly different. The business which American lines are unable to handle is likely to go to our foreign competitors. Some of it may be recovered, but much of it will never come back.[45]

To "enlighten" American seafarers and create an adequate pool of well-trained men, the commission called for increased training for both licensed and unlicensed personnel. It proposed to inaugurate a Coast Guard–supervised year-long training program for seamen on Hoffman Island, a vacant quarantine station in New York Harbor. It also recommended the creation of a Federal Maritime Academy "that would partake of the nature of the Army's West Point, the Navy's Annapolis, and the Coast Guard's New London."[46] Plans for what would become the U.S. Merchant Marine Academy at Kings Point on New York's Long Island were begun well before the United States entered World War II, but it was the urgent wartime need for thousands of licensed officers that actually brought the federal maritime academy into being.

Consideration of previous maritime policies occasioned another of the trenchant descriptions of past failures noted in *An Economic Survey of the American Merchant Marine:*

> The policy of the United States with regard to foreign-going vessels has been vague and inconsistent. As a result our shipping has been, since the decline of the clippers, an extremely unstable industry. The history of the American merchant marine for the past three-quarters of a century has been characterized by a vacillating attitude on the part of our people. We would go along for years content to entrust the carriage of our goods to foreign vessels. Then, for some reason or other, we would suddenly become conscious of the potentialities of sea-power and embark on a hastily planned and perhaps frenzied effort to establish a position at sea. In a few years public interest would die down, our vessels would be abandoned to the ravages of competition with low cost foreign operators, and the American flag would disappear from the principal trade routes.
>
> A peculiar inconsistency of the situation was found in the realm of national defense. For many years the United States had embarked on an aggressive naval program, while we did little or nothing to develop an adequate merchant marine auxiliary, even though an American, Alfred T. Mahan, propounded a philosophy of sea-power that was accepted by the entire world. Mahan developed the classic formula of maritime strength: sea-power equals naval vessels plus bases plus merchant vessels. . . .[47]

The commission concluded that American maritime policy could only achieve the steadiness and consistency it had to have if it was forthrightly accepted that subsidies were essential to the continued existence of the foreign-trading merchant marine. While some economists argued that subsidies were only needed because the operators mismanaged their companies, the commission disagreed. They tellingly pointed out that since the United States had long been protectionist in maritime affairs, the need for subsidies should hardly come as a surprise. "Were the United States a free-trade nation, the situation probably would be otherwise. Unfortunately, so far as shipping is concerned, the United States has developed as a protectionist country."[48] The long record of maritime protection has also diffused federal jurisdiction over maritime affairs to a baffling tangle of committees and agencies: "50 bureaus in the executive departments and independent agencies [exercise] some measure of jurisdiction over shipping." The commission pleaded for a consolidation of authority and responsibility, declaring that it was "desirous of calling the attention of Congress to this problem of duplication and division of functions in the hope that a program [would] be adopted whereby the costly duplication and the wasteful and inefficient separation of kindred functions may be eliminated."[49] This plea fell on deaf ears, however, and federal authority over maritime policy is still parceled out to a number of committees and agencies, a situation that makes coordination and accountability extremely difficult.

One of the purposes of the *Economic Survey* was to correct the regulatory excesses of the 1936 act. That act incorporated many of the safeguards against business exploitation that Senator Black had recommended in the final report of the investigation he chaired. Senator Black was also on record as favoring government ownership of the U.S. merchant fleet; since he did not think a subsidized merchant marine could be made commercially viable, the regulatory safeguards he recommended did nothing to encourage commercialization. But Kennedy and the other members of the Maritime Commission did place their hopes in private enterprise, as assisted by government subsidies, and their economic analysis of the industry identified ways in which the 1936 act should be changed to improve the industry's commercial prospects. Their suggestions, when compared to Black's recommendations, therefore appear remarkably pro-business. A summary of their proposed changes to the 1936 act follows:

1. That in order to compensate for the possibility of changes in government policy, subsidy contracts should be for twenty years.
2. That the requirement of a 25 percent down payment of the domestic price of new construction should be modified to require the same percentage of the foreign price only, or approximately half this amount.
3. That when the 50 percent maximum rate for construction subsidy was found to be inadequate, ship owners should be

allowed to build foreign and register under U.S. flag. Also, when domestic costs exceeded foreign costs by more than one-third and the commission considered the differential to be unreasonable, operators should be allowed to build foreign and register under U.S. flag.

4. That the cumulative period for calculating recapture of excess profits should be extended from five to ten years. (For a cyclical industry they considered the five year period too short.)

5. That in order to increase investment, contractors should be able to voluntarily increase their tax sheltered reserves. (In 1937 the corporate tax rate was 15 percent, and few would take advantage of the provision. As tax rates rose, this provision became more attractive.)

6. That the restrictions on the right for U.S. companies to have affiliations with foreign lines should be relaxed. (They reasoned "that foreign competitors are limited by no restriction whatever as regards these matters, nor would American companies be so burdened if they operated under the foreign flag.")

7. That the salary limitation of twenty-five thousand dollars should be eliminated. (The commission believed that the limitation would place the industry at a disadvantage in attracting the best talent.)[50]

The commission made these recommendations because it found the 1936 act incomplete, inadequate, and difficult to administer. Kennedy's judgment was considerably harsher; he declared the act "unworkable" and thought it was "about the worst piece of legislation" he had ever encountered, yet somehow he and his successors made it work.[51]

Some of these recommendations were accepted, others were strongly opposed. The 1936 act was amended to allow twenty-year subsidy contracts, to reduce the required down payment for new vessels, to allow voluntary tax-sheltered deposits, and to extend the excess profit recapture period. The limit on executive salaries was retained, but in later years it was easily evaded through the granting of waivers. The proposal to allow foreign building when the differential between domestic and foreign construction costs became excessive brought cries of outrage from the shipbuilders and their trade unions and was never seriously considered. The recommendation that American shipping lines be allowed to have closer working relationships with foreign lines was also opposed, although it was finally accepted in the Shipping Act of 1984. In sum, the amend-

Year	Appropriations to Shipping Board (in Thousands of Dollars)	Mail Contract Payments	Total
1917	50,075	–	50,075
1918	1,063,400	–	1,063,400
1919	1,809,477	–	1,809,477
1920	387,258	–	387,258
1921	37,281	–	37,281
1922	85,951	–	85,951
1923	20,918	–	20,918
1924	50,327	–	50,327
1925	30,307	–	30,307
1926	24,309	–	24,309
1927	14,187	–	14,187
1928	17,288	–	17,288
1929	13,687	7,612	21,300
1930	11,329	10,704	22,033
1931	6,336	16,009	22,346
1932	24,441	19,032	43,474
1933	332	22,877	23,209
1934	196	26,948	27,145
1935	219	26,326	26,545
1936	209	24,637	24,846
1937	7,421	21,762	29,183
Total	3,624,958	175,911	3,800,869

Table 7.1: U.S. Government Funding of Maritime Industry, 1917–1937.
Source: Data from the U.S. Maritime Commission, *Economic Survey of the American Merchant Marine* (Washington, DC, GPO, 1937), 85.

ments improved the business prospects of the act in several ways, but not nearly to the extent that the commissioners had hoped for.

In the final section of the *Economic Survey* the commissioners asked, "What will it cost to maintain an adequate merchant marine in foreign trade?" While acknowledging the importance of the question, they insisted that only "the roughest sort of estimate [could] be made at this time." They listed several of the variables that made it impossible to determine precisely the size and shape of a commercial shipping industry engaged in foreign trade. They also explained in detail why it was impossible to determine accurately what the subsidies based on foreign costs would be. People in government could ask such questions, but business is conducted on a day-to-day basis, and no one can predict all the costs and opportunities that will crop up. But some estimate of costs was needed, so the commissioners looked to the historical record and provided a list of annual federal expenditures to support the maritime industry since the creation of the Shipping Board in 1917. The total reached almost $4 billion; but even with that amount, the commissioners pointed out, the Shipping Board had "failed to achieve the purpose [they had] in mind, a modern efficient fleet, capable of reproducing itself in the foreign trade of the United States." Having done what they could, the commissioners guessed that on average $10 million would be needed annually for construction subsidies and $15 to $20 million per year for operating subsidies.

Three new titles were also added to the 1936 act. Title 10 and title 11 were passed in 1938 in response to recommendations made in the *Economic Survey*. To address the problem of labor relations in the shipping industry the commissioners recommended that the government create a "Board similar to the Railway Mediation Boards that [had] been conspicuously successful in minimizing labor strife in that field."[52] What was done, in title 10, was considerably less than was asked for. The new title directed the commission to encourage "the practice and procedure of collective bargaining and the prompt and orderly settlement of all disputes concerning rates of pay, hours of employment, [and] rules or working conditions," but to accomplish this task it gave the commission only a three-man Maritime Labor Board limited to fact-finding and mediation. Had the Maritime Labor Board been given the power to impose cooling-off periods, a power that had been granted to the Railway Labor Boards, it might have been able to prevent or remedy the chaotic labor conditions that developed following the war. The board that was created proved to be so weak, however, that it was not continued when its original three-year term expired.

The other new title added in 1938 addressed the need to provide mortgage assistance for companies ordering new ships. It was an issue with a tangled past. The 1920 Merchant Marine Act created a government program that provided low-cost loans to support the building of new ships. By 1935, during the depth of the depression, one-third of these mortgages were in default and the market value of the ships that served as collateral had fallen drastically. The Roosevelt administration had no desire to continue a program that created bad debts for the government, but the issue of financing ship construction still had to be faced. It therefore decided to establish a Federal Ship Mortgage Fund to insure mortgages, rather than provide low-interest loans directly, and this is what was done in title 11. The commission would charge a premium or surcharge of "not less than ½ of 1 percent on the amount of the mortgage outstanding," and this premium would be deposited in a fund managed by the U.S. Treasury. The mortgages insured by this fund were limited to a maximum of twenty years and the commission's obligation for the principal amount was not to exceed 75 percent of the cost of construction.[53] In subsequent years the limit on insurance for ships employed in the coastal and intercoastal trades was increased to 90 percent. Losses under this program were small for many years and were adequately covered by the premiums. In the 1980s, however, the Maritime Administration managed the program badly and the fund was wiped out. The title 11 program still exists, however; and although the Treasury Department has adamantly opposed any increase in the fund, it has recently been used to insure construction loans on ships being built in U.S. yards for both foreign and domestic operators.

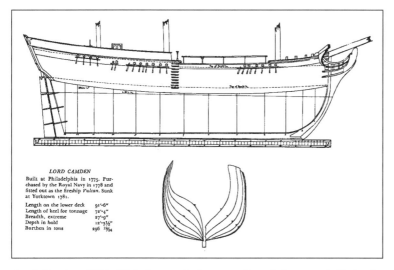

LORD CAMDEN
Built at Philadelphia in 1775. Purchased by the Royal Navy in 1778 and fitted out as the fireship *Vulcan*. Sunk at Yorktown 1781.

Length on the lower deck	91'·6"
Length of keel for tonnage	72'·4"
Breadth, extreme	27'·9"
Depth in hold	12'·3½"
Burthen in tons	296 19⁄94

Lord Camden, 1775. This is a typical cargo ship of the type built in the North American colonies for the British market. (The Mariners' Museum, Newport News, VA)

USS *Constitution* in Boston Harbor. One of the six frigates authorized by Congress in 1794, *Constitution* was completed in 1798 and remains in active commission in the U.S. Navy. (South Street Seaport Museum)

Thomas Macdonough. As a thirty-year-old lieutenant in the War of 1812, he built a fleet and defended Lake Champlain against a British attack from Canada. (U.S. Naval Institute)

Right foreground, *Washington*; left, *Hermann*. The first practical American ocean steamships, they were put into service in 1847 and 1848 by the Ocean Steam Navigation Company. They were supported in part by mail subsidies provided under the Postal Act of 1845 and competed with the Cunard Line for passengers and high-value freight. Oil painting by J. E. Buttersworth. (Mystic Seaport Museum)

Caleb Grimshaw, packet ship of 987 tons. It was built in New York City in 1848 by W. H. Webb. (Mystic Seaport Museum)

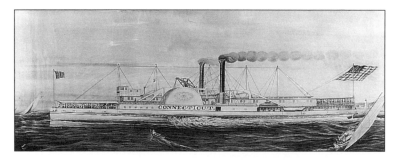

Connecticut, a Long Island Sound steamboat built in 1848. Note the boilers placed on the paddle boxes for safety and the steam engine "walking beam" that drove the paddles. (Mystic Seaport Museum)

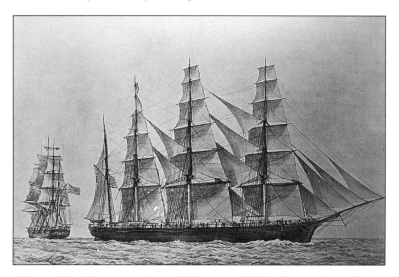

Clipper ship *Great Republic*. Built by Donald McKay in East Boston, Massachusetts, and launched in 1853, this largest of all American wooden square-riggers burned to the waterline before its first voyage but was rebuilt and sailed for many years. (South Street Seaport Museum)

Three sketches of the *Vanderbilt*, a ship with a curious career. Cornelius Vanderbilt had this wooden-hulled side-wheeler built in 1857 and operated it in the transatlantic passenger trade. After a ram was installed and the bow was iron-plated during the Civil War, it pursued the Confederate ships *Virginia* and *Alabama*. In the 1870s it was rebuilt as a three-masted clipper ship and carried freight along the Pacific coast and across the Atlantic. (Frank Braynard Collection)

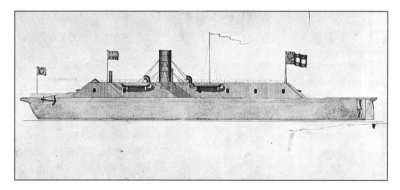

CSS *Virginia*, formerly the *Merrimack*. Originally a forty-gun steam frigate, *Merrimack* was burned and sunk by Union forces in 1861 when they abandoned Norfolk Navy Yard. Confederate forces refloated it, stripped it of its masts and superstructure, armored it with an iron casement, fitted it with a ram, and deployed it as a blockade destroyer. (U.S. Naval Institute)

Sailors aboard USS *Monitor*, James River, Virginia, July 9, 1862. (U.S. Naval Institute)

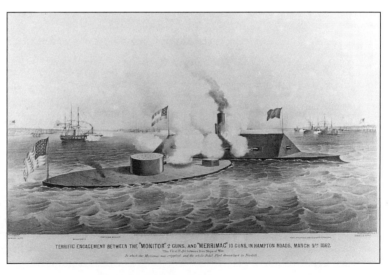

Currier and Ives lithograph of the battle between *Monitor* and *Virginia*, March 9, 1862. (Mystic Seaport Museum)

Confederate auxiliary-powered, screw-driven commerce raider *Alabama* and one of its victims, the whaler *Ocean Rover*. (U.S. Naval Institute)

SS *Ohio*. One of four iron-hulled, Pennsylvania-class passenger liners built in Philadelphia's Cramp Shipyard, it was launched in 1873 for the American Steamship Company, which was organized by the Pennsylvania Railroad and some Philadelphia investors. (South Street Seaport Museum)

USS *Trenton*. Launched in New York in 1876, *Trenton* was the most advanced of the navy's post–Civil War ships. It had a wooden hull, bronze ram, full square-rigged sail plan, a compound steam engine, and a maximum speed of twelve knots. (Mystic Seaport Museum)

USS *Chicago* underway, with other ships in company. The ABC ships, *Atlanta*, *Boston*, and *Chicago*, were the first steel ships built for the U.S. Navy. They were armed with breech-loading rifled guns and were powered by full sail plans and twin screws driven by compound steam engines. The 325-foot *Chicago* entered naval service in 1889. (Mystic Seaport Museum)

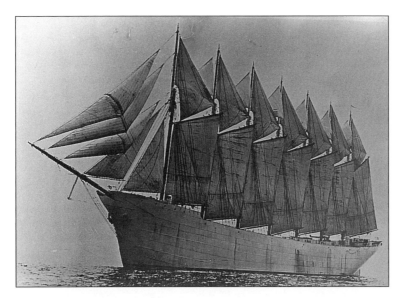

Thomas W. Lawson. The size of coastal schooners peaked in the decade 1900–1910. *Lawson,* a unique seven-masted schooner, was built of steel in Quincy, Massachusetts, in 1902 and measured 5,218 gross tons. It was not a commercial success. The largest wooden schooner, *Wyoming* (3,730 tons), was constructed in Bath, Maine, in 1910. (South Street Seaport Museum)

The battleship USS *Connecticut.* Commissioned in 1906, *Connecticut* served as the flagship of "the Great White Fleet" during its 1908–1909 cruise around the world. (Mystic Seaport Museum)

Alfred Thayer Mahan. One of the first officers appointed to the Naval War College in Newport, Rhode Island, Captain Mahan spent five years preparing and delivering lectures that were published in 1890 as his classic *The Influence of Seapower on History, 1660–1783*. (U.S. Naval Institute)

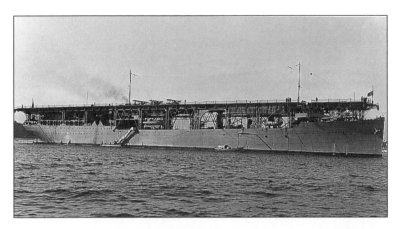

USS *Langley*. The navy converted the 542-foot collier *Jupiter* into its first aircraft carrier by installing a flight deck that bridged the hull and pilot-house. The ship was recommissioned in 1922 as *Langley*. (Rosenfeld Collection, Mystic Seaport Museum)

American Merchant, New York, 1933. One of the 7,500 gross ton freighters built in the Hog Island yard under the World War I emergency shipbuilding program. In the 1920s and 1930s these rugged ships were work horses in the American merchant marine, and many served with distinction in World War II. (Frank Braynard Collection)

Liberty ship *Robert E. Peary*, being launched on November 12, 1942, four days and fifteen hours after its keel was laid. Henry J. Kaiser, who mass-produced ships during World War II, staged this unique demonstration of his shipyards' capabilities. (Maritime Administration)

Foreground, SS *America* (1940) and, behind, SS *United States* (1952) passing in the Hudson River on the West Side of Manhattan. (Frank Braynard Collection)

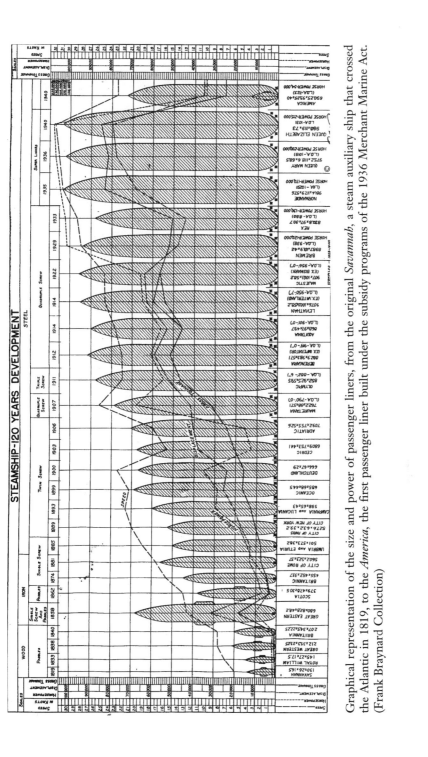

Graphical representation of the size and power of passenger liners, from the original *Savannah*, a steam auxiliary ship that crossed the Atlantic in 1819, to the *America*, the first passenger liner built under the subsidy programs of the 1936 Merchant Marine Act. (Frank Braynard Collection)

Passenger liners at Manhattan's West Side piers in the 1940s. (Frank Bray-nard Collection)

Nuclear ship *Savannah*. Proposed in 1955 as part of President Eisenhower's Atoms for Peace program and launched in 1959, *Savannah* was an engineering success, but its commercial possibilities were undermined by labor disputes, high costs, and a widespread fear of atomic power. (Frank Braynard Collection)

T-2 tanker *Ideal X*. Adapted by Malcom McLean to carry semitrailer-sized containers on its spar deck, the *Ideal X* inaugurated modern container service in April 1956. (South Street Seaport Museum)

Portrait of Malcom McLean, founder of Sea-Land Services, Incorporated, and pioneer of the modern system of containerized cargo handling. (U.S. Merchant Marine Academy)

Sea-Land vessel being loaded by a gantry crane in the Port Elizabeth, New Jersey, container terminal. (South Street Seaport Museum)

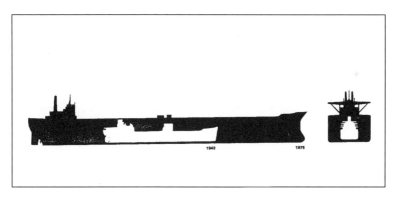

Superimposed silhouettes of a World War II–era T-2 tanker (17,000 tons) and a Very Large Crude Carrier (VLCC) (approximately 250,000 tons). The ships' dimensions are, for the T-2, length, 503 feet; beam, 68 feet; draft, 39 feet; and, for the VLCC, 1,000, 164, and 82 feet, respectively. Since doubling all dimensions increases cargo capacity more than eight times, this illustration helps explain the relationship between size and tonnage in modern tankers. (Society of Naval Architects and Marine Engineers)

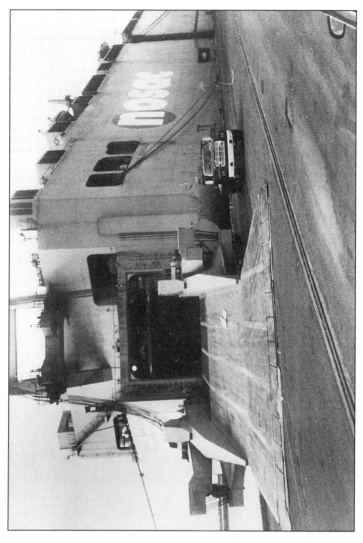

Stern view of a large Roll-on/Roll-off (Ro-Ro) vessel. The ship pictured was built primarily to transport automobiles but has specially reinforced decks and can also carry larger vehicles. Note the massive ramp skewed onto the dock alongside. (Ameri-can Automar, Inc.)

Lighter Aboard Ship (LASH) vessel. Note the shipboard gantry cranes used to lift barges over the stern and stow them aboard. The vessel was capable of carrying up to ninety barges, each loaded with 370 tons of cargo. Some LASH ships had container cranes and could hoist aboard a full deck load of containers in addition to barges stowed below deck. (American Automar, Inc.)

A C-11 containership built for American President Lines: length, 906 feet; beam, 131 feet (too wide to transit the Panama Canal); draft, 46 feet; capacity, 4,832 twenty-foot equivalent unit (TEU) containers. The size of mega containerships continues to increase; a 6,000 TEU ship was put into service in 1998. (American Ship Management, LLC)

8

THE MERCHANT MARINE
IN WORLD WAR II,
1939–1945

The Merchant Marine Act of 1936 was designed to revive the shipbuilding and liner shipping industries in a time of economic depression. The act built upon the legacy of the Merchant Marine Act of 1916 and the administrative structures it created, but its chief concern was to stimulate and promote U.S.-flag shipping, not to build a wartime merchant fleet. But pressing new concerns soon created new tasks for federal maritime policy. As Europe drifted into a second world war America found itself once again serving as the arsenal of democracy. It was soon realized that an enormous number of ships would be needed to carry the supplies produced in America to the battlefields overseas. Once again the federal government found itself organizing and funding the construction of thousands of ships to implement its military policies. Those charged with getting the job done adapted the administrative structures created by the Merchant Marine Act of 1936 to this new task. Gearing up for war achieved the act's original intentions of reviving the shipbuilding and shipping industries. When German submarines threatened to cut Allied supply lines, American shipyards addressed this ominous challenge of building merchant ships faster than the U-boats could sink them. It was a desperate time that called for heroic action; fortunately, the response by all those working in the maritime industries was equal to the challenge.

WAR IN EUROPE AND ISSUES OF NEUTRALITY

The lesson taught by history was clear: when Europe goes to war, America must defend its right to trade as a neutral, but when it does so, it risks being drawn into the war. Jefferson recognized the danger; Madison and Wilson were overwhelmed by it. In 1940 and '41 it appeared that this chain of events, which had repeatedly undermined America's belief in its exceptionalism and its policy of isolation, would once again drag the United States into a European conflict.

Germany rearmed in the 1930s. Rearmament began in secret, for it was prohibited by the Treaty of Versailles, which ended World War I, but in March 1935 Hitler felt strong enough to publicly tear up the treaty's arms provisions. He announced that the army would be expanded from one hundred thousand to five hundred thousand men, all German males would be subject to service in the military, a German air force would be formed, and a submarine fleet would be built. Germans cheered while Hitler awaited the reaction of the Great Powers. Britain and France issued mild protests; Italy under Mussolini's leadership, noting the weak response, recognized an opportune moment and invaded Ethiopia a few months later.

Hitler then launched a classic "peace" offensive; like Wilson, he had his fourteen points. Germany had no territorial ambitions, he declared, for it only wished to live in peace with its neighbors. Germany, having no desire to inaugurate an arms race, would unilaterally limit its new fleet to 35 percent of the size of the British fleet. Britain's press and government swooned with relief and in June welcomed Von Ribbentrop, formerly Germany's ambassador to London, to negotiate a new naval agreement. An Anglo-German Treaty was signed that summer, with the size of the German navy being restricted to the limits Hitler had originally stipulated.

The limits were meaningless, however, for Germany was incapable of building all the ships allowed under the treaty. Worse yet, from the British point of view, the treaty gave the Germans permission to build a submarine fleet of virtually unlimited size. The British admirals casually accepted Germany's right to rearm under the ocean as well as on its surface and agreed that it could build submarines up to 60 percent of Great Britain's strength. The Germans accepted this limit. Like the Versailles restrictions, it would be ignored at their convenience. They vowed that their submarines would never attack merchant ships. The British negotiators, in an act of forgetfulness and naivete that Churchill later termed "the acme of gullibility,"[1] accepted this promise and agreed that "if she [Germany] decided that the circumstances were exceptional, she might build up to one-hundred percent" of the British submarine fleet.[2] The British admirals, evidently having forgotten the lesson German submarines taught them in 1917, would have to relearn it again five years later.

The American government was not entirely taken in by Hitler's pretenses, yet its response to the storm clouds gathering over Europe was to withdraw rather than face the threat. In 1935 Congress passed a Neutrality

Act that was later extended; its initial provisions obliged the president to prevent the shipment of armaments to any nation designated a belligerent. Since it was widely believed that the United States got involved in World War I to protect war loans made by wealthy Americans to the Allies, a further provision was added to the Neutrality Act the next year forbidding loans to warring nations. But the opportunities for profit could not be entirely bottled up by legislation, and subsequent congressional acts permitted arms sales so long as the buyer paid cash and provided their own transport.

Public opinion in America forced President Roosevelt to play a delicate political game. Congress and the American people were in no mood for another war; they cheered in September 1938 when British prime minister Neville Chamberlain accepted Hitler's dismemberment of Czechoslovakia. Roosevelt saw more clearly what was coming and two months before Hitler invaded Poland in 1939 asked Congress to repeal the Neutrality Act, but to no avail; he could not even muster up enough votes to get his bill out of committee. An incident was bound to occur, and when it did, Roosevelt was forced to adopt the Jeffersonian strategy of withdrawal. On September 5, 1939, Roosevelt organized a "neutrality patrol" extending from Newfoundland to the Gulf of Mexico; German submarines carefully avoided challenging it. But in October, one month after the war began, the German armored ship *Deutschland* seized the American freighter *City of Flint* with a cargo bound for Manchester, England. The reaction in America indicated the public was more concerned about the danger of the United States becoming involved in war than about the violation of neutral rights. Roosevelt did what he could, barring all American ships from a war zone defined as the waters around the British Isles and the western coast of Europe from Norway to Spain.[3] As the United States withdrew from foreign trade many of its ships were put in layup; others were reflagged to avoid the neutrality laws. When Germany invaded Denmark in April 1940 Roosevelt acted again, announcing that the United States had taken over protection of Greenland.

When France fell in June 1940, Germany gained possession of submarine bases directly on the Atlantic. This improvement in its strategic situation, along with control of almost all of Western Europe, made Germany less cautious. The British position had meanwhile become extremely dire. Britain now stood alone after having suffered severe naval losses in the summer of 1940. Evacuating the British Expeditionary Force from Dunkirk had resulted in alarming losses to its destroyer fleet, leaving only sixty destroyers to protect the home waters. These losses reduced the number of ships available to escort convoys, increasing their vulnerability. At that point President Roosevelt intervened by providing fifty World War I destroyers in return for permission to build bases in eight British colonies in the western Atlantic. During the summer of 1941 American troops relieved the British force which had been in Iceland since the invasion of Denmark and the U.S. Navy assumed responsibility for patrolling the western Atlantic.

After being reelected in November 1940 Roosevelt was able to increase his efforts to aid the British. The Lend-Lease Act was signed in March 1941; it allowed the president "to sell, transfer title to, exchange, lease, lend or otherwise dispose of [war materiel to] any country whose defense the President deemed vital to the defense of the United States."[4] In May, fifty U.S. tankers were leased to the British. After the Atlantic Charter was signed by Roosevelt and Churchill in August 1941, the United States began escorting British merchant convoys as far as Iceland. American neutrality had ceased to be the central issue.

During September and October events drove America and Germany into an undeclared war. The destroyer *Greer* discovered and tracked a German submarine and broadcast its position to a British plane that attacked with depth charges. The U-boat then fired on *Greer*, which responded with depth charges. Although neither vessel sustained any damage, President Roosevelt denounced the U-boat attack as an "act of piracy" and ordered the navy to "shoot on sight."[5] Shortly thereafter he asked Congress to repeal the Neutrality Act, but opposition to doing so was still strong. While escorting a convoy a few weeks later, another destroyer, USS *Kearny*, was torpedoed and limped back into port. Another convoy escort, the destroyer *Reuben James* was not as lucky; it was torpedoed and sank with the loss of ninety-six lives. Congress was at last ready to act. The restrictive portions of the Neutrality Act were eliminated, and all pretense of neutrality vanished. The president ordered the arming of American merchant ships and allowed them to enter combat areas without restriction.

THE UNITED STATES ENTERS THE GLOBAL CONFLICT

While Roosevelt was coming to Britain's aid, Germany's attention and resources were focused on its eastern front in Russia. The confrontation in the Atlantic therefore might not have led to open war for some time had Japan not attacked Pearl Harbor on December 7, 1941. Four days after America declared war on Japan, however, Japan's allies, Germany and Italy, declared war on the United States.

While U.S. attention understandably shifted immediately to the Pacific Ocean, the unleashing of the German submarine fleet in the Atlantic brought about a catastrophe that was of far greater strategic significance than the loss of ships in Hawaii. For six months the U-boats were effectively unopposed as they preyed on shipping along the U.S. Atlantic coast and in the Caribbean. They sank nearly four hundred merchant ships, many of them tankers. The loss of tankers was especially damaging, for they were the fuel lifeline for both the U.S. war effort and the survival of Great Britain. But strategy is an abstract concept and the sudden loss of battleships and destroyers and their naval crews in Pearl Harbor shocked the public much more than the accumulating loss of merchant ships and civilian crews in the Atlantic.[6] It took time for Americans to learn that they had to darken their coastal towns and cities to avoid silhouetting ships traveling close to shore, and it took time for the navy to acknowledge the strategic importance of merchant shipping and organize for its defense. In the meantime the U-boat crews enjoyed what they called "happy times" as they picked off ships like ducks in a shooting gallery. Admiral Karl Doenitz, the commander of German submarine warfare, called this Operation Drumbeat, or *Paukenschlag*. The operation was an enormous success and inflicted six months of infamy on Atlantic shipping.

The United States had been warned that Germany was deploying its submarines off its coast, but as is so often the case in wartime, the warning was ignored. The British had broken the German secret code and informed the Americans as early as January 12, 1942, that a "concentration of German U-boats was proceeding or had already arrived on station off Canadian and Northeastern U.S. coasts"; the British also gave their approximate location.[7] The number of U-boats involved was small—only five at the outset and never more than fifteen—but they were the latest type, having great endurance, a speed of fifteen knots on the surface, and twenty-four torpedoes on board. Worldwide, during the first nine months of 1942 Axis submarines sank 520,000 tons of Allied and neutral shipping per month, which was twice the previous rate of sinkings. More than half of this total occurred in American coastal waters and the Caribbean. In May more than 85 percent of all losses occurred in waters nominally controlled by the American navy, with a large number of the ships lost being British tankers leaving Caribbean oil refineries. This caused a furious British officer to tell Admiral Ernest J. King's chief of staff, "[T]he trouble is Admiral, it is not only your bloody ships you are losing. A lot of them are ours."[8]

While the U.S. Navy eventually became concerned about the loss of merchant ships, it was initially most concerned about its public image. It asked the press to refrain from announcing the sinking of German U-boats, although in fact none had been sunk. Then on April 1 the navy itself announced that twenty-eight enemy U-boats had been "sunk or presumably sunk." This turned out to be the height of presumption, for no U-boats had been sunk off the Atlantic coast and "presumed sunk" meant nothing more than "merely sighted."[9] By the end of April the navy had finally begun to use the tactics of the British coastal command and the army was flying intensified air patrols. Results soon followed: aircraft sank two U-boats off Norfolk and the destroyer *Roper* sent a third to the bottom. In July five more U-boats were destroyed.

The navy's disregard for the merchant marine during the first six months of the war was astonishing. If, as was so often claimed, the merchant marine deserved to be sustained by federal subsidies primarily because it was important for national security, why did the navy neglect it so? Was Admiral King, who became commander in chief (CINC) of the U.S. fleet in December 1941 and chief of naval operations the following March, responsible for the navy's lack of preparedness? King, like Admiral Benson, his predecessor in World War I, was unquestionably a staunch Anglophobe. The American navy had been actively working with the British since 1939 and had full access to British antisubmarine technology and tactics. Did King reject British practices just because they were British? There may be some truth to such a supposition, but it is far too simplistic.

The real reason the Germans had six months of "happy times" was that U.S. coastal defenses were almost nonexistent. Two years of participation in mid-Atlantic submarine warfare had diverted attention away from the U.S. Atlantic coast, where peacetime conditions still prevailed. Buoys and beacons were lit, ships ran with running lights burning, and coastal cities were fully illuminated, while in Great Britain all had been darkened since September 1939. After Pearl Harbor a new operational command was created in the Atlantic: Commander, Eastern Sea Frontier. (Frontier commands were also established in the Gulf of Mexico and the Caribbean.) In March 1942, Admiral Adolphus Andrews, who was serving as commander of the Third Naval District, an administrative command based in New York, added operational command of the Eastern Sea Frontier to his responsibilities. The weapons and equipment placed at his disposal were utterly inadequate. He commanded a "sea and air force so antiquated, so unprepared . . . and so deficient in craft and weaponry that it stunned the Germans and British when they saw it go to war." Andrews lamented, "[T]here is not a vessel available that an enemy submarine could not out-distance when operating on the surface. In most cases the guns of these vessels would be outranged by those of the submarine."[10]

Huge sums had been spent on the navy before Pearl Harbor, but little had been spent on protecting merchant shipping. Although President Roosevelt had authorized the arming of merchant ships before the United States entered the war, the program was proceeding at a snail's pace. When Admiral King finally recognized the need for coastal convoys, he thought only in terms of destroyers as convoy escorts, and he knew that Admiral Andrews did not have enough. King believed that inadequately escorted convoys were worse than no convoys at all, although experience had taught the British that the opposite was true. And the British were becoming desperate. Why convoy ships safely to and from Halifax if they were going to be easily picked off when sailing unescorted in U.S. home waters? In February Britain sent over twenty-four of their best antisubmarine trawlers and ten corvettes, but they were not assigned to convoy duty and the Germans had little difficulty evading this tiny patrol force.

Historians have harshly censured the navy for its failure to defend the Atlantic merchant fleet. Samuel Eliot Morison minced no words in his definitive history of naval operations in World War II:

> This writer cannot avoid the conclusion that the United States Navy was woefully unprepared, materially and mentally, for the U-boat blitz on the Atlantic Coast that began in January 1942. He further believes that, apart from the want of airpower which was due to prewar agreements with the Army, this unpreparedness was largely the Navy's own fault. Blame cannot justly be imputed to Congress, for Congress had never been asked to provide a fleet of subchasers and small escort vessels; nor to the people at large, because they looked to the Navy for leadership. Nor can it be shifted to President Roosevelt, who on sundry occasions prompted the Bureau of Ships and the General Board of the Navy to adopt a small-craft program; but, as he once observed, "The Navy couldn't see any vessel under a thousand tons."[11]

In a detailed analysis of this strategic failure, Cohen and Gooch, both former professors at the Naval War College, concluded that the navy did not meet three measures of military performance: it failed to learn, it failed to anticipate, and it failed to adapt.[12] The navy's performance during the early months of the war certainly indicated that it failed to learn and failed to anticipate.

Eventually, however, the navy did adapt. President Roosevelt realized that control of the Atlantic was crucial both for supplying Britain and for preparing to invade the continent, and he insisted that this theater be given first priority. After months of terrifying losses, coastal convoys were at last organized; longer-range aircraft patrolled ever larger areas of the ocean; and U-boat sinkings increased. The tide was turned when small aircraft carriers, "baby flat tops," began working in hunter-killer groups with destroyers and the new destroyer-escorts to fill the gaps that land-based aircraft could not cover. In the second half of 1943 the tonnage of merchant ships sunk fell from 1,386,000 to 323,000, while improved Allied tactics and weapons rapidly raised the number of German submarines being sunk.[13] In May 1943, because of his growing losses, Admiral Doenitz ordered all U-boats out of the North Atlantic and repositioned them in the Caribbean and around the Azores. Although subsequent technological advances enabled them to return to the North Atlantic later in the war, they never regained their former dominance. The Allies were now winning the battle of the Atlantic, and the "drumbeat" that now sounded foretold the defeat of Hitler and his "Thousand Year Reich."

German submarines were not the only problem faced by those responsible for merchant shipping. The war created urgent demands that severely taxed relations between civilian and military authorities and between Allied planners and administrators. The struggles that ensued lacked the drama of clashing armies and navies, but they were intense and consequential.

The British recognized early on that the war would overtax the available shipping. When Germany occupied the entire continent, it closed off the Mediterranean, thereby forcing British ships supplying the Middle East and India to sail thousands of extra miles around Africa. More ships would clearly be needed, in addition to those needed to replace ships being sunk. In March 1941, shortly after the Lend-Lease Act was signed, the British Merchant Shipping Mission arrived in Washington to urge the United States to build more ships. At that time the war in the Atlantic had attained a kind of equilibrium, with Britain being adequately supplied and ship losses being held to a tolerable level.[14] But when the United States entered the war following Pearl Harbor, the equilibrium was upset; the United States realized that it, too, would need vast numbers of new merchant ships. At that point, with practically the entire industrialized world at war, the British and American governments took control, through purchase or charter, of most of the world's available shipping, including what remained of the Dutch and Norwegian fleets. But the inadequacies of the agencies charged with administering this vast array of ships soon became a serious problem.

The British had already placed control of their merchant shipping under a single office, and the United States moved to do the same. The U.S. Joint Chiefs of Staff had different plans, however. They wanted no civilian interference with the way they allocated and utilized merchant shipping; in their view the civilian authorities should stick to making the tools the military needed to prosecute the war. But Roosevelt was impressed by the British arrangement and became convinced that all merchant shipping should be put under civilian control. He therefore issued an executive order in February 1942 directing that the War Shipping Administration (WSA) be primarily responsible for the "operation, purchase, charter, requisition and use of all vessels under the flag or the control of the United States."[15] The WSA would be in charge of allocating shipping to the army, navy, and other government agencies.

Shortly before this executive order was issued, the British and American governments agreed on how their merchant fleets would be jointly controlled. The agreement stated in part:

> In order to adjust and concert in one harmonious policy the work of the British Ministry of War Transport and the shipping authorities of the United States Government, there will be established forthwith in Washington a Combined Shipping Adjustment Board consisting of a representative of the United States and a representative of the British Government who will represent and act under the instructions of the British Minister of War Transport.
>
> A similar Adjustment Board will be set up in London consisting of the Minister of War Transport and a representative of the United States Government.
>
> In both cases the executive power will be exercised by the appropriate shipping agency in Washington and by the Minister of War Transport in London.[16]

This agreement looked good on paper, but in the United States the War Department never acknowledged the authority that the president had bestowed on the WSA. The military repeatedly made excessive and frequently unjustified demands for tonnage and rendered civilian control and coordination with Great Britain impossible. This created an impasse that became increasingly threatening as the Allies began preparing to invade the continent.

In January 1943 President Roosevelt and Prime Minister Churchill met in Casablanca to plan the future conduct of the war. The operations to which they committed themselves all required extensive use of merchant shipping: top priority was to be given to the invasion of Sicily, a major buildup of American men and materiel would be made in Great Britain in anticipation of an invasion of France, America promised to conduct major operations against the Japanese, and Britain agreed to undertake operation "Anakim" in Burma to reopen the road to China. Yet it appears that there was no serious consideration of the shipping and landing craft needed to support these operations. The American chiefs of staff, hoping to avoid civilian meddling, brought along no representatives from the WSA. The result of this neglect was that when the Sicilian invasion was begun, all the other operations had to be postponed.[17] After Casablanca the WSA, having learned its lesson, began developing the information and skills needed to anticipate future needs and match shipping capabilities to operational requirements.

The American chiefs of staff soon found they could not simply commandeer merchant shipping. The British insisted that the first priority for merchant shipping must be to sustain the flow of vital imports to the United Kingdom, and they refused to commit additional shipping to military operations unless this condition was met. When President Roosevelt accepted their point, officials at WSA set about determining what the implications of this decision would be on the chiefs' operational plans. In the end military plans were revised to accommodate the British priority; the solution of this problem removed a major obstacle to Allied cooperation.[18]

Planning for the invasion of Europe was well underway by the beginning of 1944. The task of bringing together a vast armada of ships to carry huge quantities of men and materiels to the French beaches became paramount. Happily, by that time the British Shipping Mission in Washington and the War Shipping Administration had learned to work together in greater harmony, although a further presidential directive was needed to curb the tendency of American chiefs of staff to request extravagant shipping support.[19] By the time the war was being won on the fighting fronts, however, the bureaucratic conflicts that had impeded mobilization during the early years of the war had largely been resolved.

Shipbuilding and Maritime Training Programs

As table 8.1 indicates, America had a much larger merchant fleet when World War II began than it had at the beginning of World War I in 1914. The other shipping resources available to the Allies were also larger. Although the British merchant fleet was smaller in 1939 than it had been in 1914, the

	July 1, 1914 (1,600 DWT Minimum)	
	Ships	Thousand Gross Tons
World Total	8,445	35,145
British	4,174	18,197
German	743	3,799
U.S.*	513	2,216
French	357	1,602
Japanese	429	1,496
Italian	355	1,310
Dutch	263	1,285
Norwegian	323	1,087

	December 31, 1939 (2,000 DWT Minimum)	
	Ships	Thousand Gross Tons
World Total	9,161	51,998
British	2,529	16,321
U.S.*	1,296	7,881
Japanese	873	4,574
Norwegian	698	3,947
German	579	3,353
Italian	505	2,921
Dutch	405	2,453
French	414	2,383

* Not including the Great Lakes.

Table 8.1: World Shipping: World War I and World War II.
Source: Data from Robert G. Albion and J. B. Pope, *Sea Lanes in Wartime*, 2nd ed. (Hamden, CT: Archon Books, 1968), 309–10.

Norwegian and Dutch fleets, both of which served the Allies in World War II, were substantially larger. After World War II had begun, the Allies also made use of seized German ships and the ships of nations that Germany occupied.

The United States was also better prepared for wartime shipbuilding when it entered World War II than it had been when it became directly involved in World War I in 1917. When Joseph Kennedy left the Maritime Commission, Admiral Emory S. Land became chairman and remained in that position to the end of the war. Land's own talents and his close ties with the president enabled him to provide outstanding leadership during this time of crisis. Early in 1939 the Maritime Commission began implementing its original program to build 50 ships per year, but its target had been raised to 200 ships by December 1940. By the end of 1939, 127 dry-cargo ships were under construction, along with 12 fast tankers being built in a cooperative venture with the Standard Oil Company.[20]

By the end of 1939 the British were also building commercial ships in America. A contract was signed with the Todd Shipbuilding Corporation to build 60 ocean-class ships, 30 in the Todd-Bath yard in Maine and 30 in California. As it turned out, these ships became prototypes for America's Liberty ships. The United States adopted the British design so that production could be scaled up rapidly and with few changes. The British design fit the bill because the machinery and equipment on these ships was quite simple, could be easily manufactured, and required only limited training for the engineers who would operate it. While Roosevelt and Land kept administrative infighting to a minimum and successfully managed the flow of materials to the shipyards, the commission's shipbuilding programs grew ever more ambitious. After the attack on Pearl Harbor the goal was raised to 5 million tons of shipping in 1942, which was more than ten times the number of ships built by the commission in 1939.[21] In 1941 the United States had eight naval shipyards and nineteen private shipyards capable of building oceangoing ships. Rapid expansion programs were begun that year in all yards, with the Maritime Commission providing much of the funding for the private yards. In addition the commission began constructing the twenty-one additional yards in which the 2,708 Liberty ships were built.[22]

The first Liberty ship, *Patrick Henry*, was launched in September 1941; over a hundred more followed before the end of the year. At the source of this torrent stood Henry J. Kaiser, a successful industrialist who had never built a ship before 1941. Unburdened by the limitations of vision and practices of traditional shipbuilding, Kaiser geared up for volume production. His ships were built of prefabricated modules that were then assembled in series construction. Welding, a skill that could be learned in a month, was used extensively for the first time instead of riveting. In the first year and a half, construction time for Liberty ships was reduced from 105 to 14 days; in a much publicized stunt one ship was built and launched in less than 5 days. By the end of the war Kaiser had built one-third of the Maritime Commission's vessels and had set the standard for all other yards.

It is hardly surprising that the Germans and Japanese did not anticipate the flood of manufactured goods that poured forth from America's factories and shipyards to support its war effort. A country that had constructed only 1 million tons of merchant shipping in 1941 built more than 17 million tons by 1943. By the war's end in 1945, a workforce of 4 million men and women had built 5,000 ships at a cost of some $12 billion. The speed with which this was done and the scale of the effort still defy comprehension. When operating at their peak rate of production, America's shipyards were capable of reproducing the entire world's prewar commercial tonnage in less than three years.[23] But immense as this effort was, it was barely enough.

The demand for shipping in World War II was much greater than it had been in World War I. In fact the United States launched only a relatively few ships before the end of World War I, but that did not prove fatal. But in World War II Germany's armies were far more successful, and the Allies' positions were far more extended and difficult than they had been in World War I. Merchant shipping played a crucial role in stopping the Axis advance in North Africa and sustaining the island fortress of Great Britain. Ships lost to submarines had to be replaced, and those that escaped the submarines had to sail on supply routes whose lengths far exceeded those of World War I. Sending supplies to Murmansk in northern Russia, to the Persian Gulf via the Cape of Good Hope, and to distant islands in the western Pacific created a need for all the ships that could be built. The culminating challenge of invading Europe in June 1944, when an armada of 2,700 American and Allied ships was assembled and deployed, also required a level of shipping support unknown in World War I. And to man all these ships, thousands of officers and men had to be hurriedly trained as seafarers.

The Maritime Commission anticipated the need for training schools and met the rising demand. When the United States entered the war there were several state nautical schools already in existence, and the U.S. Merchant Marine Academy was under construction as well. As crews were lost on ships sunk by submarines and new ships slid down the ways in ever increasing numbers, new training facilities were opened to train the officers and crews needed to man the ships at sea. When the War Shipping Administration (WSA) was established in February 1942, Admiral Land was appointed as its administrator, while continuing as chairman of the Maritime Commission. The WSA, in addition to managing the operation of government-owned and chartered vessels, assumed responsibility for training seamen. Although many men who had not sailed since World War I returned to sea, there was a shortage of crews during the early days of the war. The problem began to ease by 1943, however, as training centers came up to speed and men with relevant shore jobs, men such as carpenters, riggers, machinists, and cooks, sought seagoing employment at good union wages on merchant ships. Existing regulations were modified to allow ships' officers to progress from third mate to master in two years, as many did. Third engineers were also allowed to advance to chief engineer in the same time interval. By the end of the war many American merchant ships were commanded by men in their early and mid-twenties. Today it is hard to comprehend the bravery and achievements of these young seafarers. The government gave them the minimum training required to man their ships; their youth, courage, and determination sustained them and the cause they served through their difficult and often dangerous years at sea.

9

DREAMS OF A NEW GOLDEN AGE, 1945–1960

At the end of World War II the United States was the world's most powerful nation. Preeminent on the high seas, America owned 60 percent of the world's tonnage and for the second time in thirty years had at its disposal a huge war-built fleet. At first glance it appeared that America was on the brink of a new golden age of maritime enterprise, and if this great continental power had harbored traditional global ambitions, it could have readily exploited the hegemony it enjoyed following the war. But since American tradition favored private ownership of industry and avoidance of overseas adventures, the presence of a huge fleet of government-owned merchant ships created as many problems as opportunities. In the short term many of these ships were put to use in war-related tasks, such as bringing troops home and carrying war-relief supplies to Europe and other ravaged areas. But because it took many years to revive the economies of both the victors and the vanquished, there were few commercial cargoes looking for transport in the years following the war's end. When the government-impelled demand for shipping began to abate, once again too many ships were competing for too little business. The availability of a vast war-built fleet also meant there would be a prolonged period of idleness in shipbuilding, which had expanded enormously to meet wartime needs. Clearly demobilization would require painful adaptations to postwar commercial realities.

But history still had a few surprises to unfold, for almost immediately after World War II ended the cold war began. America and the Soviet Union confronted each other around the globe at a time when the technology and strategies of intercontinental warfare were changing rapidly. Thus while in the postwar years America had to spend vast sums inventing, building, and deploying complex new weapons systems, it also had to be ready to refight the last war, complete with lines of doughty freighters bridging the oceans. So long as the cold war lasted, the merchant marine could therefore still claim its traditional place at the national security planning table. The government maintained the American merchant marine even though it remained commercially uncompetitive, and it did so by continuing and expanding cargo preference and existing subsidy programs. But these programs were costly and continued to generate domestic opposition to protection, regulation, and subsidies. Nonetheless, the cold war postponed for another generation the day on which the U.S. merchant marine would have to stand or fall on the basis of its performance in the international marketplace.

THE POSTWAR SHIPPING BOOM

World War II was truly global in ways that World War I had not been. When the war ended the American flag was prominently displayed in all the world's major ports. The established American shipping companies had profited while operating under government contract, and at war's end they were eager to buy government-owned ships with their enhanced capital reserves. The government, as at the conclusion of World War I, had to decide how to dispose of its war-built fleet. More than half of the four thousand vessels under its control were Liberty ships, which had very little commercial value. The rest of the fleet consisted of Victory ships, conventional C-2s and C-3s, T-2 tankers, and troop carriers (P-3s and converted C-4s), which American and foreign shipping companies were eager to acquire.

At the end of World War I President Wilson had insisted that the U.S. commercial fleet remain under U.S. flag so that America could become a major shipping power. After World War II the United States was more concerned with rebuilding the economies of the Allies and acted more openhandedly. The 1946 Merchant Marine Sales Act made surplus ships readily available to both American and Allied owners: by 1948 Great Britain's prewar tonnage had been completely replaced, and the fleets of Norway, Denmark, and France were within 10 percent of their former tonnage levels. By the end of 1949 American shipping companies had purchased more than 1,000 recently built vessels, and foreign owners had purchased 1,100.[1] Government-owned ships that were not sold or scrapped were placed in the newly formed National Defense Reserve Fleet (NDRF). These ships, primarily Liberty ships and the remaining Victory ships, were stored in six major anchorages, two each on the Atlantic, Gulf, and Pacific coasts.

Latin American nations that before the war had relied on foreign shipping also took advantage of the Ship Sales Act. The war had been good to several of these nations, for they not only had escaped the ravages of combat but also had exported raw materials that were vital to the war effort and had commanded high prices. The U.S. share of the carrying trade to Latin America had increased steadily during the war, but at the war's end many of these nations were determined to use their newly accumulated wealth to avoid future dependency. They therefore welcomed the opportunity to purchase U.S.-built ships.

Success in maritime commerce requires more than owning ships, however, and the newly established South American lines were seldom able to attract cargoes in open markets. Their governments therefore imposed neomercantilist trade restrictions to make sure that each nation benefited from its own trade to the greatest extent possible. These laws required that all imports and exports be carried in the nation's own ships. Like the United States in the early nineteenth century, however, the Latin American nations permitted the formation of bilateral partnerships in which cargo was shared on a fifty-fifty basis between trading partners. Since the South American lines were not as efficient as their foreign partners, they also insisted on revenue pooling agreements as well. The revenues from a specific trade would be totaled annually and the portion of the revenue that exceeded 50 percent would go to the deficient carrier, which in every case was the Latin American line.

The U.S. policy of selling surplus ships to foreign lines was generous and farsighted, but at the same time it accelerated the decline of America's position in world trade. By 1948 the U.S. share of world shipping had been reduced from 60 percent at war's end to 36 percent, which was still an impressive figure by any measure. But as the U.S. share continued to decline in later years, calls for greater protection and subsidy became ever more insistent. In the early postwar years military and government-aid cargoes had filled many U.S. ships, but as foreign shipowners got back on their feet and as the volume of government-impelled cargoes declined, U.S. carriers insisted ever more stridently on reserving U.S. cargoes for U.S. ships. The U.S. merchant marine was again retreating from the world of free trade to a mercantilist dependency on government protection and subsidy.

While the war put America on top of foreign trade for at least a few years, it precipitated an irreversible decline in coastal shipping. If one excludes ships operating on the Great Lakes and in the noncontiguous Jones Act trade to Puerto Rico, Alaska, and Hawaii, there were about four hundred ships in the coastal trade at the beginning of the war; by 1950 the number was down to one hundred. During the war many of these ships were diverted to overseas routes, and in their absence much of the cargo they had carried was shifted to rail and road transport. The Colonial

Pipeline, which connected the oil refineries around Houston to markets in the Northeast, accelerated this decline by replacing the equivalent of thirty T-2 tankers with overland carriage. As the availability of oil increased, the use of coal in home heating waned, and the large fleet of barges and colliers that had carried anthracite from Newport News, Virginia, to the ports of New England lost its markets. In the end these vessels were reduced to hauling coal for power-generating plants.[2] Coastal shipping, unlike overseas shipping, had long faced modal competition from trains, trucks, and pipelines; it did not rebound and recapture its former markets following the end of World War II.

NEW PROGRAMS FOR MARITIME PROMOTION AND PROTECTION

The federal government expanded enormously during the war, and the coming of peace brought calls for retrenchment and greater efficiency. One response to this demand was the Legislative Reorganization Act of 1946, which authorized a detailed examination of the operations of several government agencies, including the Maritime Commission. Senator George Aiken of Vermont chaired the committee that began this review in 1948, and the committee members were not pleased by what they found. The failings they identified were serious, but they were also what one would expect of an agency that had been assigned huge and ever-changing tasks during a period of national emergency. Their report stated:

1. The internal organization of the Maritime Commission was a labyrinth that hindered efficiency.
2. Authority did not follow responsibility.
3. No general policies were formulated to provide guidance for officials.
4. Routine matters were handled as special problems.
5. Lack of planning resulted in improvisation.
6. Control data and reports for management purposes were not available.
7. Legal aspects of all issues were overemphasized.
8. There was no accounting control of assets and operations.
9. Procedures were unduly complicated and were not standardized.
10. Costs of operations could be reduced.[3]

The first Maritime Commission had also wrestled with problems of this sort, but the need to mobilize for a world war had overwhelmed the administrative arrangements it had put in place. Aiken's committee therefore made a number of recommendations designed to clarify responsibility and consolidate authority. They proposed that "the direction of the operation of the agency should be vested in an administrator who would be accountable to the President, preferably through the Secretary of Commerce." They also recommended that "the quasi-judicial and regulatory functions [of the commission] should be completely separated from operations and be made the responsibility of an independent, part-time Board."⁴ President Truman welcomed these proposals and in March 1950 submitted to Congress a reorganization plan that called for the creation of a Maritime Administration with a Federal Maritime Board, both to be located in the Department of Commerce. The new maritime administrator would also serve as chairman of the Maritime Board. Congress endorsed the proposed reorganization in May, and the Maritime Commission, created in 1936, was replaced by the Maritime Administration. The separation that the Aiken committee recommended between regulation and promotion was carried a step further in 1961, when the Federal Maritime Commission (FMC) was created as an independent organization under the direction of five commissioners appointed by the president. The FMC was given responsibility for all the regulatory functions set forth in the 1916 Shipping Act, as amended, while the Maritime Administration retained as its primary function promoting the American merchant marine in accordance with section 1 of the 1936 act.

Vice Admiral E. I. (Ned) Cochrane, who as chief of the navy's Bureau of Ships had been responsible for all naval construction during the war, was appointed as the first maritime administrator. Cochrane knew shipbuilding and immediately addressed the problems of this industry. As soon as other nations revived their shipyards, the demand for construction in U.S. yards vanished, the primary reason being the high cost of U.S. shipbuilding—by 1953 costs had risen from one hundred dollars per deadweight ton (DWT) at the end of the war to five hundred dollars per DWT. Admiral Cochrane decided to initiate a thirty-five-ship "Mariner" construction program at the government's expense. The ships, classified as C-4s, were designed and built by the Maritime Administration and reflected Cochrane's wartime experience. They were 12,900 DWT in size and capable of a sustained speed of twenty knots; they were ideally suited for wartime use either as speedy transports that could sail without convoy protection or as auxiliaries capable of keeping up with naval forces. But because they were larger and faster than the C-2s and C-3s then in use, American shipowners considered them unsuitable for commercial operations. Most ship operators were unwilling to try anything new and the first five Mariners were consigned to the navy. One operator finally recognized their potential, however, and purchased several at bargain rates. The other owners then fell in line, and in the end thirty of the thirty-five ships built under this program were sold for commercial use and were operated with great success for many years.

During the first decade after the war, funds were appropriated annually to pay for the cost of transporting war-relief cargoes to foreign destinations. These cargoes were shipped on U.S.-flag vessels, as required by law, but as cheaper foreign shipping became available and American costs continued to climb, resistance to using American ships began to increase. The Department of Agriculture was especially dismayed by being required to allocate an increasing portion of its aid budget to transportation costs, an obligation that reduced the amount of money available to purchase surplus crops from U.S. farmers. In 1954 President Eisenhower's new Republican administration began looking into the problem. It concluded that the requirement that American ships be used regardless of cost should be eliminated and that if maritime subsidies were needed, they should be provided openly and directly, as Roosevelt had recommended.

This proposal caused great concern, for with the exception of a few companies serving South America and South Africa, U.S. liner companies and bulk carriers had become increasingly dependent on reserved, government-impelled cargoes. When the administration's intentions became known, the reaction in Congress was swift and sure. The labor unions and shipowners working through John Butler, chairman of the Senate Merchant Marine Subcommittee, had legislation drafted to require that all government cargoes be carried in American ships. When this bill came up for hearings the administration's witnesses, representing the Maritime Administration and the Departments of State, Agriculture, and Defense, were solidly opposed. Only the shipowners and the maritime labor unions supported the bill, but that was enough. The legislation was passed, the only significant compromise being that the requirement for shipping in U.S. vessels was reduced from 100 percent of government financed cargoes to 50 percent. The long-standing requirement that military cargoes be carried exclusively on U.S.-flag ships was left unchanged.[5]

Two bills were in fact passed. Public Law 480, the Agricultural Trade Development and Assistance Act, was approved in July 1954; a month later Congress passed a companion bill, Public Law 664—the Cargo Preference Act. Public Law 664, known as the Fifty-Fifty Act, was written as an amendment to the Merchant Marine Act of 1936 and spells out the legal basis for requiring that half of certain government-generated cargoes be carried in "privately-owned United States flag commercial vessels." Ever since the passage of these linked legislative acts, an appreciative shipping industry has referred to the cargo preference they provide as "PL 480," and the term is used this way throughout this text.

When President Eisenhower signed these bills into law, they took their places alongside the Jones Act of 1920 and the Merchant Marine Act of 1936 as legislative pillars supporting the U.S. merchant marine. But this victory for protection soon provided its critics with new ammunition. As U.S. farm surpluses grew, food shipments funded under the Agricultural Trade Development and Assistance Act expanded enormously and the demand for shipping increased accordingly. Since at least half the commodities had to be carried in U.S.-flag ships, which were in short supply, the shipowners made huge profits. Although PL 480 contained a provision requiring that rates on American carriers be "fair and reasonable," this stipulation was generally subordinated to the requirement that at least 50 percent of these cargoes be carried in American bottoms. As time passed the unsubsidized bulk carriers still operating war-built ships were squeezed out by larger bulk carriers and the subsidized liner fleet. The fact that liner companies that were already being subsidized to engage in international commerce were carrying a substantial part of the cargoes reserved for American-flag ships created a new controversy reminiscent of the double-dipping that the Black commission had uncovered in the mail subsidy system. U.S. merchant mariners, having become accustomed to thinking that the government could never do enough for them, were evidently still prepared to take whatever largess they could commandeer.

The drive for ever more protection therefore did not end with PL 480. That law did not require that ships carrying preference cargoes be American built; up to 1961 shipowners could shift ships to American registry to take advantage of the cargo preference laws. When preference cargoes were no longer available, these ships would be shifted back to foreign registry. To plug this hole in the protectionist dike, PL 480 was amended to require that a vessel operate under the U.S. flag for three years before becoming eligible for preference cargoes. This amendment had the desired effect, for no operator could afford to keep a ship idle in anticipation of preference cargoes some years hence; it effectively eliminated the use of foreign-built shipping to carry these cargoes.

In 1960 the Maritime Administration found another way to subsidize a sector of the industry; this new program, like the Mariner-building program, proved to be highly effective. Section 510 of the Merchant Marine Act of 1936 was amended to allow operators of unsubsidized ships to trade

Fiscal Year	Ships Obtained	Estimated Cost of Conversions (in Dollars)
1961	4	2,871,834
1962	4	8,241,341
1963	11	3,902,164
1964	16	39,247,609
1965	18	109,930,133
1966	13	42,334,486
1967	15	58,046,347
1968	22	89,235,866
1969	17	NA
Total	120	353,773,780

Table 9.1: Trade-In/Trade-Out Program, 1961–1969.
Source: Data from the U.S. Department of Commerce, Maritime Administration, *MARAD 1970: Year of Transition* (Washington, DC: Department of Commerce, 1971), 14.

in worn-out vessels for relatively new ships in the NDRF. This trade-in program helped modernize a segment of the fleet at little cost to either the government or the operators and was widely used by the eligible shipowners. One of the principal benefactors was Malcom McLean, the originator of modern containerization and the founder of a new company, Sea-Land Services. McLean was a newcomer to the shipping industry, having made his fortune in long-haul trucking. His decision to load cargoes aboard ships in sealed containers that were the size of truck trailers would eventually revolutionize the liner industry and lead to the construction of huge specially designed containerships. But at the outset McLean needed inexpensive ships that could be adapted to carry containers. The trade-in program enabled him to obtain at very little cost a number of T-2 tankers and C-4 cargo vessels that he then converted into container carriers by jumboizing and fitting them with container cells. It took an innovator from another sector of the transportation industry to see how the trade-in program, designed to renew the unsubsidized fleet, could be used to help launch a revolution that eventually transformed the entire industry.

WARS IN KOREA AND VIETNAM

During the closing weeks of World War II, Russia declared war on Japan and occupied the northern part of Korea as the Japanese forces withdrew. At the end of the war Korea was partitioned at the thirty-eighth parallel, with the Russians occupying the north and the United States occupying the south. By the late 1940s all foreign troops had been withdrawn. Early in 1950 the U.S. secretary of state made a speech that unfortunately omitted Korea from a list of areas considered vital to U.S. security. This oversight may have encouraged the Communist government in the north to believe it could strike southward unopposed, and on June 25, 1950, North Korean troops invaded South Korea.[6] President Truman reacted immediately; he had learned the lesson of World War II and was not about to stand aside as a totalitarian regime invaded and occupied a neutral country. On the day of the attack the United Nations, with Russia momentarily absent, called on its member states to oppose the invasion. Truman immediately ordered American troops into action.

The first task was to get U.S. troops ashore in Korea; Japanese ships, combined with a fortuitously present small naval amphibious force, did this in the summer of 1950. The task force provided support for an American armed force that held only a narrow perimeter around Pusan. In September marines and army personnel embarked in 230 naval and merchant vessels, launched a daring attack far behind enemy lines and carried out a successful landing at Inchon. The American forces and their U.N. allies, mainly South Korean troops, then drove relentlessly northward across the thirty-eighth parallel. As these troops approached the Yalu River, the border between Korea and China, they were overwhelmed by huge Chinese armies that, in the western part of the peninsula, drove them back beyond the recently recaptured city of Seoul. In the eastern part of the peninsula the U.S. Marines and other elements of the besieged Tenth Corps were evacuated in November and December by a collection of merchant and naval vessels that picked them up in the North Korean port of Hungnam. These ships made some two hundred voyages to South Korean ports and rescued some 105,000 U.S. and South Korean troops and much of their equipment. They also evacuated more that 90,000 civilian refugees.[7]

Supplies for the American and United Nations forces came initially from Japan, but it was not long before the supply lines stretched across the vast Pacific. Once again, but this time without the threat of submarine attack, a bridge of ships sustained an American army fighting abroad. During the three years the Korean War lasted, the U.S. Army of 300,000 men was annually supplied with materiel and equipment totaling 8.8 million measurement tons. Thirty-eight percent of this cargo was carried by the U.S. commercial fleet; 57 percent was carried by the navy's new Military Sea Transportation Service (MSTS) in its own ships or aboard the 180 ships it activated from the NDRF. Troopships were also used, for while after 1966 almost all personnel engaged in foreign wars would be airlifted, during the Korean War nearly all troops went by sea. This mobilization to support the war in Korea came at a time when there was also great demand for U.S. shipping elsewhere as well. The greater part of the NDRF ships activated in the early 1950s were used to transport coal to Europe during the winter of 1950–51 and to carry emergency grain shipments to India and Pakistan in the following spring.[8]

The heavy utilization of U.S. shipping resources led President Truman to establish the National Shipping Authority within the Department of Commerce in March 1951. The Shipping Authority, operating under the Maritime Administration, was assigned administrative responsibility for the reserve fleet; the following year the recently established MSTS was given operational control of the ships that had been activated for its use. This division of administrative and operational responsibility created many problems and resulted in neither the Maritime Administration nor the navy being deeply committed to maintaining the reserve fleet. This situation continued until the shipping requirements of the Gulf War in 1990–1991 demonstrated the cost of this neglect. Since then substantial programs have been instituted to ensure the improved readiness of the RRF in future conflicts.

The demand for shipping services definitely worked to the advantage of commercial carriers. When the Korean War eliminated the small surplus in world shipping, charter rates quickly doubled. The new level of rates made it possible for many unsubsidized American ships to turn a profit despite their high costs. For foreign shipowners who had lower operating costs, the war brought windfall profits. Greek and Greek-American owners referred to the period as "Santa Korea."

Other conflicts kept the demand for shipping high after the Korean War ended in June 1953. When Egypt seized the Suez Canal in 1956, Israel, with backing from France and Great Britain, invaded Egypt. By October ships sunk in the canal had rendered the vital waterway inoperable. During World War II General Eisenhower had grown increasingly disenchanted with the British government's clear intention to resurrect its empire once the war was over. As president in 1956 he was deeply disturbed by the joint invasion and firmly informed both the French and British governments that their actions would not be tolerated. As far as the United States was concerned, European imperialism was dead. Eisenhower's opposition was reinforced by both the Soviet and Indian governments, which had indicated they might intervene on the side of the Egyptians. By insisting that the British, French, and Israeli forces retire, the United States prevented further escalation and brought the war to an end. The canal remained blocked for several more years, however.

Ships sailing between Atlantic and Indian Ocean ports now had to follow the much longer route around Africa rather than the more direct route through the Mediterranean and Red Seas. The demand for ships, and especially tankers, grew enormously. Shipping rates skyrocketed, and a number of ships in the NDRF were activated in an attempt to bring about a reduction in freight rates. This was one of the few times the reserve fleet was used for purely economic reasons. When mobilized to support America's foreign wars, reserve ships were operated by private shipping companies for the government under general agency agreements, but the reserve ships that were activated following the closing of the Suez Canal were made available to private operators under bareboat charters that enabled them to use the ships for their own purposes and participate in profitable businesses. But since the reserve fleet had only a few tankers and most of the ships steaming between Indian Ocean ports and Atlantic ports carried petroleum, little could be done to offset the tripling of tanker rates. In the long run, however, the Suez Canal crisis brought about the construction of mammoth oil tankers that could not transit the Suez Canal but could nonetheless bring oil from the Mideast to the Atlantic more cheaply than the smaller tankers the canal could accommodate.

The cold war continued to generate hot spots in the 1950s. In the spring of 1958 Syria and Egypt set out to destabilize the government in Lebanon. The president of Lebanon called on the United Nations for assistance, and when none was forthcoming, he appealed directly to President Eisenhower. The Sixth Fleet was ordered into the eastern Mediterranean and fourteen thousand marines were landed to restore order. By the time the troops were withdrawn several months later, MSTS had been obliged to charter a number of foreign-flag ships to provide the sealift capability required. It was becoming clear that the ability of the U.S. merchant marine to provide the support services required by the military could no longer be taken for granted.

As early as 1961 U.S. involvement in Vietnam was beginning to escalate into a full-scale war. President Kennedy initially sent nine hundred "advisers" to South Vietnam to counter increased Communist aggression from the north. By 1963 the number of advisers had increased to seventeen thousand and the conflict was widening. It was not at all clear, however, what role the navy should play in this affair. If a blockade was imposed on North Vietnam the United States had to be prepared to confront the Soviet Union, North Vietnam's main supplier. It was decided this had to be avoided. The United States also ruled out amphibious landings in North Vietnam. While a "brown water" navy of river craft was developed and deployed, the "blue water" navy was largely restricted to using its aircraft carriers as offshore airfields for launching strikes against the North Vietnamese, even though such use ignored their chief asset, mobility.

It is ironic, in light of the navy's secondary role in the Vietnam War, that the United States seized on a naval incident as its justification for becoming openly involved. On August 2, 1964, the U.S. destroyer *Maddox*, while conducting an intelligence-gathering mission in the Tonkin Gulf, was attacked by North Vietnamese motor torpedo boats. Two nights later it and another destroyer, the *Turner Joy*, reported a second attack at night. Although serious questions were soon raised as to the reality of the alleged night attack on August 4, Lyndon Johnson, who became president following President Kennedy's assassination, authorized retaliatory raids by aircraft launched from the carriers *Ticonderoga* and *Constellation*. Three days after the alleged attack President Johnson obtained from Congress the Tonkin Gulf Resolution, authorizing him "to take all necessary measures to repel any armed attacks against the United States and to prevent further aggression."[9] By the time the war finally ended nine years later over a million Americans had been directly involved and fifty thousand of them had died. It had taken that long and losses of that magnitude to demonstrate that despite America's overwhelming military superiority, a war directed by politicians in Washington could not "repel armed attacks" or "prevent further aggression" as Congress had demanded in the Tonkin Gulf Resolution. In the end America withdrew and South Vietnam was abandoned to its fate.

Secretary of Defense Robert McNamara expected that logistical support for the war in Vietnam would be provided primarily by aircraft, but as the war expanded and dragged on, the inadequacy of airlift for such an operation became obvious. Ocean transport to the Far East increased steadily. In the early years of the war about 200,000 measurement tons per month were being shipped to Vietnam, but by the summer of 1965 that had increased to 800,000 tons per month.[10] Fifty ships were activated from the NDRF in 1965 and another 120 were activated in the following years; 1965 also saw a prolonged strike by seamen on the West Coast and by engineers on ships sailing from Atlantic and Gulf ports. These strikes halted commercial operations, but the unions did not strike ships carrying military cargoes to South Vietnam. The shipowners were therefore happy to make their ships available for military cargoes while the strikes were on.

During the early phase of the war the flood of supplies pouring into Vietnam overwhelmed the available harbor facilities. Congestion and pilferage were so widespread that the United States was in effect supplying both sides in the war. The army therefore decided to build two entirely new ports, one at "Newport," near Saigon, and the other further up the coast at Cam Ranh Bay, and to install modern container-handling equipment. The navy did much the same thing even further up the coast by improving the port of Denang for the marines. By 1969 operations at these ports had reached such a level of efficiency that the ships from the NDRF were no longer needed and were returned to layup by the end of the year.

The war in Vietnam helped sustain public awareness of the need for a U.S. merchant marine. This perception was further strengthened during the 1970s by the rapid expansion of the Soviet Union's navy. Under the leadership of Admiral Sergei Gorshkov, who was as convinced of the importance of seapower as America's famous nineteenth-century strategist Alfred T. Mahan, the Soviet Union vastly expanded its naval, merchant, and fishing fleets and began deploying them worldwide. America's position as the world's predominant seapower was being seriously challenged, and it was a challenge that had to be met. National security concerns therefore continued to justify both protection and heavy subsidies for the U.S. merchant marine, whatever its successes or failures in the world of commercial shipping.

THE RISE OF THE MARITIME TRADE UNIONS

Before America became an industrial society, control of work aboard ship and along the shore remained in the hands of shipowners and the captains and agents who served them. But as industries grew larger and managerial control increased during the final years of the nineteenth century, early labor unionists succeeded in organizing workers around shared craft skills. To secure higher wages and a degree of control in the workplace, the leaders of the early craft unions emphasized the importance of training, safety, and, where needed, government regulation and oversight. These concerns accorded well with the Progressive politics of the late nineteenth and early twentieth centuries.

Seafarers, and especially those working on steamships, were eager participants in this drive to organize workers according to their craft skills. In 1875 engineering officers sailing on Great Lake steamers formed the Marine Engineers' Beneficial Association (MEBA), a union of engineers that is still in existence.[11] In 1902 the Marine Firemen, Oilers, Watertenders, and Wipers' Union (MFOWW) was formed to represent the interests of the unlicensed workers in ships' engine rooms. In 1891 the American Brotherhood of Steamship Pilots in New York opened its membership to seagoing masters and renamed itself the American Association of Masters and Pilots of

Steam Vessels; nine years later it added mates to its membership and became the Masters, Mates, and Pilots Union (MM&P). Also in 1891 Andrew Furuseth, who had recently been elected president of the Pacific Coast Seamen's Union, merged his organization with the Steamship Sailors' Protective Union to create the Sailors' Union of the Pacific (SUP), the pioneering union for unlicensed deckhands. From these founding organizations descended a complex lineage of competing maritime unions.

The most significant legislative act of this era in the maritime field, and the crowning achievement of a generation of labor organizing, was the La Follette Seaman's Act of 1915. Passed after a decade of campaigning led by Andrew Furuseth, and with strong congressional support from Wisconsin senator Robert La Follette, this act "emancipated" seamen from longstanding legal strictures that severely limited their freedom as workers. The act also involved the federal government in setting crew qualification and manning levels for U.S.-flag ships and in ensuring that living and working arrangements aboard ship were up to American standards.

During the 1920s the presence of too many ships and too many seafarers plunged the maritime industries into a depression that rolled back the gains won by the recently established craft unions. The more general breakdown of the nation's economy in the 1930s created a larger crisis that required the construction of a new industrial order in America. President Franklin Roosevelt understood that his foremost challenge was to get America working again, and he was prepared to invoke federal power to ensure that industrial workers were given the right to organize and bargain collectively. He established the National Labor Relations Board to adjudicate labor issues. It was not long before huge industry-based rather than craft-based unions were formed under the umbrella of the new Congress of Industrial Organizations (CIO). The maritime unions again shared in this upsurge of worker organization and power.

The pivotal event for the maritime unions was the great strike that idled shipping in San Francisco Bay during the summer of 1934.[12] This violent confrontation pitted the San Francisco longshoremen, supported by the Seaman's Union, against the West Coast employers. Although the strike was over before the end of summer, a final settlement was not reached until the following April. The agreement that emerged brought about a major redistribution of power within the industry. The way in which this agreement was reached illustrates the union strategy of playing off employers against one another.

The San Francisco strike provided dramatic evidence that the reinvigorated maritime unions, operating under the benevolent oversight of the

National Labor Relations Board, could seriously disrupt the industry. In December 1934, before the issues that led to the San Francisco strike were settled, the International Seaman's Union (ISU) demanded a new contract with the Atlantic and Gulf Coast employers. The operators agreed to a contract that set a new standard for the industry. The unions were given effective control of hiring, through the use of preferential hiring halls; a three-watch system; and an eight-hour workday aboard ship. There was also a significant increase in wages.[13] This agreement then provided the basis for settling the dispute in San Francisco in the following April. The Pacific contract in fact contained even better terms for the West Coast workers, and after it was signed the ISU was elected as the bargaining agent for the Pacific Coast sailors as well.

It was not long, however, before the strategy of pitting one region against another generated irreversible dissension within the maritime unions. In September 1936 the ISU again went on strike against the Pacific Coast employers. The union, hoping to avoid a potentially dangerous political reaction, sought to restrict the strike to the West Coast; nonetheless, some East Coast sailors joined in a sympathy strike both to support their brothers on the West Coast and to obtain the better contract terms they had secured. These East Coast dissidents refused to follow the orders of their ISU leaders and in May 1937 withdrew to form their own organization, called the National Maritime Union (NMU). They then demanded new representational elections, which they won overwhelmingly. Dissent was then compounded when the NMU affiliated with the recently formed Congress of Industrial Organizations (CIO), which was locked in a power struggle with the older craft-union–based American Federation of Labor (AFL). By 1938 the NMU was the exclusive bargaining agent for the overwhelming majority of the fifty-two Atlantic Coast steamship companies.[14]

This split between coasts and between branches of the labor movement permanently shattered the recently achieved solidarity of maritime labor. The NMU consolidated its position as the representative of unlicensed personnel employed by the major subsidized lines on the Atlantic and Gulf Coasts and dealt with the bargaining unit these firms established. The remnant of the International Seamen's Union (ISU) on the Gulf Coast formed themselves into the Seafarers International Union (SIU) and represented employees of a few companies in the Gulf of Mexico. On the West Coast the ISU dissolved into three separate unions that represented unlicensed personnel. Although the officers' unions remained intact, they established bargaining units that conformed to the divisions among the unlicensed unions.

This disaggregated and intensely competitive array of maritime labor unions existed only in America. When coupled with the federal government's reluctant willingness to pay whatever it cost to sustain a U.S.-flag merchant fleet, it fatally distorted labor costs and labor relations in the U.S. maritime industry. In most other maritime nations the unions spoke with one voice or at most with two, one for the officers and the other for the unlicensed seamen. But on some U.S.-flag ships on the West Coast crews were represented by six different unions—a situation that has become increasingly dysfunctional as crews have been reduced to twenty-one or less. The intense rivalry for power that occupied the NMU and the SIU made unified action impossible. The unions succeeded in continually obtaining more costly contracts from the employers, and the employers, not encouraged by the government to force the unions to modify their demands, passed their skyrocketing labor costs on to the subsidy system and to customers obliged to use U.S.-flag shipping. No other feature of the twentieth-century American maritime industry has been nearly as effective in bringing about its demise.

Following World War II, unlike the period following World War I, the maritime unions were able to maintain their power within the industry. When the Merchant Marine Act of 1936 was passed, an able-bodied seaman earned fifty to sixty dollars per month and lived aboard ship in conditions that the Maritime Commission described as "crowded, unsanitary and poorly ventilated."[15] Federal intervention and unionization gradually changed this, and at the end of World War II the size and strength of the maritime unions matched that of the war-built fleet itself. American ships employed two hundred thousand seamen, and the unions to which they paid dues had enormous resources and power. Demobilization was bound to create problems, yet the unions were well positioned to prevent a repetition of the collapse of wages that had followed World War I. Labor had power; the cold war guaranteed that the government would not permit the political arrangements that sustained that power to be disrupted.

By 1964 able-bodied seamen on U.S. ships earned an average of $825 per month, including overtime payments and benefits, and lived in quarters that the Maritime Administration required be equal to or exceed those on any other ships in the world. By skillful use of strikes and the threat of strikes, the unions had achieved a sevenfold increase in the purchasing power of a seaman's wage, an increase that far exceeded the average for comparable members of the American workforce. In the 1960s wages on American ships were three to five times higher than those on the ships of leading European maritime nations.[16] Such a situation could only be maintained so long as the government continued to serve as paymaster.

The pattern of labor relations that emerged following the 1934 strike, and the subsidies provided under the 1936 Merchant Marine Act, dramatically shifted decision-making power within the maritime industry and created an unlikely community of interest between the shipowners and the unions that represented their seafaring employees. U.S.-flag operators had little inclination and even less ability to control the steady escalation of wages, one consequence being that those of their ships not covered by subsidy agreements were slowly priced out of the world market. Those with subsidy contracts were increasingly dependent on the government both for operating funds to meet their high wage costs and for cargoes. Union leaders and ship operators formed joint lobbying organizations that pressured the government to provide the freight and funds they needed to survive. They could usually count on support from the Department of Commerce and the Maritime Administration, but other governmental departments opposed their requests. The State Department pointed out that friendly nations objected to cargo reservations, and the Department of Agriculture complained of the high cost of shipping grain on U.S.-flag ships. The last issue came to a head in 1963 after large grain sales to the Soviet Union had been negotiated.

The Russian grain sales were private transactions and hence, it was assumed, not subject to existing cargo preference requirements. The grain had been grown under a government agricultural subsidy program, however, and the unions and shipowners claimed this made the shipments subject to PL 480. President Kennedy, seeking to avoid controversy, decided to impose the condition that American ships be employed "to the extent that they were available." Kennedy's action satisfied one highly vocal domestic interest group, but it angered others. When the Russians refused to pay any shipping premium, the grain dealers involved, Continental Grain and Cargill, claimed there were no American ships available that could meet the particular conditions at the Black Sea ports where the grain was to be discharged. The companies therefore applied for waivers to PL 480 requirements, and the government approved. The seamen's unions objected and, in alliance with the longshoremen, began a boycott of all Russian ships and cargoes. George Meany, president of the now merged AFL/CIO and an ardent anti-Communist, applauded their action. The administration opposed it, but Congress was unwilling to consider legislation that would have ended the boycott. The stalemate killed the grain deal and postponed for another decade the opening of regular trade with the Soviet Union.

In the postwar era ship operators formed a number of organizations to engage in collective bargaining. Subsidized liner companies having contracts with the National Maritime Union (NMU) created the American Merchant Marine Institute (AMMI), while employers in the tanker industry whose workers were represented by the NMU bargained through the Tanker Service Committee. The Pacific Maritime Association (PMA) included all West Coast employers and bargained with the six unions that represented their shipboard employees. In general the AMMI and NMU took the lead, and once they had reached an agreement the Seaman's International Union (SIU) would sign similar agreements with its companies. Paul Hall, the president of the SIU, favored this approach because it helped limit the number of strikes for his member companies and the loss of employment for his union's members.

In addition to forming bargaining organizations, the operators formed lobbying organizations that frequently included union representatives as well. The Labor-Management Maritime Committee (LMC) represented the NMU and several subsidized employers, while the Committee of American Steamship Lines (CASL) included members from most of the subsidized lines, including some that were also in the LMC. The American Maritime Association (AMA) was oriented toward the SIU and included representatives from the unsubsidized lines primarily concerned with protecting PL 480 programs. The Transportation Institute included companies having SIU contracts, although the union was not directly represented. The Marine Engineers' Beneficial Association (MEBA) formed its own lobbying group, the Joint Maritime Congress (JMC). Thanks to the high wages subsidized by the American taxpayer, each of these groups was well funded by member contributions and in the 1960s and '70s exercised power that vastly exceeded the relative size and importance of the American maritime industry. Given the dynamics of the industry, it is not surprising that all the principal liner companies and maritime unions maintained headquarters or large offices in Washington, D.C., as this was the location of power and money. In Europe and the Far East, where the shipping industry was less politicized, maritime union headquarters were normally located near seaports where the members were employed.

The implications of management's loss of control over wage costs and of the industry's reliance on third party subsidies, either through protection or direct federal payment, became clear in the 1960s. The world maritime industry was changing, but the American industry remained hobbled by dependency and self-interest. The war-built ships were nearing the end of their useful lives in the 1960s, and the subsidy contracts under which the liner companies had operated required them to buy new ships. The new ships would be larger and faster and would incorporate new control technologies. Main engines could now be controlled from the bridge, an innovation somewhat misleadingly called "automation." The new technology had been proven effective and reliable. An adjustment in union work rules would be needed, but the engineers, represented by MEBA, and the deck officers, represented by MM&P, continued to exhibit the narrow focus of craft unions. Each union claimed jurisdiction over the new control equipment. Two ships with the new equipment were delivered in 1965, the Moore-McCormack Line's *Mormacargo* and the Gulf and South American Steamship Company's *Gulftrader*. Both were struck before making their maiden voyages; when they did sail, they were operating under an interim agreement that settled nothing.

The expiration of the contract between eight subsidized lines represented by AMMI and MEBA in June 1965 marked the beginning of one of the longest and most destructive strikes in U.S. maritime history. The companies, struggling to regain some control over their operations, were determined to settle three key issues:

1. Craft jurisdictions and manning scales for the new, "automated" ships had to be clearly defined.
2. The permanent arbitrator, who had decided many of the issues arising under the old contract, had to be replaced. The companies believed his record indicated he was totally biased in favor of the union.
3. The guarantee of welfare and pension benefits for all of the union's members was to be avoided. Up to this time the contract had called for a specific percentage of wage contributions to be used to pay for benefits. With a declining industry and a reduced number of companies to fund these benefits, the owners were fearful that by guaranteeing a specific level of benefits, the surviving companies would be heavily burdened and thus possibly jeopardized.

The attitudes revealed during the strike that ensued were all too characteristic. Management proved to be inept and disorganized. The unions involved were in no hurry to settle, for the war in Vietnam was creating a demand for shipping that meant there were plenty of jobs available on government-owned ships and with the companies that were still sailing. The maritime administrator, Nicholas Johnson, muddied the waters by attempting, at this delicate moment, to enforce the "fair and reasonable" restrictions on subsidy payments that were written into the 1936 act. Existing agreements called for certain wage increases during the life of the contract. President Kennedy, in an attempt to control inflation, had called for a limit of 3.2 percent on wage settlements. Johnson therefore pressured the Subsidy Board into limiting subsidy increases to this figure for ships operating under labor agreements signed two years earlier. The board also agreed that it was unreasonable to pay pensions of three hundred dollars per month to engineers and that subsidy funds should not be paid into an "automation" fund.[17]

Johnson's purpose was clear but his timing was terrible. The constraints he sought to impose made administrative sense, but the way in which he imposed them enraged the unions whose appetites he was attempting to curb. His proposals also utterly disoriented management in the midst of difficult negotiations by forcing them to give all their attention to ensuring that any increase in wage costs resulting from the current negotiations would be fully covered by the subsidy program. Johnson would provide no such guarantee, but the secretary of commerce eventually overturned most of the Subsidy Board's disallowances.

Thirty days after the strike began Secretary of Labor Willard Wirtz intervened, but to no avail. After another thirty days President Johnson offered to use his office to bring the parties together and help resolve the remaining issues, so long as the strike ended and work resumed. His proposal was accepted, but the ships remained idle. The MM&P, having remained on the sidelines to that point, decided it was time to fish in these troubled waters, and the strike continued for another two weeks. In all, the strike lasted seventy-four days, and in the end the MM&P and MEBA had to accept the same terms. These were as follows:

1. The automation issue was to be resolved on a company-by-company basis. Crew sizes were eventually reduced slightly, to an average of thirty-five to thirty-seven men; this was still at least ten more men that the companies needed. The engine-room control equipment on the bridge was never used.
2. The biased arbitrator was replaced. Even the union felt embarrassed once his record of decisions was examined.
3. The companies' worst fears were realized: pension benefit levels were specified, and the costs to the company per indi-

vidual worker increased rapidly. But they need not have worried, for once again the companies were able to pass the cost on to the taxpayer through increased subsidies or higher payments for preference cargoes.

Once the strike was over the president's offer to facilitate resolution of the fundamental issues was forgotten. The eight companies involved had lost over $50 million but had learned an important, if dispiriting, lesson—no one in government had any interest in backing their effort to confront the unions and control costs. In the future they should make the best deal they could while avoiding a strike.[18] The companies also suffered longer-term damage that for many proved fatal. Although no future strike lasted as long as the 1965 strike, that stoppage convinced shippers throughout the world that American lines provided less reliable service than their foreign competitors. The Bloomfield Steamship Company never resumed operations after the strike. Four years after the end of the strike the Grace Line was sold, its parent company, W. R. Grace, having had enough. Six years after the end of the strike, all the passenger ships owned by the companies that had been struck, including the magnificent *United States*, were in layup. Eventually four of the remaining six lines in the AMMI were sold or disappeared: Moore-McCormack Lines, United States Lines, Prudential Line, and the Gulf and South American Steamship Company. Only the Lykes and Farrell lines survived. The progressive failure of the American merchant marine was undeniable; the cause of its failure was hardly a mystery.

The postwar self-destruction of the U.S. merchant marine could have been avoided, but it would have required greater industrial statesmanship and wiser governmental action than anyone was prepared to champion. In 1937 the Maritime Commission had blamed the industry's chaotic labor relations on the "unenlightened attitude . . . of both employer and employee"; it had also clearly warned that relations would have to improve if the industry was to survive. In an age of industrial consolidation, however, the shipping industry was still burdened with the problems of regionalism and small-scale operations. In the 1930s and '40s powerful industrial unions were organized in highly concentrated sectors of the economy, such as the automobile and steel industries; they helped bring a degree of order to labor-management relations in these industries. But in the maritime industry competing unions continued to fight each other by dividing and conquering management.

Such warfare would have quickly destroyed the industry or brought both management and labor to their senses had the government not, for its own reasons, absorbed the cost of escalating wages. In wartime, of course, high wages were paid for extraordinarily hard work and for taking huge personal risks to sustain the war effort. But in peacetime one would have expected the government to insist on reasonable labor relations in return for operating subsidies. In 1937 the Maritime Commission had suggested that a law similar to the Railway Labor Act be written for the maritime industry, and such a law might well have been as beneficial for the maritime industry as the Railway Labor Act was for the railroads. But the suggestion was ignored, and the government remained indifferent, except as a paymaster.

By the 1970s the long cycle of self-destruction was coming to an end. A National Maritime Council was formed in 1971, its purpose being to encourage American shippers to use U.S.-flag carriers. The council included representatives from government, labor, and management; the government's participation was authorized by the section of the Merchant Marine Act of 1936 that directed the Maritime Administration to promote the commercial use of the U.S. merchant marine. When the council's representatives made their appeal to shippers, they were told again and again that foreign carriers were preferred because they were more dependable. Shippers told the council members what Paul Hall, the president of the SIU, had been saying for years: labor and management in foreign countries resolve their differences without abusing their customers by tying up cargoes or canceling passengers' long-planned vacations. The message finally got through; there has been only one major strike since 1970. But great damage had been done, and labor attitudes and relations in the American maritime industry were all too often a source of insupportable cost and inefficiency.

PART III

THE APPROACHING END

10

ATTEMPTS TO AVOID
A SECOND DECLINE,
1960–1980

Public attitudes shifted dramatically between the election of President Kennedy in 1960 and the election of President Reagan in 1980. The cold war continued, but the experience of waging war in Vietnam called into question America's responsibilities as a superpower. The Soviet Union invested enormous sums in its navy, merchant marine, and fishing fleet, but the U.S. response was muted until President Reagan committed his administration to a six-hundred-ship navy.

Those charged with administering the laws that governed regulated industries in the United States, especially in transportation, communications, and energy, began to implement deregulatory policies that eventually would profoundly alter the relationship between those industries and the government. And in some major industries that were not directly regulated, such as the steel and automobile industries, fierce international competition forced drastic restructuring and considerable dislocation. It was a period of crosscurrents, one which might have but did not occasion a radical rethinking of U.S. maritime policy. Some new initiatives were taken, to be sure, most notably in the Merchant Marine Act of 1970, but at the end of these two decades the basic structure of the U.S. maritime industry remained fundamentally as it had been, with the pressure exerted by increasing foreign competition still being addressed through protective legislation rather than economic adaptation.

INCREASING GOVERNMENT-INDUSTRY DISARRAY

In the 1960s relations between the maritime industry and the federal government became increasingly contentious. The industry, while becoming ever more dependent upon federal protection and subsidy, was also exhibiting the irritating attitude that it was entitled to such funding and to shelter from competition. Savvy administrators tried to steer clear of the maritime policy quagmire, but the issues could not be entirely ignored. The entire generation of ships built during World War II was approaching obsolescence; if the federal government did not take the lead in arranging for their replacement, it was likely that the U.S.-flag fleet would quite suddenly experience a precipitous decline, a denouement no one was prepared to accept. But it was only after years of halfhearted bureaucratic efforts that an ambitious new initiative was launched at the end of the decade. This effort proved only partially successful, however, and at the end of the 1970s the question of how the federal government could effectively marry the promotion of maritime commerce and public investment for national security remained largely unanswered.

President Johnson made several stabs at resolving maritime policy problems, but the industry's representatives were unmoved by his famous powers of persuasion. After the Soviet grain boycott began in 1964 the president created a grievance committee composed of labor and management representatives and charged it with advising the maritime administrator when industry disputes arose. But the unions distrusted the committee, which in any case had no power, and it was soon disbanded. The grain boycott was resumed shortly thereafter.

President Johnson also created a more prestigious Maritime Advisory Committee (MAC) chaired by the secretary of commerce. Its membership was carefully structured: in addition to the chairman, the government was represented by the secretary of labor and five other members; organized labor also had five representatives; and five members represented management and the general public. The committee's purpose was to provide "a forum within which the recommendations of the public, labor and management members . . . shall be represented to, and discussed with, the Secretaries of Labor and Commerce."[1] As the committee soon learned, however, getting the parties into one room was easier than establishing fruitful dialogue, and allegiance to special interests continued to dominate.

The MAC chairman, Secretary of Commerce John Connor, promised that no new maritime program would be proposed until it had been reviewed by the committee. Subcommittees were then established to develop policy proposals. Theodore Kheel, a prominent labor lawyer, chaired a group composed of the public representatives. They drafted a long-term maritime program, known as the MAC Report, that called for continuing the existing maritime policies and substantially increasing commercial shipbuilding. A majority of the full committee approved the MAC

Report, but the government representatives, under the leadership of Alan Boyd, the under secretary of commerce for transportation, ignored it while working up their own proposals. The report produced by Boyd's group was titled the Interagency Task Force Report but was commonly called the Boyd Report. It was released to the press at the same time it was submitted to the full committee for their consideration.

The Interagency Task Force Report pulled together a number of ideas that Alan Boyd had been advocating for some time, and he clearly saw his chairmanship of the government subcommittee as an opportunity to stage a policy coup. His report called for phasing out the cargo preference program, for allowing the use of foreign-built ships in subsidized foreign trade and the Jones Act trades, for extending subsidies to the bulk trades, and for modifying the existing operating subsidy program. These proposals, and their premature disclosure to the press, angered the other members of the Maritime Advisory Committee, yet when the secretaries of commerce and labor forwarded the committee's recommendations to the president, it was the Boyd Report they sent forward. In the absence of broad support, however, these proposals were not sent on to Congress for further consideration.

This was not the end of Alan Boyd's encounter with maritime policy, however. President Johnson wanted to pull all the government's transportation activities into a single office, and in October 1966 Congress approved the creation of a new Department of Transportation (DOT). The department was to take charge of all the transportation functions previously exercised by the Department of Commerce; it was also assigned the duties of some independent agencies and responsibility for the U.S. Coast Guard, which had been a part of the Treasury Department. Alan Boyd was confirmed by the Senate as the first secretary of transportation.

Although it was originally intended that the Maritime Administration (MARAD) would be relocated in the new DOT, the maritime industry opposed the move, and it remained in the Department of Commerce. This opposition was in part a reaction to the proposals Alan Boyd had foisted on the MAC the preceding year. But there was also an administrative rationale for keeping MARAD in the Department of Commerce. The functions of the new DOT were primarily regulatory, whereas the primary functions of the Maritime Administration were promotional. The Department of Commerce was created to promote American business; there was little reason to think DOT would understand or strongly support such programs. And so, although Boyd wanted the Maritime Administration in DOT, it remained in the Department of Commerce.

Since Boyd could not command the maritime industry, he began to court it. He let it be known he was prepared to modify his previously announced policy proposals and would support continuation of existing subsidy programs. This move skillfully pitted two factions within the industry against one another. One group immediately sought to put MARAD beyond Boyd's reach. The American Maritime Association (AMA), a trade association of unsubsidized lines that carried PL 480 cargoes and was allied with the Seaman's International Union (SIU), worked with SIU president Paul Hall to have legislation introduced that would make MARAD an independent agency. In short order one hundred congressmen pledged to support this move. The other group was ready to work with Boyd. The subsidized lines and the unions that manned their ships opposed Hall's initiative and made their position known through the Committee of American Steamship Lines (CASL) and the American Merchant Marine Institute (AMMI).

The issue was debated in July 1967 at hearings held by a subcommittee of the House Merchant Marine and Fisheries Committee. The subcommittee chairman had introduced a bill to create an independent maritime agency, and he pressed the representatives of CASL and AMMI to explain why they now wished to see MARAD shifted to DOT, whereas they had originally opposed such action. It soon became clear that Boyd's change of heart on subsidies satisfied their main concern. Privately they welcomed his proposal to allow the use of foreign-built ships in the subsidy programs and Jones Act trades, and they did not oppose his proposal to end cargo preference. Indeed, the end of cargo preference legislation as it was currently structured well might mean more cargoes and higher freight rates for the subsidized carriers. Boyd had also shown a little steel to keep his supporters in line. At a breakfast meeting two of his senior staff members told Frank Nemec, executive vice president of the Lykes Steamship Company and president of CASL, that the subsidized lines had better endorse the transfer of MARAD to DOT or they would be "budgeted into oblivion."[2]

But Boyd had no comparable way to force the shipbuilders and the unsubsidized operators to bend to his will. In his opening testimony Paul Hall, the spokesman for these interests, laid out why they opposed Boyd's proposal:

> In the consideration of this legislation we are confronted on the one hand with a political theory: and on the other hand with history, experience, and facts. And the theory and the facts are in complete conflict. The theory is a very good one. It is that all executive agencies concerned with transportation

should be combined in the Department of Transportation, presided over by the Secretary, who will keep all modes of transportation healthy and prosperous, and correlate their programs and activities in a national transportation industry.

History, experience, and the facts prove beyond doubt that the Maritime Administration, if included in an executive department having other duties and responsibilities, becomes submerged, is largely ignored and languishes from neglect. Broadly speaking, these are the reasons why we have opposed and still oppose the transfer of the Maritime Administration to the Department of Transportation.[3]

As is so often the case in government, uncompromising opposition prevented any action; both proposals, one being to make MARAD an independent agency, the other being to shift it to DOT, died. Boyd, having been hammered in the hearings for his stand on foreign shipbuilding, let it be known that if he did not prevail there would be no maritime program at all, and he was good to his word. As the Johnson administration wound down, the estrangement between the maritime industry and the government increased. At one point members of the National Maritime Union paraded around the White House carrying mock coffins to dramatize the decline of the merchant marine. When they disbanded they threw the coffins over the fence onto the White House lawn, a gesture President Johnson found deeply offensive. In 1966 Nicholas Johnson, who was then serving as maritime administrator, was appointed chairman of the Federal Communications Commission. President Johnson not only did not submit new maritime legislation, he pointedly left the office of maritime administrator vacant for the remainder of his term.

THE MERCHANT MARINE ACT OF 1970

Johnson's maritime policy gridlock presented Richard Nixon with an opening that he exploited during the presidential campaign in 1968 and during the early years of his administration. While on the campaign trail Nixon gave a speech in Seattle titled "Restoring the U.S. to the Role of a First-Rate Maritime Power." He noted that the Soviet Union was rapidly expanding its merchant marine as part of its program to become a dominant maritime power; spurred on by his well-known anti-Communist sentiment, he insisted the United States must respond. His vision was comprehensive, addressing the needs of both shipowners and shipbuilders, and the industry found his proposals noteworthy. "We shall," Nixon declared, adopt a policy that will enable American-flag ships to carry much more American trade at competitive world prices. The old ways have failed, to the detriment of the seamen, the businessmen, the balance of payments and the national defense.

The time has come for new departures, new solutions and new vitality for American ships and American crews on the high seas of the world.[4]

More surprising than these oratorical flourishes was Nixon's determination to carry out what he had promised to do. Shortly after taking office he told a gathering of subcabinet officials that while he had made many promises during the campaign, some of which perhaps had been ill-advised, he had been elected to fulfill the program he had run on, and that was what he intended to do. Revising and reinvigorating the nation's maritime programs was one of the promises he meant to keep.

Soon after Nixon was inaugurated, Maurice Stans, his new secretary of commerce, recruited Andrew Gibson, the senior author of this study, as maritime administrator and directed him to develop a program that would fulfill the president's campaign commitments to the maritime industry.[5] After being confirmed in mid-March Gibson recruited a staff and began work on the new program. Robert Blackwell was recruited as his deputy, and Roy G. Bowman was chosen as general counsel. Both of these lawyers brought considerable government experience to the task of putting together the new program, and both were appointed despite the fact they were registered Democrats. The president had made it clear that while he preferred Republican appointments, his first priority was to hire talented personnel.

By May a program had been formulated and was being reviewed in a series of meetings chaired by an executive office staff member. Representatives of the Maritime Administration attended these meetings, as did representatives from the navy, the Departments of Defense, State, and Transportation, and the Bureau of the Budget (the forerunner of OMB). Representatives of other departments, including Labor, Treasury, and Agriculture, and of the Council of Economic Advisors also occasionally attended. The reception given to the proposed program varied from negative to openly hostile.

Debate within the administration continued into the early fall of 1969. Had an unauthorized account of the proposed program been leaked to the press, the internal policy debate well might have been aborted, but the firing of Secretary of the Interior Walter Hickle and some of his staff a few months earlier for such indiscretion had demonstrated that such a move would cost the individual involved his job. The public therefore knew nothing about the program being hammered out until shortly before the president presented it. The maritime administrator's proposed program was opposed by nearly the entire Nixon Cabinet, but the president supported it, and that was enough.

President Nixon sent his new maritime program to Congress on October 23, 1969. In the accompanying message he said:

> Both government and industry share responsibility for the recent decline in American shipping and shipbuilding. Both government and industry must now make a substantial effort to reverse that record. We must begin immediately to rebuild our merchant fleet and make it more competitive. . . . Our program is one of challenge and opportunity. We will challenge the American shipbuilding industry to show that it can rebuild our Merchant Marine at reasonable expense. We will challenge American ship operators and seamen to move toward less dependence on government subsidy. And, through a substantially revised and better administered government program, we will create the opportunity to meet that challenge.[6]

To revive the shipbuilding industry the president proposed coupling new shipbuilding initiatives with a rollback of construction subsidies. Title 5 of the 1936 Merchant Marine Act set the normal rate for construction differential subsidies at 33 1/3 percent, but payments up to 50 percent were authorized in special circumstances. The higher rate had quickly become the norm, yet the industry was still hobbled by "low production rates and high production costs." As Nixon noted, the aging American merchant fleet could only be replaced by American-built ships "if our builders are able to improve their efficiency and cut costs." To create work for the shipyards, Nixon proposed a ten year program of thirty ships each year. To force improvements in efficiency, he proposed reducing the shipbuilding subsidy in the first year from 50 to 45 percent and by 2 percent every year thereafter until a maximum of 35 percent had been reached. He also proposed that construction subsidies be made available for building bulk carriers as well as the liners authorized by the 1936 act, and that the subsidies be paid directly to the shipbuilders rather than channeled through the shipowners. To help finance new construction, he proposed that the level of federal mortgage insurance be raised from $1 billion to $3 billion. These changes, the president believed, would encourage the builders to "improve designs, reduce delays and minimize costs." "We are confident," he said, "that the shipbuilding industry can meet this challenge. If the challenge is not met, however, then the Administration's commitment to this part of our program will not be continued." In other words, if the proposed shipbuilding program did not produce the expected increases in efficiency, the option of allowing foreign building of U.S.-flag vessels would be realistically considered.

The new program also proposed changes in the Operating Differential Subsidy system. Operating subsidies were to be made available to bulk carriers, the expectation being that the premium rates being paid for cargo preference shipments would be eliminated. The idea of providing direct subsidies instead of cargo preference for ships eligible to carry cargoes that had to be shipped on U.S.-flag vessels had been around for years. In 1950 Gordon Gray, a former secretary of the army, had made the argument in a memorandum to President Truman: "Cargo preference is, first of all, a concealed subsidy, and thus not subject to the scrutiny and supervision which is accorded to open subsidies. . . . Second, cargo preference is a blunt and capricious instrument for maintaining a fleet of security size. The volume of shipping kept in operation through cargo preference . . . is completely unrelated to estimated security requirements. . . . Finally, the cargo preference policy tends to relieve some of the pressure on ship operators to compete in service and rates."[7] President Eisenhower had also tried to substitute direct subsidies for premium payments on preference cargo when Nixon was vice president.

Other changes in operating subsidies that President Nixon proposed in 1969 included eliminating subsidies for maintenance and repair, the expectation being that the requirement that such yard work be done in the United States would be eliminated. Indexing increases in subsidized wages to a scale reflecting wage increases received by comparable American workers was also proposed as a way of bringing such subsidies within reasonable limits. Finally, all shipowners in foreign trade, and not just those receiving subsidies, would be allowed to defer payment of federal taxes on reserve funds set aside for new construction.

Hearings on the proposed maritime legislation were held during the following year. Despite earlier opposition, the bill finally passed through Congress with only two dissenting votes, one in the House and one in the Senate; it became law in October 1970. In its final form it contained all the essential elements of the president's program, but with several modifications to the original proposals. The subsidy for maintenance and repair was eliminated, but the 50 percent ad valorem tax on repairs done in foreign yards was not repealed. The result was an increased cost for the operators. The extension of tax payment deferral on funds placed in reserve was broadened to include all shipowners, that is to say those in domestic as well as foreign trade. The indexing of subsidy increases to cover wage increases was passed, but since the new scale was applied on top of an already extremely high base, it had little practical effect.

The 1970 Merchant Marine Act also elevated the maritime administrator to the level of assistant secretary, as had been recommended by the Intergovernmental Committee that investigated postal contracts in 1935. This was an important change because government policies are worked out in detail at the assistant secretary level. Department secretaries generally offer overall guidance and occasionally take issues to the president for final adjudication; it is assistant secretaries who develop, promote, and administer programs within the department. Giving the maritime administrator the additional title of assistant secretary of commerce for Maritime Affairs identified him as the administration's chief spokesman for maritime policy. With so many constituencies having an interest in maritime policy, the responsible individual in this area needed considerable support in order to develop and administer a coherent program.

The new law stimulated the largest peacetime private shipbuilding program in U.S. history. Shipbuilders spent over a billion dollars on capital improvements to their yards between 1970 and 1975. The maximum rate for shipbuilding subsidies was reduced for several years until it reached 41 percent. A number of liquified natural gas (LNG) carriers were built, some without subsidy payment, others at rates below 20 percent. While on average only twelve ships were built each year, rather than the intended thirty, the ships that were produced were considerably larger that those that had been projected; thus annual tonnage expectations in fact were met. The stimulus provided by the act continued into the administrations of Presidents Ford and Carter. In her end-of-the-fiscal-year report in June 1979, the secretary of commerce told the president:

> The cargo-carrying capacity of the privately owned deepdraft U.S. merchant fleet reached a record 21.6 million deadweight tons.
>
> New merchant vessels under construction or on order at private American shipyards on September 30, 1978, totaled 48 (with a contract value of $3 billion), . . . second only to Japan among the world's shipbuilding nations. Twenty new vessels were delivered by private U.S. yards in this reporting period.[8]

The 1970 Merchant Marine Act, passed twenty-five years after the end of World War II, provided a new burst of energy for the U.S. merchant marine. The Maritime Administration was reinvigorated, new shipbuilding programs were inaugurated, and new marketing programs designed to produce cargoes for the new fleet were begun. The spirit of optimism created by the act softened the antagonisms that had plagued the industry and its relations with the government and laid the basis for what promised to be a new era of cooperation. Yet the long-term results of the new programs were disappointing. Why was this so?

Even after passage of the 1970 act the U.S. merchant marine had to contend with two closely related facts that worked to its disadvantage. One of these was rooted in the long history of federal maritime policy. While the new act made significant changes in existing maritime policies, it left in place a number of restrictions that continued to make U.S. shipping a high-cost venture in international trade. The other fact was that the effects of new U.S. policies designed to promote foreign shipping had to be measured in an international market that was intensely competitive and changing rapidly. The continuing requirement that U.S.-flag ships had to be U.S. built, for example, meant that if American shipbuilders did not succeed in lowering their costs to competitive world levels, U.S. carriers would remain dependent on government construction subsidies and would have to deal with all the restrictions such dependency entailed. In fact U.S. shipbuilders did not increase their efficiency to the point that they could compete internationally, and when construction subsidies were eliminated, U.S. shipowners engaged in foreign trade had no choice but to buy foreign, flag out, or go out of business.

Foreign shipowners were also prepared to compete vigorously to defend their shares of the international shipping market. European and Japanese shipowners successfully resisted American efforts to acquire more cargoes, and U.S. ship owners were only able to attain the levels of market share called for in the preamble of the 1936 act in the South American and African trades. The results of extending operating subsidies to the bulk trades was also disappointing. Competition was fierce, and U.S. operators were burdened with many restrictions and trade-route concepts carried over from the liner trade. Obliged to employ U.S. crews and repair their ships in U.S. yards, American bulkers, unlike tramp ships operating under less restrictive flags, were not really free to roam the world in search of cargoes. The American owners who succeeded in the bulk trades operated under foreign flags.

The results of the 1970 Merchant Marine Act were, therefore, mixed. Since the Operating Differential Subsidy program of the 1936 act required subsidized shipowners to replace their ships on schedule, many liner companies had already committed themselves to new building programs before 1970. Some of these initiatives were aided by the new law. Two subsidized companies, Waterman and Delta, built a series of large Lighter Aboard Ship (LASH) vessels under the new program, and several existing ships were enlarged and converted into container carriers. A number of large tankers were also built under the new program, but most of them quickly retreated from foreign competition and ended up carrying crude oil under Jones Act protection from Alaska to California or through the Panama Canal to the Texas refineries.

Events beyond American control soon brought worldwide tanker production to a halt, however. The war that broke out in 1973 between Israel and the surrounding Arab states was accompanied by an oil-producers' embargo that sent petroleum prices skyrocketing. International demand plunged as prices soared, and tankers were laid up all over the world. New ships delivered by foreign yards went directly into layup and stayed there for years before being operated. This collapse of the market stopped the fledgling U.S. tanker-building program dead in its tracks. Without series production, construction subsidies quickly returned to 50 percent. Many established European shipbuilders went out of business at this time, and the center of world shipbuilding subsequently shifted to the efficient, low-cost yards in Japan, Korea, and Taiwan.

LATER ATTEMPTS TO PROVIDE ASSISTANCE

In addition to modifying the subsidy programs of the 1936 Merchant Marine Act so as to promote U.S. participation in world shipping, President Nixon also sought to increase the number of cargoes shipped on U.S.-flag vessels. As always, domestic and international affairs were closely linked.

By 1970 the United States had accumulated a huge grain surplus, and the Soviet Union was the only customer prepared to buy it. But no grain bound for Russia could be loaded at U.S. docks so long as the longshoremen sustained their boycott. The president's emissaries therefore approached the seagoing unions with an offer that they hoped would break the stalemate. If the longshoremen could be convinced to handle the grain, the administration would negotiate a bilateral agreement with the Soviet Union guaranteeing that a significant portion of the cargoes would move in American ships. The key figure in selling this bargain to the union leadership was Paul Hall, president of the Seafarers' International Union (SIU). Teddy Gleason, leader of the International Longshoremen's Association (ILA), was willing to go along with the deal so long as George Meany, president of the AFL-CIO, approved. Meany, however, being an implacable anti-Communist, would never have agreed had Hall not already established a close relationship with both Meany and his lieutenant, Lane Kirkland. In the end Hall convinced Meany to approve, and the deal went forward.

The Russians also had their needs and strategies. Leonid Brezhnev, the Soviet leader, had staked his political career on reaching detente with the West. Reopening trade with the United States was crucially important to him, and he was therefore prepared, although reluctantly, to pay subsidies in the form of higher freight differentials to ship part of the grain on U.S.-flag ships. The agreement that was signed in October 1972 stipulated that one-third of all cargoes would be reserved for American ships while the remaining two-thirds was to be divided between Soviet and third-flag ships. This agreement also opened all the major ports in the United States and the Soviet Union to each other's shipping. The Russians quickly exploited this feature of the deal by using selective rate cutting to take over a substantial share of the U.S.–northern European trade. Their penetration of the American market continued until 1979, when, following the Soviet invasion of Afghanistan, President Carter cut off all further grain shipments to the Soviet Union and the longshoremen renewed their boycott.

Other bilateral treaties were also negotiated as a way of enlarging U.S. trade and making cargoes available to U.S. carriers. During the Carter administration an agreement was signed with the People's Republic of China (PRC), but unlike the Russian agreement it did not require that the U.S. trading partner subsidize the use of American-flag ships; as a result, no grain was moved in U.S. ships. Some liner trade with China was begun, however, once its ports were open to international commerce.

The situation in South America was complicated, but new bilateral agreements were reached there, too. Although U.S. companies had previously entered into bilateral agreements with the Latin American carriers serving the Caribbean and the Pacific Ocean, such agreements had not been reached in either Brazil or Argentina. By 1970 the U.S. share of the Brazilian coffee trade, coffee being that country's chief export, had been heavily eroded by foreign shipowners using massive rebating to capture the export of coffee. The federal government entered the picture to help the two U.S. lines that had been affected, Moore-McCormack and Delta Lines. A bilateral maritime agreement was negotiated with Brazil reserving 40 percent of cargoes for each nation's carriers and leaving 20 percent for third-flag ships; under such an arrangement there was in fact very little third-flag participation. A similar agreement was negotiated with Argentina several years later. This 40-40-20 formula appealed to other less-developed commodity-exporting nations as well, for it assured them a large degree of participation in the carrying trade generated by their exports. In time this formula was temporarily adopted as the standard for all developing nations by the United Nations Committee for Trade and Development (UNCTAD). Opposition by American and some European governments prevented its implementation.

Rapid fluctuations in the global petroleum trade, and America's growing dependence on imported oil, also created new opportunities for those who wished to use legal restrictions rather than open competition to secure cargoes for U.S. ships. The 1973 oil embargo that caused sudden shortages and rapid price increases made oil policy a hot topic in America. In addition to promoting energy conservation and the development of alternative sources of energy, the government decided to create a Strategic Petroleum Reserve by storing large quantities of oil in underground salt domes located offshore in Louisiana and Texas. The 1977 law that authorized this program stipulated that 50 percent of all oil purchased for the reserve had to be transported in American ships.

Four years earlier the U.S. maritime labor unions had begun another concerted effort to turn the energy crisis to their advantage. Their goal was to require that a large percentage of the oil imported into the country be carried in U.S.-built, U.S.-flagged, and U.S.-crewed tankers. When a bill proposing that 30 percent of petroleum imports be shipped on U.S. carriers was introduced, lobbyists quickly dug in on both sides of the issue. The unions donated more than three hundred thousand dollars to 141 members of Congress to persuade them to support what they termed a "cargo equity" bill.[9] The coalition supporting the bill had a familiar look to it; it included the Marine Engineers' Beneficial Association (MEBA) and the SIU, the shipbuilders, and those shipowners most likely to benefit from the new law. In May 1974, the House quietly passed the legislation, and in September it was successfully voted out in the Senate.

The bill generated fierce opposition, for the stakes were high. The major oil, steel, and aluminum companies all had large proprietary fleets sailing under foreign flags, and they quickly united to oppose this threat. The national press, responding to public sensitivity to fuel prices, was solidly opposed to the bill. An editorial in the *New York Times* on October 12, 1974, called on President Ford to veto the bill when it was forwarded to him.[10] Japan and the European nations whose tanker fleets would be adversely affected had their ambassadors make their opposition known to Secretary of State Henry Kissinger.

Since Congress had already acted, it was left for President Ford to determine the fate of this contentious bill. Paul Hall was optimistic, for he had become a good friend of Ford's when he was Republican leader in the House. It therefore appeared that the shipbuilders, shipowners, and seafarers were about to receive a bonanza. But Ford, for reasons that have never been made clear, did nothing, and the legislation died by year-end through a "pocket veto." Many concluded that foreign protests, forcibly represented by the secretary of state, convinced Ford that he should not sign the proposed legislation.

Angry and disappointed, the labor unions stepped up to the plate for one more time at bat. Hall felt that Ford had betrayed him, and together with Jesse Calhoun, president of MEBA, he threw his full support behind Jimmy Carter, who was challenging Ford in the next election. When Carter was elected, a bill that would reserve a more modest 9.5 percent of oil imports for U.S. ships was prepared. While hearings on this bill were being held, a damaging internal White House memorandum was made public. The memo urged the president to support the legislation as a way of repaying the maritime unions for the extensive support they gave him during the 1976 election campaign. This was the kiss of death. Thundering editorials accused the maritime industry of using its power selfishly to obtain benefits at the expense of the American taxpayer and consumer. Many members of Congress who had supported the bill abandoned it, and it died after being defeated in the House.

President Carter, thoroughly embarrassed by this episode, was clearly not going to go to bat again for the maritime industry. Realizing this, some industry leaders and key members of the House Merchant Marine and Fisheries Committee in 1979 put together a massive proposal called the Omnibus Bill. It attempted to solve all the industry's problems and succeeded in antagonizing nearly everybody in some way. The bill proposed that foreign-built ships be allowed to register under U.S. flag so long as a matching vessel was built in an American yard; neither the shipbuilders nor the shipowners found this acceptable. Peter McClosky, an influential Republican member of the House Committee, hoped to win bipartisan support by limiting the size of crews on ships eligible for subsidy, but the labor unions would have none of it. The prospects for the bill took a turn for the worse when John Murphy, chairman of the Merchant Marine Sub-Committee, was indicted for bribery. When he was forced to surrender the chairmanship, his successor declared the Omnibus Bill dead.[11]

By 1980 the U.S. merchant marine had surrendered the dominant position in world trade that it had occupied at the end of World War II. Few doubted that the United States should have a merchant marine, yet none in the administration or Congress was willing to absorb the political heat that would have been generated by an attempt at real change. Those who worked in the maritime industry realized how dependent they were on public subsidy and protection, and they clung desperately to the programs that sustained them. From an industry point of view, laws that required the use of U.S.-flag ships and passed the higher costs directly on to consumers, such as the Jones Act, were the best form of support. The unions knew this, as they demonstrated when they sought to capture a share of the oil trade for U.S. ships, but this strategy alone was not enough. Direct subsidies and inflated shipping costs that had to be paid out of agency budgets were more visible and vulnerable forms of support, yet they remained indispensable to the industry.

During the cold war the military called on the nation's maritime resources frequently enough to render the naval auxiliary argument for funding the merchant marine credible. In any case, the cost of maintaining the merchant marine was insignificant when compared to military budgets. The economic arguments for using public funds to sustain the merchant marine were less compelling. As a generator of tax revenues and a contributor to the balance-of-payments account, the maritime industry was too small to be of much significance. Furthermore, the funds spent sustaining it could have been far more profitably employed elsewhere. Economically, U.S.-flag shipping was a marginal activity at best.

When Ronald Reagan was elected president in 1980, the position of the U.S. merchant marine engaged in international trade had slipped back to the level it had been at before World War I. Ships flying the Stars and Stripes were carrying less than 5 percent of the nation's foreign commerce. America's overseas trade was booming, and President Reagan was committed to making sure that the United States had the world's leading navy, but commercially the nation had ceased to be a maritime power of any consequence.

11

THE RAPIDLY CHANGING
MARITIME WORLD

The changes that transformed the maritime industry in the decades following World War II were astonishing. During the war Henry Kaiser had made modular construction techniques central to America's enormous shipbuilding program. After the war, however, minimal use of these new techniques was made in the United States and Europe, where labor unions were strong. As Kaiser's government-owned yards were closed, the remaining yards reverted to using many of the construction methods employed before the war. In postwar Japan, however, Kaiser's new construction techniques were welcomed; by combining efficient methods of production with low labor costs, the Japanese soon captured the dominant position that Great Britain had held in shipbuilding before the war.

Equally dramatic changes were introduced in cargo handling. A few years after the war ended, ships began to carry loaded truck trailers either on Roll-on/Roll-off ships (Ro-Ros) or as trailer-sized cargo boxes on containerships. Cargo-handling productivity soared, and the quick turnaround time containerization made possible increased vessel productivity as well. Efficiency in the bulk trades also rose dramatically as ships, and especially tankers, grew to enormous proportions. By the early 1980s behemoths of five hundred thousand deadweight tons (DWT) were being built, and although ships this size proved to be too large for efficient operation, tankers fifteen times larger than the T-2 tanker of World War II became the norm.

In the 1970s, as the various provisions of the Merchant Marine Act of 1970 were implemented, there was considerable optimism that the American maritime industry might be revived. A massive influx of new orders gave U.S. shipbuilders a new lease on life, and they invested heavily in equipment and new facilities in an effort to increase productivity. Containerization, an American innovation, offered such order-of-magnitude increases in efficiency, security, and dependability in cargo handling that many new importers and exporters were attracted to the world market. It appeared for a while that American ships would again carry a significant share of world trade, but such optimism would not last.

The Japanese were quick to adapt to the new circumstances, and Korea and other Asian nations followed close behind them. European shipbuilders and shipowners, who for many decades had dominated international shipbuilding and ocean transport, were slower to respond, and many fell by the wayside. The few that survived, to their credit, learned to play by a whole new set of rules. U.S. shipbuilders counted on the navy to keep them in business. They also welcomed environmental laws that mandated double hulls for tankers trading in U.S. coastal waters. Since vessels engaged in Jones Act trade had to be U.S. built, the shipbuilders looked forward to eventually replacing all the Jones Act fleet.

Unlike the shipbuilders, the shipowners had the option of escaping the costs and constraints associated with U.S. registry by sailing under a foreign flag of convenience (FOC), as they were called by their detractors. Many former colonial nations established "open registries" that imposed no restrictions on the nationality of the officers and crew employed and did not tax the income earned by ships or their crews. Several European nations responded to the "flight from the flag" by setting up their own "international" registries that granted nearly as much freedom of operation as the FOCs; the most successful of these was the Norwegian International Ship Registry (NIS).

Containerization and Intermodalism

The container revolution began on an April day in 1956 in Port Newark, New Jersey. Fifty-eight specially strengthened truck trailers bound for Houston, Texas, were loaded on the spar deck, which ran above the main-deck cargo-piping system, of *Ideal X*, a T-2 tanker built during World War II. During the war the spar deck of *Ideal X* had been used to carry military aircraft; in 1956 it carried the first modern containers.

The man behind this novel experiment was Malcom McLean, the founder and owner of one of America's largest long-haul trucking companies. McLean wasn't trying to do anything complicated; he simply realized that it would be cheaper to move loaded trailers between port cities on ships than to haul them over the road. When not aboard ship, each box could be attached to a trailer chassis and driven from its point of origin to the ship and from the ship to its destination. McLean called *Ideal X* a "trailership." But while the central idea was not complicated, its implementation proved to be profoundly consequential. The container revolution transformed surface-freight transportation throughout the world. McLean didn't set out to make a revolution, but his innovation had that effect. The shipping industry was never to be the same again.

Not satisfied with having built up a major trucking company, McLean bought the Pan-Atlantic Company from its parent company, the Waterman Steamship Company of Mobile, Alabama. Pan-Atlantic operated in the Gulf and Atlantic Coast trades and therefore already had the Interstate Commerce Commission (ICC) certificates that McLean needed to try out his new idea. On the return trip from Texas, he originally intended to have *Ideal X* carry petroleum, as it had done during the war, as well as deck cargo; but for reasons best known to themselves, the U.S. Coast Guard would not allow this. The prohibition made his initial venture unprofitable, but McLean pressed on. Unlike members of the shipping fraternity, who rarely took risks not covered by insurance, McLean was a classic entrepreneur. He bet his fortune and all the money he could borrow on containerization, an idea most shipowners considered an utter folly.

McLean was practicing what came to be called "intermodalism" before the term was coined. Ship operators traditionally began thinking about their cargoes only when they arrived on the dock in barrels, bags, and boxes; they stopped worrying about them when they had been off-loaded and placed on a distant shore. McLean simply played across the ship-dock interface while looking for a more efficient way to load and move cargo. Today everyone in the transportation industry sees the problem of moving freight the way McLean did in 1956; but back then ship operators, and particularly those who eventually had to compete with him, considered McLean a meddlesome outsider. McLean stuck to his goal of reducing the costs of loading and unloading cargoes, however, and the way he found to do that radically altered the economics of shipping.

Before containerization, cargo was loaded and stowed in the hold piece by piece, or pallet load by pallet load, in a system called break-bulk shipping. On a typical break-bulk voyage of two to three thousand miles, cargo-handling costs represented about half of the total voyage costs. To understand how containerization changed these cost calculations, one must note that break-bulk cargo-handling costs were variable; if no cargo was being moved, no dock workers were hired and no expense was incurred. Shipowners in the break-bulk era could afford to be easy going, with prolonged stays in port. But the introduction of containers radically altered the calculation of cargo-handling costs. Containerization substitutes expensive capital equipment (a fixed cost) for manual labor (a variable cost). Capital equipment, such as gantry cranes, tractors, trailer chassis, container facilities, and the containers themselves, must be paid for whether they are producing revenue or not. In container shipping, fixed costs represent approximately 90 percent of total voyage costs; equipment use therefore must be maximized if a profit is to be realized.

Container technology is comparatively simple, more *Popular Mechanics* than quantum mechanics, yet like railroads and electric power production it is a technology that by its very nature forces system building and standardization. The dimensions of McLean's original containers were, naturally enough, determined by the state highway codes that stipulated the maximum size of truck trailers. The first containers were thirty-five feet long, eight feet wide, and eight feet high, but highway codes changed as new roads were built, and it was not long before larger boxes were being introduced. An international standard of twenty- and forty-foot containers was soon established, but boxes of different dimensions have subsequently been employed, and the tension between responding to specific needs and opportunities and conforming to the standards of a universal system remains.

Containerships can be loaded and unloaded so quickly that new systems for locating, moving, and positioning containers also had to be developed. A gang of longshoremen could typically handle ten to fifteen tons of break-bulk cargo per hour; a considerably smaller number of men can load containerized cargo at the rate of six to seven hundred tons per hour. Development of the support systems needed to realize the full potential of containerization would have been stunted had computers not become available to record and manage tracking, loading, and discharging programs. And as the ability to locate and move containers improved, customer expectations rose accordingly. The railroads had grown accustomed to losing railcars for weeks at a time in switching yards, although customers served by trucks were more demanding. As pickup and delivery times became more reliable, inventories could be reduced and production lines and marketing schedules could be geared to just-in-time arrivals. In brief, the introduction of a truck-trailer-sized standard cargo container gave rise to an intermodal freight system that possessed an internal dynamic for greater coordination, higher levels of utilization, and steadily falling cost. No one, not even Malcom McLean, had anticipated that the larger consequences of this innovation might be so vast.

Containerization also provides greater cargo security. Goods can be packed in a container in less time and with much lighter packing materials than are needed in break-bulk shipping, and they are much less liable to be damaged. Thus, casual pilferage of high-priced cargoes was virtually eliminated and cargo insurance rates fell accordingly.

Goods flowed through containerports without being broken out and repacked, but moving and storing these huge boxes requires open space of a kind that was seldom available near the piers that lined urban waterfronts. As the container lines developed hub-and-spoke distribution systems, many ports that had previously been served directly by transoceanic liners found themselves receiving shipments in smaller lots brought by feeder ships distributing cargoes off-loaded in major containerports. The few hubs where the largest containerships call are located where suitable land, deep water, and rapid road and rail connections can be found in conjunction. In New York Harbor containerships call at the docks in Newark and Elizabeth, New Jersey, instead of Manhattan and Brooklyn; in England, Liverpool and Southampton declined as Felixstowe expanded; and San Francisco watched its piers rot away as ships went across the bay to Oakland. Some older ports, such as California's Los Angeles/Long Beach, Washington's Seattle/Tacoma, and Rotterdam and Yokohama, were able to add sufficient land to meet the demands of their booming container business. The container has been and remains an incredible engine of urban change in traditional port cities.

But these larger consequences only became evident over time. McLean's immediate problems, when he first introduced containers, were to establish reliable service and to increase the market share of his innovative maritime venture. He added additional ships to his coastal fleet and in August 1958 inaugurated container service to Puerto Rico. Carrying cargoes between Puerto Rico and the mainland was, like the coastal trade, restricted by the Jones Act to U.S.-flag ships. American ships receiving subsidies were also excluded from Jones Act routes. The Bull Line had enjoyed a virtual monopoly for many years in the Puerto Rico trade, but in 1958 the Bull Line had been shut down by a jurisdictional strike between maritime unions. When the strike ended Bull Line acquired two ships partially adapted to carry containers, but by then McLean's Pan-Atlantic line was firmly established and was providing superior service to Puerto Rico. The Bull Line eventually went bankrupt after having served the island since 1910.

On the U.S. West Coast another set of forces eventually drove Matson Navigation in a similar direction, although the system it developed was in many ways unique. In the 1950s Matson's owners, the "Big Five" merchant families of Hawaii, were pressing the company's president, Randolf "Joe" Sevier, to increase the company's profitability. Matson Navigation, operating as a domestic carrier under the jurisdiction of the Interstate Commerce Commission, was finding it exceedingly difficult to obtain rate increases large enough to offset the huge cargo-handling costs that were consuming most of its revenue. In 1953 Sevier pulled Stanley Powell out of the sales department and put him in charge of the company's faltering research operation. Powell soon recognized that the savings realized in industries that made use of assembly-line technology could be realized in liner shipping. He and others at Matson saw that McLean's container system could be used to establish an assembly line between San Francisco and Honolulu. To implement this vision, they studied the cargoes Matson carried and determined that their needs would be best served by a twenty-four-foot-long container. Unlike the designers of the Sea-Land system, the Matson executives did not begin with the size limits on trucks set by East Coast states. The twenty-four-foot container had an added advantage—West Coast highway regulations allowed two such units to be towed in tandem on public roads.

There were other differences as well. Malcom McLean insisted that there be a trailer chassis in the port terminal for every container that was off-loaded, a stipulation that clearly required very large dock areas. But because terminal facilities on the West Coast and in Hawaii were more expensive than on the East Coast, Matson elected to stack their containers three high when storing them and only placed them on trailers when they were being readied for highway transport. To stack and unstack containers, Matson adapted the straddle carriers commonly used to move stacked lumber. And because Matson would be operating its containerships between only two ports, the company elected to use shore-based cranes to lift boxes on and off its ships. McLean, who planned to serve a wide variety of ports, fitted out his original containerships with onboard cranes to load and discharge their containers. Sea-Land later followed Matson in relying on shoreside cranes and only put lifting gear on ships opening new trades to container service and calling at ports without adequate shoreside facilities.

McLean, being the sole owner of his company, could move swiftly and decisively when acquiring ships and inaugurating new services. Sevier could only move as quickly as Matson's very conservative board of directors would allow. Nonetheless, by 1960 Matson was able to inaugurate the first container service in the Pacific. When Sevier retired, Powell became Matson's president and pushed hard to extend an unsubsidized container service to Japan, Korea, Hong Kong, and the Philippines. These countries resisted introduction of the new technology, causing delays and large operating losses.

Although the losses were not out of line with Powell's original projections, the Matson directors became increasingly uneasy and eventually insisted that the Far East service be discontinued. The four containerships that Matson had ordered, two of which were being built in the United

States and two in Germany, were eventually acquired by Sea-Land and became the most modern ships in that company's fleet. Meanwhile Matson concentrated on containerizing its West Coast–Hawaii service, which under the presidency of Robert Pfeiffer was brought to a high level of profitability.

While both of these developments were unfolding, Grace Line, which served ports in the Caribbean, Central America, and the west coast of South America, decided in November 1957 to convert two of its C-3s, *Santa Eliana* and *Santa Leonore*, to full containerships. Two years later, in January 1960, the *Eliana* became the first full containership to enter foreign trade when it sailed for La Guaira, Venezuela. This was a daring venture beyond U.S. protection, and the risk involved immediately became apparent. The Venezuelan longshoremen quickly realized that this new way of handling cargo would undermine their archaic but firmly established work rules. Determined to defend their jobs, they struck *Eliana* as soon as it arrived. Not a single container was moved for eleven months as the ship sat at its berth in La Guaira while efforts were made to settle the dispute. In the end the longshoremen agreed to discharge the ship so long as it never returned.

Bloodied by unbowed, Grace Line suggested to U.S. Lines that they open a joint container service in the North Atlantic. Grace Line would provide the ships, three full sets of containers and equipment, and the U.S. terminal; U.S. Lines was to provide marketing and a terminal in the United Kingdom. The response to this suggestion was not unexpected. What, the U.S. Lines officials politely asked, would be put in the containers? When it was suggested that they be loaded with whiskey, high-priced woolens, china, and other high-value items, the U.S. Lines executives were not impressed. Who ever heard of such a ridiculous idea as putting whiskey in containers! And that was how the first experiment in international container shipping ended. Grace Line sold its containerships to McLean, who renamed them *Ponce* and *Mayaguaz*. An incident involving the latter ship brought it a degree of fame during the war in Vietnam.

The Grace Line's dismal experience dampened interest in investing in container shipping and secretly delighted those operators who considered McLean's innovation imprudent. Yet a few lines continued to experiment with the new system. In 1966 Moore-McCormack introduced the first of its four combination Ro-Ro and container ships into its North Atlantic service. An intermediate deck that could be adjusted to accommodate containers or Ro-Ro cargo made these ships unusually adaptable. McLean, whose Pan-Atlantic Line by this time had been renamed Sea-Land, also entered the North Atlantic trade in 1966.

Moore-McCormack, being restricted essentially to its subsidized trade route to Scandanavia, was unable to generate sufficient cargoes to profitably fill its ships and was forced to withdraw. Sea-Land, operating without any trade route restrictions, was able to continually expand, eventually covering much of Western Europe via the ports of Rotterdam and Bremen. In time it penetrated the British market by using Felixstowe and Grangemouth as its ports of entry. By utilizing two previously underdeveloped facilities, it successfully avoided possible confrontation with militant British trade unions.

While European and Japanese shipowners began planning how they would enter the new world of containerization, American shipping companies continued to evaluate their options. In the watershed year of 1966 a new management team at U.S. Lines became convinced that there might be some benefit in "carrying whiskey in containers" and began converting their American Racer-class to full containerships. It was not long before they also converted their American Lancer-class ships as well. Shortly thereafter U.S. Lines ordered an entirely new class of large containerships, the first of which, *American Astronaut*, could carry thirteen hundred twenty-foot-container equivalent units (TEUs). U.S. Lines was now in a position to challenge McLean.

But in 1969, just as the rush to containerization was reaching its height, McLean sold Sea-Land to the tobacco conglomerate R. J. Reynolds. In return for this shrewdly timed transaction McLean received a top-of-the-market price and a seat on the R. J. Reynolds Board of Directors. Other American liner companies did not share his good fortune. After being merged into the Isbrandsten Steamship Company, American Export Line inaugurated an unsubsidized and unsuccessful North Atlantic container service. It had better luck with a subsequent Mediterranean container service, which was continued after Isbrandsten sold his company to the Farrell Line. In 1969 Grace Line was sold to Prudential Lines, which was committed to a LASH barge system for its Mediterranean service. Prudential later sold the former Grace Line service to the Delta Line of New Orleans and converted its remaining LASH ships into full containerships. The company's management failed to make a success of it, however, and within a few years Prudential Lines ceased operations.

Like the Farrell Line, the Moore-McCormack company served ports in South and East Africa as well as the east coast of South America. Although initially containerization had had little impact on these routes, the president of Moore-McCormack, James Barker, realized that eventually the company would have to take the costly risk of introducing container service. Barker decided that the company's assets could be better employed elsewhere and sold its two passenger ships to Holland-American Line. He then bought Pickens-Mather, a coal and iron-ore resource company that also operated a fleet of ore carriers in the Great Lakes, and changed its name to Moore-McCormack Resources. The new company's headquarters was moved from New York City to Stamford, Connecticut, a move which turned out to be the beginning of a general exodus from New York City. Its remaining liner service was sold to U.S. Lines.

The exports and imports that moved through the Gulf of Mexico ports of Houston, New Orleans, and Mobile were largely bulk commodities, many of which were reserved for U.S. ships by PL 480, and they did not lend themselves to containerization. After buying the Pan-Atlantic line, McLean bought its parent, the Waterman Steamship Corporation of Mobile; he later sold Waterman to the Walsh family, who in turn sold it to Central Gulf, which is headquartered in New Orleans. Waterman continued to operate three LASH ships in the Far East trade under federal subsidy. Delta, another Gulf Coast company, operated at a substantial level of profit in 1980 and 1981 after acquiring the Grace Line's ships and trade routes. To a large extent these profits were generated by the high cargo rates paid by Latin American countries that were in the process of accumulating the massive foreign debts that nearly drove some of them into bankruptcy. Holiday Inns, the parent company of Delta, realized this could not continue, and, instead of undertaking a major ship-replacement program, sold the company to Crowley Maritime in December 1983. Crowley sustained two years of substantial losses before selling what remained of Delta to Malcom McLean, who by that time had reentered the shipping business by purchasing U.S. Lines.

Lykes Brothers Steamship Company, the foremost U.S. player in the Gulf of Mexico and the largest subsidized carrier, was slow to react to the container revolution, and when it did, it took on massive debts while purchasing new vessels. This burden eventually overwhelmed Lykes, and in October 1995 the company was forced to file for protection under chapter 11 of the U.S. bankruptcy code.

Other American-flag lines on the West Coast were also in difficulty. States Line, a proud company with a long record of service, bet its future on a new fleet of Ro-Ros just as the Pacific Basin was becoming containerized. It lost its gamble and quietly disappeared. The Alioto family of San Francisco acquired the Pacific Far East Line from American President Lines (APL) and tried to run LASH ships and Ro-Ros in the Pacific trades. It too failed.

APL, under the leadership of Bruce Seaton, looked to the future and invested heavily in containerization. It merged its subsidiary American Mail Line into the parent company and became the dominant container carrier in the Pacific when it put its fully containerized vessels into service in 1973. A few years later it decided to concentrate all its resources in the Pacific Basin and discontinued its around-the-world container service. To serve its East Coast customers, APL developed its intermodal rail connections and introduced the first express container-train service connecting the West Coast to New York. The company survived the transition and has been a leading innovator in containerized shipping ever since.

Back in the Atlantic, container service to Puerto Rico continued to evolve following Sea-Land's elimination of the Bull Line. During the 1960s the island was served by three companies: Sea-Land, Seatrain, and Trans American. When Seatrain announced that it was going to discontinue its service, the Puerto Rican government, in an ill-considered political move, decided to control rising freight rates by acquiring Seatrain's ships. The government soon realized that to effectively control rates it would have to acquire all the ships serving the island. In 1973 it formed a new company, Navieras de Puerto Rico, and purchased the ships of Sea-Land and Trans American. A few years later, however, Crowley Maritime developed a highly efficient fleet of towed Ro-Ro barges to transport fully mounted containers to the island, and shortly thereafter Sea-Land again began operating to Puerto Rico. These challenges kept freight rates low and prevented Navieras from showing a profit. The company constituted a considerable drain on the island's Treasury, and when a new governor was elected, he acknowledged the venture had failed and the company was sold.

The introduction of containerization has continued to take its toll on American shipping companies. Being a member of the R. J. Reynolds Board of Directors was not enough of a challenge for Malcom McLean, so when the opportunity arose, he cashed in much of his stock and purchased U.S. Lines, his goal being to recapture first place in the U.S. container trade. Focusing as always on reducing costs as much as possible, McLean decided to build twelve of the largest containerships in the world in Korea. Because he accepted the common assumption that fuel costs would continue to rise, he decided to sacrifice speed for fuel efficiency, and his sixteen-knot ships were to be the slowest of all the major container carriers. He planned to operate them in an eastbound around-the-world service at a time when freight rates on several of the connecting links had begun to deteriorate. This put him in direct competition with the well-established Taiwanese company Evergreen Line. Construction contracts were let in

1983, and the first sailing of *American New York* from the Far East began in July of the following year. Because these huge ships had to pass through the Panama Canal, the beam restriction added not only to their length but also to their draft. Their much deeper draft prevented them from calling at many ports, particularly in the United States, when fully loaded. But their fatal flaw, as it turned out, was probably their slow speed, for fuel costs fell instead of continuing to rise, and the trade-off that made them fuel efficient became a serious liability.

McLean had previously financed his acquisitions primarily through bank financing, but to pay for his new ships he not only loaded U.S. Lines with debt, he also sold a considerable amount of stock in the company. In the past he had been able to meet his debt payments because of his highly profitable service. But in the 1980s McLean was competing against other equally efficient container companies, not old-line, break-bulk companies, and Sea-Land and the recently established foreign container lines were not about to yield market share without a fight. In this instance the father of containerization guessed wrong, and by November 1986 U.S. Lines was bankrupt. Another company with a proud lineage, in this case stretching back nearly a century to J.P. Morgan's International Mercantile Marine, was gone. McLean's new ships were laid up and Sea-Land purchased eight of them at bargain prices.

Containerization, an innovation that began in the United States, continues to evolve worldwide. The pressure for system-building and increased efficiency that drives this evolution has led to the construction of ever larger containerships and ever more tightly scheduled intermodal linkages. In 1987 American President Lines (APL) added five huge ships of a new L-9 class to its Pacific fleet. Rather than restrict the beam and increase the draft of these ships, as McLean had done, APL decided to forego the possibility of sending them through the Panama Canal. A wider beam gives these ships greatly improved sea-keeping ability, but it also required that the gantry cranes used to load and unload them be rebuilt and extended. The APL system was stretched in another direction when forty-five- and forty-eight-foot containers were introduced. Boxes of these lengths are especially efficient on the double-stacked railcars that APL introduced on its unit trains to inland and Atlantic coast destinations. These innovations have helped create new markets. Japanese car assembly plants in Tennessee now know exactly when they will be receiving parts manufactured in Japan, Taiwan, and elsewhere and are able to schedule their production on a "just in time" basis. Sea-Land, meanwhile, continues to grow and has become not only the largest American shipping company,

Top Ten (First Quarter, 1996)	TEUs
1. Sea-Land	308,835
2. Evergreen	254,480
3. Maersk Line	206,376
4. Hanjin	171,682
5. American President Lines	143,906
6. COSCO	122,599
7. NYK	118,864
8. OOCL	112,892
9. Hyundai	108,303
10. "K" Line	100,458

Top Ten (First Quarter, 1993)*	TEUs
1. Evergreen	219,207
2. Sea-Land	206,342
3. Maersk	175,668
4. American President Lines	147,301
5. Hanjin	122,381
6. NYK	105,546
7. OOCL	89,947
8. Mitsui	86,918
9. Hyundai	84,906
10. "K" Line	81,578

Table 11.1: International Container Traffic through U.S. Ports.
*Comparative figures for the same period in 1993. The dramatic emergence of COSCO, the shipping line of the People's Republic of China (PRC), must be noted.
Source: Data from the *Journal of Commerce*, May 29, 1996.

but also, primarily because it moves five hundred thousand containers a year through its Rotterdam Terminal, the largest U.S. transport firm in Europe.

SHIPS DESIGNED FOR THE TRADE

With the coming of containerization the five-hatch break-bulk freighter that had been the standard cargo-ship design up through World War II no longer predominated. McLean had introduced one new specialized-ship type in 1958 when he put into service the first fully containerized vessel, the converted cargo ship *Gateway City*. But ships built to carry containers were only one type among a number of new specialized-ship types that appeared in the 1960s and afterwards, and as new types of ships were built and put into service, maritime commerce became increasingly segregated into distinct markets. These new ships were highly efficient when serving the trades for which they were designed, but most of them could not easily be shifted among markets, nor could they be readily adapted to military use. Thus as these ships increased in size and cost, stable financing and a steady steam of appropriate cargoes became ever more important. These needs in turn imposed ever greater demands on management and intensified competition in every sector of the industry.

Although Grace Line had been thwarted in its first attempts at containerization, it did not turn its back on the new technology. Rather than commit itself again to fully containerized ships, however, the company ordered four ships of fourteen thousand DWT designed specifically for the trades it served. The first of these ships, *Santa Magdalena*, entered service in 1963 on the New York–South American west coast run. The ship had accommodations for 125 passengers. It also successfully combined Lift-on/Lift-off (Lo-Lo) container service with Ro-Ro capability. Since none of the ports it served at that time had dockside container gantries, the ship had its own cranes to handle containers stowed on deck and in below-deck compartments. It also had large side ports through which automobiles could be driven and cargo could be loaded by forklift trucks. The lower decks were served by elevators. On northbound voyages the side-port compartments were loaded with one hundred thousand boxes of refrigerated bananas, making these ships the largest fruit carriers operating at the time. These ships were highly profitable when operated by Grace Line, but the banana contracts were lost when the ships were sold to Prudential, and they were not successful when operated merely as general-cargo freighters. The ships were later transferred to San Francisco and assigned to a route that circumnavigated South America, a service that was continued when they were sold to Delta. When Delta was in turn sold in 1984, the Santa Magdalena–class ships were retired, marking the end of America's two-hundred-year history of providing first-class international passenger transportation. Today only one fifty-year-old former American Export Line passenger ship continues to sail under the U.S. flag in service among the Hawaiian Islands.

The Lighter Aboard Ship (LASH) system of cargo handling that was developed at about the same time provides a high degree of carrying flexibility. The Delta Line built three LASH ships, *Delta Norte, Delta Sud,* and *Delta Mar*, and used them successfully for several years. The ships carried eighty-nine barges, each of which could be loaded with 370 tons of cargo. Once the barges had been loaded they could be brought aboard at the rate of 1,200 tons of cargo per hour. LASH ships were especially useful when serving ports connected to large river systems or coastal waterways, over which the barges could be operated intermodally. Delta regularly gathered and distributed cargoes along the Mississippi river system and the gulf coast inland waterway from Texas to Alabama. It made similar use of the waterways attached to its principal South American ports of Santos, Brazil, and Buenos Aires. These ships were also fitted with a container crane and could carry a large number of twenty-foot containers on deck.

The LASH design has proven useful in military as well as in commercial service. When Delta was sold to Crowley Maritime, the government purchased its three LASH ships for use in the Ready Reserve Fleet. Six LASH ships are still in commercial service; three are being operated under U.S. flag by Waterman, and three under foreign flag by its parent company Central Gulf. Several other LASH ships were bought by APL and Sea-Land and converted to full containerships. The remaining U.S. LASH ships on charter to the Department of Defense are loaded with ordnance and are prepositioned to provide quick response to overseas emergencies.

Lykes Line experimented with Seabee ships, a variant of the LASH design. On LASH ships the barges are lifted aboard by a gantry crane that is cantilevered out over the stern of the ship. The crane then lowers the barges into slots on the main deck, as is done with containers. Seabee ships used a synchro lift to raise the barges. This was similar to the shipyard systems used to lift small craft. The barges were then transferred to their stowage positions by power-driven chain conveyors. Lykes built three Seabees of 37,000 DWT, each capable of carrying thirty-eight barges of 850 ton capacity or thirteen hundred containers or a combination of the two. The company found it difficult to operate these ships profitably in the northern European service, and the ships were eventually laid up and then sold to the Department of Defense.

While American innovators were developing the containership and LASH and Seabee ships, European operators were developing new types of Ro-Ros and bulk carriers. The Ro-Ro ship combines the advantages of the short sea ferry with those of large amphibious-landing craft, such as the World War II Landing Ship Tank (LST) and Landing Craft Infantry (LCI). High loading and discharge rates can be achieved with these ships with simple terminal facilities when their cargoes move under their own power. These ships are also highly adaptable and can carry special cargoes of almost any size and weight that will fit within their cargo openings.

Scandinavian shippers took the lead in developing deep-sea Ro-Ros. These ships were originally loaded through side-port ramps, but slewing stern ramps were later developed to make them independent of any special docking facilities. The cargo spaces on these ships are huge and highly adaptable; the decks are strengthened to carry the heaviest loads. Ramps to upper and lower decks are accessible from the entry ramp, and many of these ships also have large elevators to service the lower decks. Such ships have proved especially useful when servicing undeveloped ports in Africa and the Middle East. During the past four decades the size of Ro-Ros has increased dramatically. Some have now reached the maximum size that can pass through the Panama Canal and have operational speeds of twenty-three knots.

As the international automobile trade grew, specialized Ro-Ros were developed to carry cars. Immediately after the war, cars were treated like any other freight, being loaded and discharged by means of ship's cargo gear and stored below decks in general cargo compartments. Such handling frequently caused damage, and as more and more cars were exported, a way had to be found to load and transport them more quickly and inexpensively. In the late 1960s the Volkswagen Company and several Scandinavian shipowners addressed this problem and created the Pure Car Carrier (PCC). The first of these ships could carry 1,000 to 2,000 cars, but PCCs that can carry 4,000 to 5,000 cars are now common, and vessels capable of carrying 6,000 units are being built. Flexibility has also been increased by strengthening fixed decks and making intermediate decks adjustable, the result being that many of these ships are now described as Pure Car/Truck Carriers (PCTC). The Japanese quickly became important PCC customers and soon moved to take over the business of shipping cars as well. By 1982 Japanese interests owned 145 PCCs and controlled two-thirds of the world market. U.S. protests against this near-monopoly position persuaded the Japanese to award a few charters at competitive rates to American PCC owners.[1]

Although the size of oil tankers began to increase shortly after the war, the explosion in size occurred following the closing of the Suez Canal. Europe and the United States had become dependent on Middle Eastern oil. Severe restrictions on supply or sharp increases in cost would have profoundly disrupted their economies. Increasing the size of tankers was thus a strategy for limiting the rise in the cost of transporting oil around Africa once the more direct Suez route had been closed. Construction costs per ton of cargo capacity decreased steadily as tanker size increased, and crew costs for operating a mammoth ship were nearly the same as for a much smaller ship. Very Large Crude Carriers (VLCCs) of between 200,000 and 300,000 tons were soon followed by Ultra Large Crude Carriers (ULCCs) of over 300,000 tons.

Dry-bulk carriers have also increased in size, but not nearly as much as tankers, their average deadweight capacity being in the 100,000-ton range. Because it is inefficient to have a ship haul cargo in only one direction, a new vessel type called the Oil-Bulk-Ore (OBO) ship was developed. The cargo holds of these ships are subdivided so that ore and grain can be carried in the center compartments on one leg of a voyage and petroleum cargoes can be carried in wing tanks on the return passage.

The 1973 war in the Middle East and the oil crisis it created also directed attention to the possibilities of transporting natural gas by ship. Natural gas has long been used in petroleum-producing nations, and the technology of sending it long distances via pipeline was well developed. But the economics of carrying natural gas aboard ship required that it be liquified, which meant that it had to be kept at minus eighty-two degrees centigrade throughout the voyage. Shipping liquified natural gas (LNG) therefore called for the development of new and highly sophisticated shipboard systems for handling and storing cryogenic cargoes.

The U.S. government encouraged three American shipyards to begin constructing LNG tankers using three different designs. Six ships were built for a Texas company, El Paso Natural Gas. Three of these were constructed by Avondale in New Orleans employing a design utilizing rectangular aluminum tanks entirely encased in a new insulating material. When tested before delivery, this insulating system failed, and since the problem could not be corrected, the insurance underwriters declared the ships a constructive total loss. El Paso's three other ships were built by the Newport News Shipyard in Virginia using a proven but costly design. These ships were later sold to Shell and continue to be successfully operated. The third series of LNG ships was built to a Kaverner-Moss design by General Dynamics in Quincy, Massachusetts. Each of these 63,000-DWT ships has five spherical tanks, giving it a total cargo capacity of 125,000 cubic meters. The ships were built for Christopher Chen, owner of Energy Transport Corporation, and were financed by a group of American banks whose risk was underwritten by federal Title XI mortgage guarantees and a twenty-five-year time charter provided by Burmah Oil. The first of these ship, LNG *Aquarius*, was delivered in 1977; today these ships maintain an overseas "pipeline" between Indonesia and Japan. They still fly the American flag and employ American crews.

The 1970 Merchant Marine Act provided American shipbuilders with a new burst of optimism, but apparently the LNG ships were its last gasp. These high-tech ships called for the use of sophisticated construction techniques that American yards had developed when building ships for the navy. The yards made substantial investments in new facilities, and the LNG ships were built with little or no subsidy. The hoped-for steady flow of orders did not follow, however, since oil prices soon fell to their precrisis level. As the commercial market dried up, the shipbuilders were forced to turn back to the navy for future business.

The decline of orders for cargo ships was compounded by the collapse of the passenger liner trade in the postwar era. This demise was not immediately anticipated, however, and shortly after the end of the war the government underwrote the building of *United States*, perhaps the finest commercial ship ever produced in America. When its keel was laid in 1950, passenger liners still carried three times more travelers across the Atlantic than airplanes, but this ratio would soon shift dramatically.

United States was a high-tech marvel that was enormously expensive to build and operate; its $70 million cost was heavily underwritten by subsidies from both the Maritime Administration and the navy. Designed to also serve as a troop carrier, the ship could accommodate two thousand passengers when serving as a liner and fourteen thousand troops in time of war. On its maiden voyage *United States* broke the eastbound and the westbound Atlantic speed records with an average speed of thirty-five knots. But by the mid-1960s it was all over. Passengers and troops were increasingly being flown to their destinations, and speed offered no advantage to cruise ships. In 1969 *United States* was laid up, and in 1992 it was sold for $5 million and towed to Turkey. Although it has since been returned to America, its fate remains uncertain, the ship's glory long since lost.

The fifty years following the end of World War II saw greater changes in shipping than had occurred since sail gave way to steam. The multiplication of ship types and the enormous growth in ship size that took place in this period is continuing. A new generation of containerships, more than 1,100 feet long and having a capacity in excess of 6,000 TEUs, is now being placed in operation. However, few ports are currently equipped to handle ships of this size and the growth of a fewer number of megaports is indicated, with most other ports being served by feeders. It is also likely that in the future continuing attention will be given to improving fuel efficiency, lowering operating costs by reducing crew size, and improving maintenance. Built-in safeguards to minimize the environmental damage of oil spills will command a high priority. The national character of ships, owners, and companies is also becoming blurred as globalization of standards, markets, and intermodal freight operations increases. In one long generation shipping passed through technical changes that in turn completely transformed the industry.

OPEN REGISTRIES: FLAGS OF CONVENIENCE OR NECESSITY?

Shipowners engaged in international trade must adjust their operations to specific political and economic contexts, and each of these contexts imposes distinctive constraints on the operator. The political context is defined by the rules associated with the flag under which the ship is registered; the economic context is defined by competitive world costs for operational components of the ship, such as construction, financing, insurance, fuel, crew, taxation, maintenance, and so forth. The evolution of world trade and the shipping industry since the middle of the nineteenth century has brought these two systems into increasing conflict. The struggle over which flag ships should sail under continues to distort consideration of maritime policy in the United States.

The flag flying on a ship's stern, like the flag flying over an embassy, is a symbol indicating that this vessel or piece of real estate is governed, by virtue of extraterritorial agreements, by the laws of the nation whose flag it flies. There are several circumstances in which ship registry can be a matter of considerable importance. Before the age of instantaneous communication by cable and wireless, a ship arriving in a foreign port relied on the country whose flag it flew for diplomatic protection and representation when engaged in trade or involved in local disputes. Such diplomatic representation is still called upon today, but it is much less crucial when one can talk directly with headquarters from any place on the globe. Similarly, in time of war or when threatened by pirates and other outlaws, ships relied on the protection provided by their country's navy. But providing such protection entailed considerable cost and reinforced the insistence that merchant ships fly the flag of the nation called upon to defend them. So long as no war is in progress and international commercial cooperation is flourishing, however, shipowners and national governments have little reason to insist on this coupling of flag and ownership.

Indeed, when a registry's rules impose costly demands on those enrolled, it makes flying that flag economically disadvantageous and leaves shipowners operating in international trade with only a few options. They can get out of the business, they can seek government assistance in the form of cargo protection or subsidies, or they can find ways to reduce their costs so that they can continue to compete. During the past century all three options have been utilized by American shipowners. Many left the business. Others have continued to operate under U.S. registry while relying on cargo protection and federal subsidies. The third option has also been increasingly used, as American shipowners have reflagged their vessels to registries that do not impose crewing, manning, and taxation burdens. The major registries also provide for international standards and inspection services more liberal than those imposed by most of the developed maritime nations. Open registries, i.e., national registries open to nonnationals, were created and have flourished as legal arrangements that enable shipowners to cope with economically unrealistic political demands.

In the twentieth century Panama provided many American operators with a convenient open registry. Early in the century many considered Panama virtually a colony of the United States, and it was relatively easy to exert U.S. influence on its shipping policy. W. Averell Harriman, who operated the passenger ships *Reliance* and *Resolute* as part of the International Mercantile Marine, was one of the first to use the Panamanian registry. When he purchased these ex-German vessels from the Shipping Board he secured an agreement that he could transfer them to foreign registry should he need to. The need that drove him to do so was created by the Volstead Act, which Congress passed in 1920 to enforce the Eighteenth Amendment to the Constitution, an amendment that outlawed the sale of alcoholic beverages in the United States. The law applied to all ships within U.S. territorial waters, whatever their nationalities, but as soon as foreign ships crossed the three-mile limit, they could open their bars and get on with this important aspect of shipboard life. But according to a determination by the U.S. attorney general, U.S.-flag ships were American territory, regardless of location, and hence could not serve alcoholic drinks anymore. Harriman insisted that the ruling forced his hand, and in 1922 his request to transfer his ships to Panamanian registry was granted. It appears likely that Harriman was at least equally concerned with the relief that reflagging provided on matters of inspections, taxes, and crewing.[2]

The political significance of reflagging depends on the circumstances in which it occurs. Before the rise of Nazism in Germany, Standard Oil had registered twenty-five of its tankers in the free city of Danzig; in 1935, as the European political situation became more ominous, the company transferred their registry to the Panamanian flag. When World War II began in 1939, Standard Oil replaced the German crews on these ships with other European and American seafarers. After 1939 the U.S. government actively encouraged American owners to reflag some of their vessels to avoid the restrictions imposed by the U.S. Neutrality Act. High freight rates tempted U.S. owners to reflag, and the federal government assisted them by insuring them against loss. The International Mercantile Marine Company (IMM) reflagged seven of its combination cargo-passenger liners built during World War I, transferring them to the Belgian flag. All except the *Ville d'Anvers*, the former *American Banker*, were lost during the war. This lone survivor was incredibly lucky; it made thirty-eight wartime crossings without a scratch and ended its career under the Honduran flag.

International tensions following World War II continued to encourage some owners to register their ships under foreign flags. When the war was over the United States sold many of the ships in its war-built merchant fleet to foreigners, and most of these were registered in the owners' nations. Many of these ships were purchased by Greeks. But in 1946 Greece was in the throes of a civil war, and, to avoid being embroiled in this conflict and to avoid Greek taxes, many Greek and Greek-American shipowners chose to register their ships in the Panamanian registry. A few years later, when the flow of relief supplies to Europe began to slow down, a number of American freighters that could not obtain cargoes while operating under the restrictions imposed by U.S. registry transferred to foreign registry. The American owners favored three foreign registries that were historically close to the United States, those of Panama, Liberia, and Honduras (PANLIBHON). Much new construction was also placed under PANLIBHON registry, and between 1950 and 1963 the tonnage registered to these flags increased from 6.1 to 24.2 million DWT and constituted 12.8 percent of the active world fleet.[3] The United Fruit Company had been the principal user of the Honduran flag, but when it shifted to Liberian registry, the Honduran flag ceased to be an important registry of any major owner.

Today flags of convenience (FOC), or "open registries" as the United Nations prefers to call them, provide shipowners with low operating costs and maximum operating flexibility. The registries offered by such countries as Liberia, Panama, the Bahamas, and a long list of former colonial nations have a number of common characteristics. They are available to noncitizens in return for a registration fee and an annual payment based on the tonnage of the vessel. Owners are free to transfer in and out of the registry at will, no tax is imposed on income, there are no nationality requirements for crewing, and ships can be built and repaired anywhere in the world. The nations that operate these registries provide no government inspection oversight, and responsibility for insuring vessel safety remains with the insurance underwriters.

Flag of Registry	Number of Ships	Deadweight Tons
Liberia	214	16,493,694
Bahamas	61	4,775,003
Marshall Islands	42	3,512,945
Panama	81	3,268,327
Greece	28	1,260,946
Hong Kong	7	891,475
Kerguelen Island	5	850,950
Bermuda	8	757,414
Cyprus	36	684,851
France	2	551,446
Norway (including NIS)	6	547,821
Malta	8	395,204
Singapore	7	335,177
Vanuatu	8	270,244
St. Vincent	6	239,476
Australia	2	184,602
United Kingdom	7	129,620
Myanmar	3	79,565
Barbados	1	44,999
Canada	4	44,048
Cayman Islands	2	24,404
South Korea	1	23,507
Philippines	1	22,133
Japan	1	8,788
Indonesia	1	6,412
Totals:	542	35,401,051

Table 11.2: Foreign-Flag Vessels Owned by American Companies.
Source: Data from *Lloyd's Register*, November 15, 1996.

The U.S. government is of two minds on the issue of flags of convenience. The State Department, which negotiates with foreign nations, and those agencies that seek to minimize their shipping costs, such as the Departments of Defense and Agriculture, generally oppose economic and legislative pressure to require the use of U.S.-flag vessels. The Department of Defense position is complicated by its need to be able to mobilize American merchant shipping in time of war. The department has addressed this need by asserting that U.S.-owned but foreign-flagged ships are under "effective control" of their American owners and are therefore available in time of emergency. Opposition to the use of flags of convenience has long been centered in the American maritime labor unions, and their position is occasionally supported by the Department of Labor, the International Transport Workers Federation (ITF), and the United Nations International Labor Organization (ILO).

American owners were not the only ones who made use of open registries. Greeks, as we have noted, also sailed under foreign flags, and so did many others from nations with high wages and taxes. As the national fleets of older maritime powers declined, governmental investigations into why the decline had occurred were undertaken. One of the more notable of these was the 1970 British Committee of Inquiry into Shipping, chaired by Viscount Rochdale.

Although the Rochdale Report was written following a comprehensive examination of Great Britain's shipping problems, it failed to address directly the problem posed by flags of convenience. The report concluded that given the existing level of investment grants provided by the government for the purchase of new ships and the free depreciation of ships allowed under the British tax code, British owners were favorably positioned when compared to their European competitors. This well may have been true, but the report ignored the challenge of non-European registries and did nothing to solve the problems that were driving ships away from British registry.[4]

Another investigation sponsored by the British government was conducted in 1990, this one being cochaired by Sir Jeffrey Sterling, chairman of the P&O Group, president of the General Council of British Shipping, and secretary of state for the Department of Transport. Once again the report that followed avoided the central issue. The report did not claim that supporting the shipping industry would contribute substantially to Great Britain's balance of payments account, although obviously some benefit would have been realized. Nor did it address the issue of maritime support for national defense. By 1990 the Falklands War was a fading memory and the report chose not to note that while it was being prepared, a number of foreign-flag ships had been chartered to move the British army and its equipment to Saudi Arabia for Operation Desert Storm. The report did acknowledge that a ship flying the Liberian flag had a distinct financial advantage when compared to a ship flying the "red ensign," yet it made no recommendations for bringing ships back under the British flag. It appears that the loss of British-flag shipping was accepted as inevitable. The report therefore limited itself to recommending that certain Department of Transport rules that restricted shipping be liberalized, that other countries be persuaded to reduce their cabotage restrictions, and that increased training be provided for maritime personnel.[5]

Several European nations have introduced one or another of two kinds of registries to compete with open registries. A few nations have established second, "international" registries that offer many of the advantages provided by open registries yet still make it possible for ships to sail under the national flag. The most successful of these, the Norwegian International Ship Registry (NIS), was established in 1987. Denmark later founded a similar Danish International Ship Registry (DIS), and Germany did the same. Norway's success can be measured directly. Prior to the establishment of NIS only 9.7 million DWT of shipping remained under the Norwegian flag; after 1987 the fleet grew rapidly as ships returned from FOC registry and new tonnage was added. By 1990 the entire Norwegian-owned fleet stood at 41.2 million DWT, which approached its previous high point, and of this total, 722 ships, or 36.3 million DWT, were registered under NIS. Although the Danish International Registry has also enjoyed some success, no other country has done as well as Norway. In 1996 Hapag-Lloyd A. G., after transferring some of its ships to Singapore registry, agreed with its maritime union to keep its remaining ships in the German International Registry. Under this agreement seven of each ship's personnel will be German, the remainder, regardless of nationality, will receive wages according to ITF agreements.

The second kind of registry established by European nations are located in the countries' offshore islands, which have a degree of local autonomy. British-dependency registries have been established in Bermuda, the Cayman Islands, and the Isle of Man; the French have established an offshore registry on isolated Kerguelan Island in the Indian Ocean. The new British offshore registries have enrolled fleets of considerable size; the French government-owned line, Companie Generale Maritime (CGM), began to reflag its vessels to the Kerguelan flag. Yet it is widely recognized that these registries are at best half-way houses. The European Union has attempted to carry registry reform further by creating a European ship register, to be called EUROS, which it is hoped will replace the existing international registries. While the proposal may have a certain logic, it has not attracted much support beyond the Brussels bureaucracy.[6]

The availability of open registries means that the flag under which a ship sails tells the observer nothing about the citizenship of its owner, and yet ownership is for most purposes far more significant than registry. As one would expect, ownership of the world's merchant fleet has traditionally been concentrated in the hands of the most highly industrialized

Country of Ownership	Number of Ships	Deadweight Tons (in Millions)
Japan	1,956	86.2
Greece	1,374	57.0
USA	1,017	52.6
Norway	732	37.3
UK	576	34.7
People's Republic of China	1,295	30.3
Hong Kong	563	22.9
South Korea	446	20.9
Singapore	543	20.8
Liberia	315	16.5
Germany	675	14.6
Republic of China	357	13.7
Bermuda	114	12.4
Sweden	199	12.4
Russia	626	12.4
India	283	11.9
Denmark	284	11.3
Saudi Arabia	90	10.2
Italy	318	10.2
World Totals	17,344	660

Table 11.3: World Merchant Fleets.
Note: The nationality is of the vessel owner; this list includes vessels of greater than 5,000 DWT and nations having over 10 million DWT.
Source: Data from *Lloyd's Register*, November 15, 1996.

nations. In the 1970s the United States, Japan, and the nations of Western Europe owned 87 percent of the world's tonnage, 22 percent of which they operated under flags of convenience. Since by the 1970s all leading industrial nations welcomed multinational corporations and were beginning to deregulate their economies, there was good reason to think that this concentration of ownership would persist. Yet by 1990 the percent of the world merchant fleet owned by these nations had dropped to 67 percent. This redistribution occurred because certain developing nations, most notably the so-called "little tigers" of South Korea, Hong Kong, Taiwan, and Singapore, had rapidly developed national fleets of their own.[7] They did so partly as a natural extension of the industrial policies that produced their remarkable export-led economic growth, but there was an element of good fortune turned to advantage as well.

Over the years the nations of the Pacific Rim gained considerable maritime experience by supplying crews for open registry ships. Then in 1977, when the bottom dropped out of the tanker market, many European owners lost their ships when they were unable to keep up their mortgage payments. Excellent ships were available at bargain prices, and many of the most active buyers lived in Hong Kong. Indeed, for a while Sir Y. K. Pao of Hong Kong was the world's largest shipowner.

Having entered the big leagues via tankers, the new Far East ship owners moved aggressively into the container-carrying trades. Their initiative did not depend on government regulation or subsidies; they competed instead as efficient, low-cost operators and they have been remarkably successful. The Evergreen Line of Taiwan was the world's largest container-shipping company; it gained that position by providing reliable service and by consistently quoting freight rates below those established by the rate conferences for the routes it serves. As table 11.1 (p. 219) indicates, if one includes the mainland Chinese lines among the companies based in the newly industrialized Pacific Rim countries, then seven of the top ten container lines that ship to U.S. ports belong to this group. This represents a sea change in ownership of the world commercial fleet.

The U.S. maritime unions have continually fought American ownership of foreign-flag shipping, and their repeated attempts to organize the crews of these ships have been unsuccessful. When they have organized pickets to coerce these companies, their acts have been found to be illegal. In 1986 the AFL-CIO Maritime Committee believed that the Tax Reform Bill offered it a chance to retaliate. Previously, American owners of ships sailing under FOCs had paid no taxes on their foreign earnings so long as they reinvested them in qualified shipping assets. Subpart F of the U.S. Revenue Code allowed such deferral so that companies could accumulate reserves to maintain modern fleets. While this did not exempt them from taxes, an exemption many other governments provided to their shipowners, it was the next best thing. But in 1986 the Maritime Committee persuaded Congress to cancel the use of subpart F for this purpose. No public hearings were held and the only rationale offered was that shipping should be taxed like other sectors of American industry. It is a fine principle in the abstract, but it ignores the fact that international shipping is not like domestic industries.

The outcome of this change was predictable: since being an American company now entailed a tax burden that put one at a disadvantage in international commerce, it no longer paid to be an American company. Between

1986 and 1991 the U.S.-owned foreign-flag fleet dropped from 17 million to 12 million DWT, and the decline has continued. American owners simply sold their ships outright or retained no more than 50 percent ownership. Since the ships were no longer American owned, they paid no American taxes and in any case created no jobs for U.S. mariners. It is clear that unless the tax-exempt provision is restored, few Americans engaged in international trade will own ships of U.S. registry in the future.

A similar situation arose in the liner trades when President Reagan terminated further funding for construction differential subsidies and stipulated that no new operating differential subsidy contracts would be awarded once those already in existence had expired. Since this termination of subsidy programs was not accompanied by a relaxation of U.S. registry requirements, it effectively sounded the death knell for the U.S.-flag fleet engaged in international commerce. The liner companies that had received subsidies to operate ships under the American flag in international commerce suddenly faced an impossible situation. Early in 1992 the two largest American liner companies, Sea-Land and American President Lines, announced that if the regulations were not changed or further subsidies provided, they had no alternative and would follow their international brethren in transferring their remaining U.S.-flag ships to open registries. During the final months of his administration President Bush tried to address this problem, but without success.

Sea-Land, sensing that no subsidies would be forthcoming, applied for permission to transfer registry of the newest portion of its fleet to the flag of the Republic of the Marshall Islands (RMI). This was an interesting choice, for it approximates closely the offshore registries established by Great Britain and France. The Marshall Islands had been part of the Japanese empire and were placed under U.S. trusteeship at the end of World War II. They became a self-governing republic in October 1986, at which time they entered into a Compact of Free Association with the United States. Under this compact the United States agreed to provide full consular services to citizens of the Republic and to defend its people from military attack. The issue of naval protection was tested in 1987 during the tanker war in the Persian Gulf. In that incident the U.S. Navy initially defended only U.S.-flag vessels against Iranian attacks; this included a number of reflagged Kuwaiti tankers as well as one ship flying the RMI flag. Thus a ship flying the flag of the Marshall Islands is not only free from U.S. taxes and regulations, but also receives the protection of the U.S. Navy and any help it requires from the U.S. consul in a foreign port.

In the 1990s Liberia was ravaged by years of civil war. The resulting political and economic anarchy made owners who had registered their ships under the Liberian flag understandably nervous, and some began switching their ships to other registries. When the Liberian faction that held power at the time of the armistice insisted that the roughly $20 million generated annually by the nineteen-hundred ship registry be turned over to it, the flight from the Liberian flag accelerated. The company that operates the Liberian registry, International Registries, Incorporated, refused to comply. It could do this because its offices are located in Reston, Virginia, where the same company also operates the Marshall Islands Registry. In 1998 the Liberian government withdrew supervision of the registry from the Reston group and placed it with a new Washington, D.C. company organized for that purpose. By the end of the twentieth century the evolving practices of global capitalism have brought about as complete a separation between registry and political power as that which has long existed between registry and ownership.

ANTITRUST REGULATION AND SHIPPING CONFERENCES

Steamship conferences, which are essentially cooperative rate-making bodies, were first established in the later part of the nineteenth century to reconcile two conflicting features of industrialized shipping. On the one hand, steamships were enormously expensive, at least when compared to the cost of building wooden sailing vessels, and the capital investment required to enter the steamship business could only be justified if it seemed likely that a steady and adequate rate of return on investment would be realized. But on the other hand, the new iron-hulled and steam-driven ships could be built to a very large scale and could be operated on more frequent and reliable schedules than sailing ships, which meant that excess capacity was likely and that competition for cargoes would be intense. Shipping conferences were established to alleviate this conflict between the need for steady, adequate revenues and a systematic inclination to build excess tonnage and engage in ruinous rate cutting.

Shipping conferences are cartels, that is to say, associations of business firms that agree to restrict trade and set price levels. Their agreements are consensual and not legally enforceable, but a variety of commercial tactics can be employed to enforce their policies. European nations normally have not been opposed to the formation of cartels, possibly because their experience with strong central governments has reassured them that their political institutions will not be overwhelmed by the new power of industry. In America the new captains of industry appeared to threaten, and indeed often did work against, the economic interests of small farmers, merchants and individual consumers, the citizens who constitute the backbone of the American

Republic. The antitrust movement of the late nineteenth and early twentieth centuries was a political response to this behavior, and it opposed in principle all forms of economic concentration and collusion. But as was shown in chapter 6 above, the Sherman Anti-Trust Act of 1890 and the Clayton Act of 1914 were only effective against horizontal marketing arrangements designed to maintain prices and allocate market shares. Thus in antitrust policy, as in so many other areas, U.S. legislation and the practices of international maritime commerce were from the outset at loggerheads.

In some ways the policy debate over shipping conferences pitted the natural characteristics of industry against the political dreams of agrarian America, although the contest was not described in these terms at the time. The failures of Louis Nixon's attempt to consolidate U.S. shipbuilding, of Charles Morse's attempt to monopolize shipping along the Gulf and Atlantic Coasts, and J. P. Morgan's attempt to dominate North Atlantic liner shipping in the early twentieth century all indicated that the maritime industries have certain characteristics which make it difficult if not impossible to organize them into effective monopolies. The shipping industry, denied the internalization of control and the protection against new entrants that comes with vertical consolidation, had no choice other than to engage in collective market and rate manipulation to counter its "natural" inclination to excess capacity and bankruptcy. But in America the swelling tide of political reaction to the concentration of industry reached its peak in 1914, just when the "natural" limits and consequences of such concentration was becoming clear. The Clayton Act of 1914 was followed by the establishment of the Federal Trade Commission, which was charged with investigating unfair competition and violations of antitrust laws. The antitrust movement became firmly entrenched in the Department of Justice; American liner operators engaged in international trade, which was then and still is largely coordinated through conferences, have been wrestling with the burden of its regulations ever since.

In 1914 Congressman John Alexander, chairman of the House Merchant Marine and Fisheries Committee, was conducting an investigation into current shipping practices, which had already attracted the attention of antitrust lawyers in the Justice Department. Alexander understood the need for shipping conferences, but he also had to devise a compromise that the antitrust lawyers would find acceptable.

The results of his efforts were embodied in the Shipping Act of 1916 and were not seriously challenged until after World War II. The antitrust provisions of the act granted the liner companies engaged in international trade immunity from antitrust prosecution, which enabled them to be active members in shipping conferences, but inevitably certain conditions were attached. Some of the new restrictions, such as the ban on the use of "fighting ships" to enforce conference agreements, did not strike at the heart of the system. But two other conditions that were imposed on American members of conferences threatened to compromise the entire system. The first of these required that all conference agreements and freight rates be filed with and accepted by the Maritime Commission; the second required that all conferences, most of which had a closed membership, be open to all who wished to enter the trade. The Wilsonian diplomatic credo of "open agreements openly arrived at" was to pertain in the shipping industry as well.

Officials in the Anti-Trust Division of the U.S. Department of Justice have always opposed the antitrust immunity granted by the 1916 act, just as federal tax officials in the Treasury Department have long been opposed to special tax provisions for shipowners engaged in international commerce. In the 1950s this opposition led to a series of legal actions that attempted not only to apply normal antitrust policy to American shipowners, but also to extend the enforcement of these policies to foreign carriers. The results significantly increased the financial risk of participating in the international carrying trade and generated a political backlash that further isolated the United States from the world shipping community. By the 1970s foreign carriers found themselves legally liable for acts that were condoned in their own countries and only a few years earlier had been legal in the United States as well. Such was the unwelcome outcome of a well-intentioned but ill-considered attempt to apply American law to an international industry.

One of the first goals in this new antitrust offensive was to extend the application of U.S. antitrust laws to foreign corporations in general. A Canadian subsidiary of the Alcoa Aluminum Company had reached an agreement with several European firms to limit the sale of foreign aluminum in the United States. No Americans were involved in the agreement and it was executed outside the United States. Nonetheless, it was held that because these actions had the intent of limiting U.S. imports, they had the effect of contravening U.S. antitrust law. This ruling became known as the "effects doctrine" and made foreign companies doing business in America liable to U.S. antitrust law wherever the agreements in question may have been made.

A second goal was to give U.S. antitrust laws precedence over the antitrust provisions of the 1916 Shipping Act. In what became known as the *Carnation* case, it was determined that a shipping conference had implemented a secret price-fixing agreement that went beyond the terms approved by the Federal Maritime Commission (FMC). Because this secret agreement did not conform to the law, the court held that the shipper, the Carnation Milk Company, was not protected by the immunity provisions of the 1916 act, and it imposed triple damages, as provided by general antitrust law.

This antitrust offensive led Congress in 1961 to amend the antitrust provisions of the 1916 act and assign regulatory responsibility to a newly created independent agency, the Federal Maritime Commission (FMC). This was a clarifying move, for while the Maritime Administration is fundamentally committed to promoting the American-flag maritime industry, the FMC is, in principle, flag-neutral, its function being to protect the public interest whatever the registry of the ships involved. The legislation that established the FMC gave it the authority to disapprove any agreement that was found to be contrary to the public interest. Justice Department lawyers made extensive use of this power by appearing regularly before the FMC to argue that agreements and price adjustments that formerly would have been routinely approved now had to be justified as not being "contrary to the public interest."

The effects of this shift to a presumption of guilt until proven innocent were demonstrated in what came to be known as the *Svenska* case. The FMC initiated this action against the Swedish-American Line, and the Supreme Court decided that the burden lay with the conference to demonstrate that the agreement under question was not contrary to the public interest. The conference, the Court said, must "bring forth such facts as would demonstrate that the [agreement] . . . was required by a serious transportation need, necessary to secure important public benefits or in furtherance of a valid regulatory purpose of the Shipping Act."[8] This "public test" requirement greatly increased the risk of antitrust liability and threatened to make it impossible to operate a conference system on routes serving United States ports.

The FMC then carried its campaign against foreign carriers one step further by demanding that they produce documents located in their home offices whenever they were called for. Other nations were understandably not happy with this attempt to extend to foreign corporations, by virtue of a dubious assumption of extraterritorial jurisdiction, the discovery procedures of U.S. law. They considered it an invasion of their sovereignty and blocked all attempts to gain documents in this manner. In 1980 the British Parliament passed the Protection of Trading Interests Act, which was specifically designed to foil any foreign discovery requests. Among the act's interesting features is a "claw-back" provision that allows "qualified British companies to recover in British courts the 'punitive' portion of any foreign multiple damages judgment entered against them."[9] The United States was having to learn again, as it had learned after World War I, that it could not unilaterally set the rules for international commerce.

The FMC's "public test" requirement and the systematic attacks by the Anti-Trust Division of the Justice Department succeeded in effectively destroying the immunity from prosecution originally provided by the Shipping Act of 1916. Indisputable evidence of this outcome appeared in 1979 when the Justice Department obtained an indictment charging several North Atlantic conferences and the executives of twelve shipping companies with entering into agreements "to fix, raise, stabilize and maintain price levels for the shipment of freight in the United States/European trade."[10] Faced with a charge which accused them of doing precisely what the conferences were designed to do and aware of the antitrust hostility of their accusers, the defendants pleaded "no contest" and were fined $6.1 million, the largest antitrust fine ever assessed to that time.

Although the Justice Department lawyers who zealously pursued all cartels were unburdened by any knowledge of the shipping industry, important members of Congress were becoming interested in reestablishing the intent of the 1916 law. The vehicle they used to achieve this end was the Shipping Act of 1984. The overall goal of this act was to bring U.S. shipping policy into harmony with accepted international practices. The law strengthened the FMC's right to regulate shipping conferences without interference from the Department of Justice, and it stipulated that the FMC could only reject a conference rate-adjustment application for stated cause. This meant that conferences no longer had to prove that their applications were in the public interest.

The 1984 law also addressed a number of problems that had arisen after 1916. The FMC's power to fight foreign discrimination when U.S.-owned ships were operating as third-flag carriers was increased. Conference members were also guaranteed the right to take independent rate action so long as they gave the conference a twenty-four-hour notice and filed the change with the FMC. The use of service contracts, which in some respects replaced the loyalty contracts that were no longer allowed, was authorized. In one type of service contract a shipper guarantees a minimum quantity of cargo over a fixed period of time and the ocean carrier commits to providing a defined level of service at a specific rate or rate schedule.

In what may turn out to be the most consequential provision of the 1984 act, Americans were permitted, for the first time, to join foreign-flag consortia. This meant they could at last become fully fledged members of the international shipping community, and in recent years American membership in such consortia has become commonplace.

In 1996 some of America's largest shippers and its largest container-ship operators tentatively agreed to a proposal to fundamentally alter certain features of the 1916 Merchant Marine Act, as amended in 1984. Their proposal, while retaining shipping conference antitrust immunity, would eliminate the requirement that tariffs be filed with the Federal Maritime Commission. It would also require that conferences allow member lines to enter into service contracts with individual shippers and that contracts would be confidential. After two years of bitter controversy, the Ocean Shipping Reform Act was signed into law; it became effective May 1, 1999. The Ocean Shipping Reform Act completes the transition begun in 1984 from essentially one of common carriage to contract carriage. These changes, in time, could prove fatal to the conference system under which the liner industry had been operating for almost a century.

Half a century after the end of World War II and forty years after the introduction of modern containerization, the world maritime industry is still being transformed by technological and organizational change. Most of the once proud American shipping firms have disappeared, and at the end of the twentieth century the U.S. fleet engaged in world commerce is fast approaching the low level to which it had descended by 1914. But even this figure is misleading, for the large domestic fleet that serviced the coastal trade in 1914 is now all but gone as well.

The world maritime industry increasingly operates without reference to specific national interests. In 1950 only 6.4 percent of world tonnage was registered under flags of convenience. By 1970 this had grown to 22.9 percent, and by 1991 the total had reached 44.6 percent. If one adds to this 1991 percentage the tonnage registered to the "international" registries that some industrial nations have created, more than half of the world's tonnage is sailing under flags of convenience.[11]

During the years immediately following 1970 the future of the U.S. maritime industry looked promising. Applications for operating subsidies were routinely approved and ample funds were available to support ship construction through differential subsidies and Title XI ship mortgage insurance. Many companies therefore invested heavily in new ships, but had they looked more closely they might have detected signs that would have tempered their optimism.

The Vietnam War, which provided a steady demand for American-flag shipping, was winding down. Traditional break-bulk ships had handled most of the cargo for that war, but toward its end the containerports built by the army and the containerships operated by Sea-Land were providing superior service. As it turned out, Vietnam was the last war in which break-bulk shipping played a major role.

It was expected that by extending full subsidy support to American owners of bulk carriers, the Merchant Marine Act of 1970 would attract these ships to American registry, but none successfully made the transition. Subsidies for ship construction and operation might have made sailing under the American flag attractive had the law removed other well-known, costly impediments, but it did not do so. A history of maritime labor unrest also deterred investors. A vice president of a major oil company observed that his American-flag Jones Act ships represented only 10 percent of his fleet but created 90 percent of his headaches.

One by one the familiar company names in American shipping disappeared. U.S. Lines, Grace Line, Moore-McCormack, American Export, Delta, States Lines, and Pacific Far-East Line were no longer seen in the world's seaports. Most of the well-established shipbuilders survived on orders provided by the navy; shutting down many of the navy shipyards helped reduce competition for navy work. Venerable navy yards such as those in Boston, Brooklyn, Philadelphia, and Charleston, as well as several on the Pacific Coast, were no more.

The maritime labor unions were also losing power, although it was far from evident in the 1970s. The death of Paul Hall in 1980 deprived them of one of their most effective spokesmen, and no one of comparable stature has emerged since to champion their cause.

12

MILITARY SEALIFT
AFTER WORLD WAR II

While the maritime industry was being transformed by new types of ships and new cargo-handling techniques, the U.S. military was reorganizing its transport and logistic services. The relationship between the commercial fleet and military preparedness has always been problematic and the way the industry has evolved recently has made it even more so. The containerships and huge crude carriers being built for the U.S. merchant fleet had little value as auxiliary transports. Since the war-built break-bulk ships in the National Defense Reserve Fleet had rusted away and could no longer be relied on, the armed forces increasingly depended on the Ready Reserve Force (RRF), which had been established in 1984. By the time Iraq invaded Kuwait in August 1990, ninety-eight ships had been acquired for the RRF. Forty-five of these were Ro-Ros that were especially useful for transporting mobile military equipment. These ships were deployed, along with a large number of chartered foreign-flag Ro-Ros, as soon as the United States decided to put an army ashore. Some U.S.-flagged containerships were used in the sustainment phase of the Gulf War, but not in the initial surge phase.

The Gulf War provided the most recent full-scale test of U.S. sealift capability and left the military with a number of pressing questions. In a time of emergency, where will the necessary ships be found? Are there enough qualified officers and unlicensed personnel to man the naval auxiliary fleet? And more generally, how useful is the existing U.S.-flag fleet to the armed forces? And depending on how this question is answered, what policy should the military pursue to ensure it has adequate sealift capability? Since the middle of the nineteenth century the subsidized U.S. merchant marine has cited its usefulness to the military in times of national emergency as an important justification for continued federal support. How the armed forces decide to acquire and sustain the sealift capability

241

they need will therefore be a crucially important factor in determining any future federal maritime policy.

THE U.S. MILITARY TRANSPORTATION SYSTEM

At the end of World War II the United States had millions of men and women on active duty in its army, navy and air force. While peace was inevitably followed by massive demobilization, it brought calls for the reorganization of the armed forces as well. The separate military services, each of which had its own traditions and strategic imperatives, had long competed with each other for control of resources and operations, but rationalization and efficiency were the watchwords once the war was over. The possibility of creating a single unified service was being discussed well before the war ended, and a few years later the National Security Act of 1947 in fact created a Department of Defense that subordinated the separate services to the overall control of a single civilian administrator.

In August 1949 the secretary of defense directed that the various units of the military involved in providing ocean transportation be consolidated into a single command. The secretary of the navy was designated the manager for Department of Defense ocean transportation and was told to establish an appropriate agency within the navy for this purpose. He did so in October 1949 by creating the Military Sea Transportation Service (MSTS). The new service's mission was to provide U.S. military forces with point-to-point sealift of passengers, fuel, materiel, and supplies. The vessels and equipment with which it was to fulfill this mission were transferred from the Naval Transportation Service and the Army Transportation Service (ATS) and included government-owned troopships, tankers, and cargo ships.

At first glance the creation of MSTS appeared eminently sensible, a triumph of efficiency over partisan interest, but the terms under which it was created included a budgetary arrangement to which the army strenuously objected. The Army Transportation Service had been an integral part of the army, but the budget of the new MSTS was set up as an industrial fund, with each branch of the government being billed for services provided. Thus the army, being the largest user of ocean transportation services, found itself paying for 85 percent of the MSTS budget but having no operational control. While naval personnel controlled the new command and operated much of the navy's noncombatant fleet, the navy paid for only a small part of the new service.

By 1969, when the Nixon administration assumed office, it was realized that the new transportation command had merely added another layer of bureaucracy to the previously existing systems. After a period of study the secretary of defense ordered that many of its original sealift functions be returned to the army. David Packard, who was then serving as deputy secretary of defense, supported this change, but the navy, led by Admiral Thomas Moorer, chairman of the Joint Chiefs of Staff (JCS), strenuously opposed it. This battle within the Pentagon soon attracted congressional attention; the navy, which had many powerful friends in Congress, arranged to have a rider attached to the Defense Appropriations Bill barring the use of any funds to bring about the change in MSTS ordered by the secretary.

While this episode in interservice rivalry was being played out, the organization and capabilities of airlift were also being considered. Before the escalation of U.S. involvement in Vietnam, enthusiasts for airlift had argued that it could replace much of the sealift that was needed to support foreign intervention, but the experience of actually landing and supplying an army in Vietnam demonstrated that, except for the movement of the troops themselves, sealift was still essential. By this time a specified Military Airlift Command (MAC) had been created and was working well, and the Department of Defense (DOD) once again began planning for comprehensive military transportation coordination. To make sure that the Joint Chiefs of Staff would be onboard before a new plan was announced, Deputy Secretary of Defense Frank Carlucci commissioned a year-long study of the issue. But before that study was completed the JCS, responding to an army request, persuaded Carlucci to approve the immediate transfer of cargo booking and contract administration from the Military Sealift Command (MSC), the new name of the MSTS, to the army's Military Traffic Management Command (MTMC). This transfer was made, and the JCS study, when completed, recommended that a unified surface command including MSC and MTMC be created and that the army and navy be given the authority to designate its commander in chief (CINC) on a rotating basis. But once again the navy objected, this time under the leadership of President Reagan's feisty Secretary of the Navy John Lehman. Lehman persuaded Senator John Tower, chairman of the Armed Services Committee, to again amend the DOD Authorization Bill so as to prevent the proposed reorganization.

As the cold war continued and defense expenditures skyrocketed, the weaknesses and inefficiency of the military contracting system attracted increasing attention. The House and the Senate held hearings, and widely publicized cases of exorbitant and careless procurements called into question the Defense Department's ability to manage the vast sums it was being allocated. A commission was therefore named to review DOD procurement practices; its chairman was David Packard. Even before the commission was formed, Packard, still stung by the navy's opposition to his earlier proposals, vowed that one result of the commission's work would most certainly be the creation of a joint transportation command.

As it turned out, a number of events strengthened Packard's hand. While his commission was conducting its investigation, the House and Senate, under the leadership of Congressman William Nichols and Senator Barry Goldwater, were holding far-reaching hearings on the breakdown of operations involving different branches of the armed services. The tragedy of the second U.S. entry into Lebanon and the confusions experienced in the invasion of Grenada were repeatedly cited as evidence that the services needed to coordinate their activities much more closely. The Goldwater-Nichols Act of 1986 in fact incorporated many of the recommendations made by the Packard Commission. The commission also recommended that a unified transportation command (USTRANSCOM) including surface units and the Military Airlift Command (MAC) be created.

In due course Admiral William Crowe, chairman of the Joint Chiefs of Staff, informed the secretary of defense that he would fully support the new transportation command; his endorsement, along with that of the secretary, was passed along to Frank Carlucci, the new national security adviser. Carlucci hurried these recommendations to the president, and the necessary legislation was soon moving forward with little opposition. In April 1987 General Duane Cassidy, of the U.S. Air Force, was confirmed as the first commander in chief (CINC) of USTRANSCOM, with Vice Admiral Albert J. Herberger, of the U.S. Navy, as his deputy; in October USTRANSCOM stood up as an operational command. The CINC of USTRANSCOM replaced the navy's chief of naval operations as overall commander of the Military Sealift Command (MSC). One year later General Cassidy was able to notify Admiral Crowe, chairman of the JCS, that USTRANSCOM was fully operational.[1]

The navy's long resistance to sharing control of its auxiliary fleet was rooted in an attitude that also made it very difficult to coordinate relations between the navy and the merchant marine. While admirals were unwilling to relinquish any of their autonomy, the army, while maintaining its own fleet of transports, had generally been cooperative in its dealings with the shipping industry and had used the merchant marine for most of its postwar sealift. The army recognized that merchant ships had a vital role to play in fighting wars overseas; the navy, concerned primarily with supporting its fleets at sea, looked upon the merchant marine as just another set of ships under its command and control.

244

The navy's reluctance to make use of commercial ships both aggravated and threatened U.S.-flag shipowners. Working with the commercial industry would have required considerable cooperation and coordination, and the navy simply did not welcome what it viewed as outside interference. It defended its position by citing the need to guarantee the security of military cargoes. The navy said that the long record of strikes in the maritime industry demonstrated its unreliability, this despite the fact that military cargoes were routinely exempted from work stoppages. The navy also insisted that using its ships first, before utilizing commercial services, saved taxpayer funds, but it failed to note that its stock of ships could have been reduced if commercial vessels were used instead. This unfriendly debate continued for years. The Korean War provided a period of respite, the demand for shipping being so great that all available ships and crews found work, but after the war tensions increased again. The shipowners knew that they were not only losing cargoes to naval vessels, they were also losing their main claim for continued political support, the claim that the merchant marine is essential to the national defense in time of war.

This issue was finally addressed directly by Charles E. Wilson, the secretary of defense, and Sinclair Weeks, the secretary of commerce, who together drew up an agreement to govern the relationship between MSTS and the merchant marine. The Wilson-Weeks Agreement, as it came to be known, was signed on July 1, 1954; its preamble states that "the Department of Defense and the Department of Commerce agree that, for the national defense, they have complementary interests in the control and utilization of ocean-going merchant ships." The agreement acknowledged that MSTS must maintain a large, dedicated fleet for specific DOD purposes, yet it required that available ships in liner service be used to the fullest possible extent to move DOD cargoes. Its stipulations were unambiguous and to the point:

> All merchant shipping capability required by the Department of Defense, in addition to that provided by the MSTS nucleus fleet, will be obtained, consistent with military requirements and prudent management, in the following order of priority:
>
> (1) Maximum utilization of available U.S. flag berth space.
> (2) Time or voyage charter of suitable privately-owned U.S. flag merchant ships to the extent these are voluntarily made available by the maritime industry. Such time or voyage charters will be kept to the minimum necessary to meet

requirements which foresight indicates cannot be met by
U.S. flag berth operators.
(3) Shipping provided by National Shipping Authority under
General Agency Agreement or other arrangement.[2]

Over time this agreement has been largely ignored; Congress, at the urging of shipowners, found itself repeatedly reminding the Military Sealift Command of the requirement that it give priority to using available commercial shipping.

During the war in Vietnam 170 ships were activated from the National Defense Reserve Fleet, and at the height of the war in 1967 over 500 ships, representing 40 percent of the country's commercial sealift capacity, were involved. This flood of ships and supplies created massive congestion, but by 1969 the efficient use of containers and new port construction had caught up with the flow of war materiel and the supply system was working efficiently. Since the number of commercial ships required to sustain the war effort declined as the navy increasingly relied on government ships, the maritime administrator suggested that the ships activated from the NDRF be deactivated. The admiral in charge of MSC rejected this suggestion and insisted on retaining what he called "his" fleet. When these ships returned to the United States the maritime administrator canceled the general operating agreements with the companies operating them for the government, and the ships were returned to the NDRF. This indicated unambiguously whose fleet it was in actuality. Within six months the entire fleet of 170 ships was gone, and privately owned American merchant ships were sustaining the war effort without a hitch. Needless to say, this action further embittered relations between the navy and the Maritime Administration.

Although the 600 Liberty ships in the National Defense Reserve Fleet were essentially worthless, DOD planners continued to count them as military assets when asked if the country had adequate emergency shipping resources. While the newer, faster, and larger Victory ships had been activated in the Korean and Vietnam Wars, none of the Liberty ships were used. Once the Maritime Administration became aware of the situation, it began scrapping the Liberty ships wholesale, and in a few years time the ships were gone and three of the six NDRF anchorages were closed. The Department of Defense responded not by acquiring additional sealift, but by reducing its estimates of the amount of sealift required so that the available resources still appeared adequate.

A turning point was reached in 1978 when the Soviet Union invaded Afghanistan and President Carter, concerned about possible Soviet aggression in the Middle East, ordered a buildup of DOD's overseas logistical support capabilities. A year later funds were provided for an Afloat Prepositioning Force (APF) consisting of chartered merchant ships loaded with military equipment and supplies and anchored in the Indian Ocean island of Diego Garcia. The Marine Corps moved in a parallel direction by acquiring a fleet of Maritime Prepositioning Ships (MPS) that it located in Diego Garcia, Guam, and the United States. In 1981 the navy purchased the eight SL-7s that Sea-Land had built and then put in layup. Sea-Land built these thirty-three-knot ships when fuel oil costs were between one and two dollars per barrel, but when oil prices soared to thirty dollars per barrel they could not afford to operate them. These ships were ideal for military rapid response, however, and the navy converted them to partial Ro-Ros and designated them Fast Sealift Ships (FSS). These ships were then assigned to a new group within the NDRF called the Ready Reserve Force (RRF). Since only ships that could be activated on five-, ten-, or fifteen-day schedules were included in the RRF, the old Victory and other worn-out ships in the NDRF were effectively dropped from the list of ships that could be activated in time of need.

The election of President Reagan marked the beginning of a massive expansion of the U.S. military. The budget for sealift, which had been $40 million in 1979, increased to over $1 billion by 1983, and over $8 billion was eventually spent acquiring and modernizing ships for the RRF. By the time Iraq invaded Kuwait, the RRF had 96 of the 142 ships it was scheduled to contain, with most of the new ships being relatively modern foreign-built Ro-Ros. And as the RRF expanded, the gulf separating MSC and the U.S. shipping industry continued to widen.

With the creation of USTRANSCOM a new attempt was made to establish a workable sealift policy. In June 1987 President Reagan had signed a National Airlift Policy, which defined in general terms the relations between the U.S. Air Force and the airline industry. Something similar was sought for shipping, and General Cassidy labored hard to formulate such a policy. His efforts were finally rewarded, and President Bush signed a new National Sealift Policy in October 1989. It was a comprehensive and enlightened set of guidelines, but its implementation remained a problem. The policy states that:

1. The goal of the U.S. Government is to ensure that adequate sealift resources are available to meet national economic and security requirements during time of war or national emergency.
2. Development and implementation of specific sealift and sup-

porting programs to meet this goal will be made with full consideration of the costs and benefits involved.

3. Although U.S. policy is to rely on both U.S. privately-owned, U.S. flag and Effective U.S. Controlled (EUSC) and allied shipping resources to meet strategic commitments to our established alliances, the U.S. will be prepared to respond unilaterally to security threats in geographic areas not covered by these alliance commitments. Sufficient U.S. privately-owned sealift with reliable crews must be available to meet these requirements and the U.S. Government will initiate actions to achieve this objective.

4. The U.S. flag commercial ocean carrier industry, to the fullest extent possible, will be relied upon to provide sealift resources for national security in peace, crisis, and war. This capability must be strengthened and further augmented by reserve fleets comprised of ships with military capabilities and trained nucleus crews that cannot be provided in sufficient numbers or types from active U.S. commercial shipping sources. In peacetime the Department of Defense will operate the minimum number of sealift ships needed to meet Joint Chiefs-of-Staff exercise and other unique military requirements that cannot be accommodated by scheduled commercial ocean carriers.

5. The Department of Defense will determine the requirements for sealift of deploying forces, follow-on supply and sustainment in contingency, crisis, or war, as well as shipyard and other required support for their mobilization. The Department of Transportation will determine the economic sealift and shipyard requirements for mobilization. Both Departments will promote the incorporation of national defense features in new and existing ships.

6. During peacetime, federal agencies will promote, through efficient application of existing laws and regulations (e.g., cargo preference laws and acquisition policy) the readiness of the U.S. Merchant Marine and supporting industries to respond to critical national security requirements.

7. The Departments of State and Treasury, the U.S. Trade Representative and other appropriate departments and agencies shall ensure that the international agreements and federal policies and regulations, including, but not limited to govern-

ing use of foreign flag carriers; protect our national security
maritime interests—consistent with national sealift policy.

8. To the extent deemed necessary to protect our security interests,
U.S. Government policies and programs should provide for an
environment which fosters the international competitiveness
and industrial preparedness of the U.S. maritime industries.

9. Government policies should also support research programs
which promote the development of technologically advanced
sealift ships and supporting industries.

10. The domestic fleet and shipyard mobilization base should be
enhanced by rigorous enforcement of the Jones Act.

11. Troop carrier capability should be encouraged by the promotion of the U.S.-flag passenger trades.[3]

Implementation of this new policy was to be directed by a senior interagency
group operating under the auspices of the National Security Council; the
administration would request new legislation as required. But these fine
plans did not survive General Cassidy's retirement in 1990. The United
States was clearly not "prepared to respond unilaterally to security threats in
geographic areas not covered by alliance commitments" (paragraph 3), and
as Desert Shield/Desert Storm demonstrated, "the U.S. flag commercial
ocean carrier industry" could no longer be "relied upon to provide sealift
resources for national security in peace, crisis, and war" (paragraph 4). A
plausible policy had been formulated, but little else had changed.

The Department of Defense did not really take this exercise seriously
and in fact obstructed the adoption of such a policy. Although it was
"directed to determine the requirements for sealift for deploying forces,
follow-on supply and sustainment in contingency, crisis or war, as well as
shipyard and other required support for mobilization," little meaningful
information was forthcoming. Instead, logistical requirements that would
cost billions were advanced and published without careful scrutiny. Had
the National Sealift Policy been properly implemented, it might have provided a reasonable basis for future maritime programs. But in the absence
of dedicated and successful implementation, it became just another piece
of government paper gathering dust.

OPERATION DESERT SHIELD/DESERT STORM

When Iraqi tanks rumbled across the Kuwait border in August 1990, the
United States suddenly found itself in a new strategic situation.
Throughout the cold war America had relied on garrisons and bases lo-

cated near its adversaries for rapid response and containment. Large concentrations of U.S. Army units in Germany and Japan helped stabilize contested borders in Eastern Europe and Korea, and naval and air force bases in the Philippines and Okinawa provided regional support for the nation's military presence in the Far East. But to counter Iraq's threat to Middle Eastern oil fields that were essential to the economies of Europe, the United States, and Japan, America had to expand and project its power in a region where none of its forward-deployed forces were based. The invasion of Kuwait thus became a test of U.S. responsiveness in a novel strategic setting, a setting in which sealift was crucial. As the United States began assembling for Operation Desert Shield the forces it would send into battle in Operation Desert Storm, it learned once again that it had made strategic commitments that required adequate sealift.

As it turned out, the United States was extraordinarily fortunate in the Gulf War. The great danger was that Saddam Hussein, having occupied Kuwait, would press on into Saudi Arabia and conquer the entire oil-rich Arabian Peninsula. Had he chosen to do so, the United States could have done little to stop him short of using nuclear weapons. But Hussein, perhaps thinking that he might get away with seizing Kuwait alone, held his army in check for week after week as the United States assembled its forces and organized its response. The modern ports that Saudi Arabia made available played a vital role in this buildup. This interval also gave the United States time to mobilize its collection of diverse forces under a United Nations banner. When the allies finally launched their counteroffensive in Operation Desert Storm, it was brilliantly executed and swiftly victorious. But the Gulf War was far more than "the Hundred Hours War," as it has been called. It was a seven-month logistical buildup called Desert Shield. Even though the allies enjoyed undisputed control of the air, the sea, and the neighboring territories, it took seven months to assemble a force capable of responding to Iraq's aggression. Under less favorable circumstances the inadequacies of America's sealift capability would have been disastrous.

The military commanders should have anticipated the sealift shortfalls they encountered. Force projection plans for Southwest Asia had clearly identified the inadequacy of existing sealift capability, but the navy had once again failed to learn its lesson. While some efforts had been made to address the problem of initial surge buildup by expanding the fleet of government transports, the problem of long-term sustainment, which was to be provided by ships drawn from the U.S.-flag commercial fleet, was ignored. When planning Desert Shield, the military allowed three weeks for arrival of the initial heavy combat forces and eight weeks for five divisions to be in place with their equipment. Yet after the first month only the marines and the army's lightly equipped Eighty-second Airborne Division had taken up positions. The schedule for arrivals of other forces had gone from weeks to months. Had the Iraqi army not stayed in a defense posture, the military commanders would have had a great deal of explaining to do.

And yet, as the commander of the Military Sealift Command, Vice Admiral F. R. Donovan, reported after the war, the sealift forces that were available and were deployed generally performed well. Indeed, the Department of Defense had every reason to be satisfied with the performance of the ships it had acquired in the 1980s for its Ready Reserve Fleet. The four Maritime Prepositioning Ships (MPS) based in Diego Garcia arrived in Saudi Arabia ten days after call-up to deliver their Marine Corps equipment, and all nine of the activated MPS had off-loaded by the first week of September. Eight of the twelve Afloat Prepositioned Ships located at Diego Garcia and loaded with army and air force equipment and ammunition had also off-loaded by September 6. Eight Fast Sealift Ships (FSS) were loaded in the United States and had sailed by August 22; the first, USNS *Capella*, was activated, moved to Savannah, loaded with twenty-four thousand tons of equipment, and sailed within six days after activation. The first two FSS arrived in Saudi Arabia on August 27; of the seventeen ships initially requisitioned from the RRF, all but one completed the transit after activation.

This record of successful activation was achieved despite inadequate funding for maintenance and little training. For the fiscal year 1990 the Maritime Administration requested $239 million to maintain the Ready Reserve Force, but Congress approved only $89 million. As Secretary of Transportation Samuel Skinner pointed out, the RRF had been "short-changed by the Congress in the appropriation process for a number of years." "Funding has been kept so low," he continued, "that the readiness status of many Ready Reserve Force ships is not realistic."[4] Many of the ships were not undergoing periodic test activations and sea trials, and more than half of the RRF ships that were activated for the Gulf War had not been tested since becoming part of the reserve fleet. But it was not just a lack of funds that hampered maintenance. In 1986 the navy had directed that management contracts for RRF ships be shifted from the cost-plus general agency agreements customarily awarded by MARAD to low-bid, fixed-cost ship management contracts. The structure of these new contracts, and the evaluation process used to award them to the lowest bidder, encouraged contractors to bet that the ships would not be activated, one consequence being that maintenance and readiness suffered severely on some ships.

Only forty-two of the ninety-eight ships in the RRF were activated to support Desert Shield/Desert Storm, but even with this limited mobilization the Maritime Administration found itself facing a serious shortage of qualified seafarers. No reserve system comparable to the National Guard had been established for mariners, and those who responded to urgent calls by leaving their jobs had no guaranteed reemployment rights. MARAD was forced to plead with the maritime unions to call members out of retirement and cancel all leave so that the ships could be manned. A traditional reason to support a national merchant marine, which in British history goes back to at least the time of Queen Elizabeth I, was to maintain a sufficient number of qualified seamen who could be called into service in time of war; the Gulf War made the continued relevance of this reason evident once again. As the U.S. merchant marine shrank during its long postwar decline, the average age of mariners increased to fifty-five years. And as diesel-engined ships replaced steam-driven ships, the skills required to reactivate and operate the steam plants of ships retained in the RRF were lost. The average age of the Ready Reserve Force ships was thirty years and over 70 percent had steam propulsion systems. It takes both ships and operating personnel to support an army fighting abroad, and in the 1990s the constraint that most severely limited U.S. military response when sealift was required was a lack of trained personnel.

To augment the sealift provided by ships from the RRF, the navy chartered nearly all the useful vessels it could find. Three months after the invasion of Kuwait seventy-three ships had been chartered, with over half of them coming from foreign fleets. Before the allies launched their counteroffensive, the United States was using just about every available vessel in the world capable of moving heavy equipment, including more than one hundred foreign-flag ships. It is instructive to note, however, that America's allies in this war, who were far more dependent on Middle Eastern oil than the United States, mobilized few of the ships sailing under their national flags. Not one ship flying the Japanese or German flag was made available to move war materiels; none of these ships were directed by their government to support the war, this despite the fact that the Japanese have many Ro-Ros available in their car-carrying fleet. The United States had hoped for a multilateral response to Iraq's aggression, and in some areas it was achieved, but when it came to sealift the United States was forced to virtually go it alone. Clearly even in the era of global trade and multinational firms an adequate National Sealift Policy is still needed.[5]

The argument that the merchant marine must be supported so that it can provide a naval auxiliary in time of national emergency is now taken less seriously. The containerships and large bulk carriers remaining under U.S. ownership are of limited use to the armed forces when they need to transport military supplies to undeveloped areas. The navy relies increasingly on the foreign-built Ro-Ros it has acquired for the RRF. It has also embarked on a $6 billion building program to create its own specialized fleet. It is revealing that during the initial buildup prior to Desert Storm, not a single American ship was pulled off a trade route to support the war. It was only several months later, when the sustainment phase was reached, that U.S. containerships could be utilized, and then only because they could discharge their cargoes at excellent port facilities in Saudi Arabia.

The disappearing U.S. maritime industry gained well-earned glory while contributing to victory in two world wars. Since then it has attempted to survive by using political means to sustain the forms of support on which it has relied for so many years. But in the age of nuclear weapons and space vehicles, the ways in which conventional wars begin, are waged, and are terminated have changed dramatically. The merchant marine may still have a role to play in supporting the U.S. military, but it is hardly the role imagined by the authors of the Merchant Marine Act of 1920.

13

The Search for
a Workable Maritime Program,
1980–1992

The preceding chapters have described the sweeping changes that fundamentally transformed the international maritime industry in the fifty years after World War II. Before the war shipbuilding and international shipping were largely European enterprises, with the United States playing the role of a recently admitted member of the club. But after the war developments that were not anticipated and only belatedly perceived brought the nations of the Far East into dominate positions in the industry. Japan and then South Korea began by carving out commanding positions for themselves in shipbuilding. They then moved into the carrying trade as well, as did Taiwan, Hong Kong, and Singapore. The formerly dominant players in Europe and America were caught by surprise. Hobbled by institutional conservatism and outdated policies, they responded slowly and inadequately to the new challenges. Many world-famous shipyards and shipping firms fell by the wayside when forced to compete in the new world of maritime commerce.

In shipbuilding, as in other areas of their export-led economies, the Asian nations initially won market share by emphasizing low-cost production and then consolidated their positions by producing superior goods. Japanese shipyards were world leaders in labor productivity and built some of the world's finest ships. Service, rather than price, was becoming the dominant concern in the maritime industry, for competitors quickly matched price cuts, and in the competition to provide superior service the Asian nations more than held their own.

The United States resisted facing up to many of these changes and maintained its merchant marine with subsidies and protection long after most modern maritime nations had largely abandoned them. U.S. shipbuilders and shipowners, together with their labor unions, laid great weight on the national defense argument to ensure continuation of the federal policies and subsidies they needed to survive. But when the Berlin Wall came crashing down in November 1989, the national defense argument became more difficult to sustain. The East-West tensions that had dominated U.S. defense strategy for almost fifty years suddenly dissipated with a speed that no one had anticipated and few could believe. It was far from clear what would happen next. But while aggressive nationalism soon surfaced in many of the smaller countries freed from Communist rule, it rapidly became evident that the threat of a third world war had passed.

In this changed environment U.S. shipbuilders and shipowners began to look for new solutions to their problems. At first the shipbuilders focused their policy concerns on the subsidies that other nations were providing to their shipyards. Since the U.S. government was no longer providing direct subsidies to its shipyards, the shipbuilding industry called on the government to insist that other nations stop subsidizing their builders. The assumption was that given a level playing field, the U.S. shipbuilding industry could compete successfully in the world market. Since operating subsidy contracts were also scheduled to expire soon, the two leading carriers, Sea-Land and American President Lines (APL), urged that archaic U.S. government regulations be eliminated, that maritime tax policies be revised, and that a new program be developed to help offset the higher cost of employing American crews. If such support was not provided, they saw no alternative to abandoning the American flag and operating their ships under foreign registries. In the 1990s congressional discussion of maritime affairs largely revolved around these two sets of proposals.

Changes in Maritime Policy during the Reagan Administration, 1980–1988

Governor Ronald Reagan, while campaigning for the presidency in 1980, announced to an enthusiastic audience of maritime industry executives his "Program for the Development of an Effective Maritime Strategy." His proposals, Reagan said, would

1. Provide a unified direction for all government programs affecting maritime interests of the United States.
2. Ensure that the vital shipbuilding mobilization base was preserved.
3. Improve utilization of military resources by increasing commercial participation in support functions.

4. Recognize the challenges created by cargo policies of other nations.
5. Restore the cost competitiveness of U.S. flag operators in the international marketplace.
6. Revitalize the domestic water transportation system.
7. Reduce the severe regulatory environment that inhibited American competitiveness.[1]

This list of promises, which touches on practically every vital issue in U.S. maritime policy, was vintage Reagan—visionary in its reach but silent on the means and costs of implementation. The governor's staff had been deeply divided over this set of campaign promises. His political advisers reminded the candidate that the maritime industry and labor unions had given President Nixon vital support in 1972, and they argued that Reagan would need every vote he could attract in his close race against an incumbent president. But Martin Anderson, one of Reagan's longtime economic advisers and a classic free trader, categorically opposed all government subsidies to industry. Anderson did not prevail on the maritime issue during the campaign, but after Reagan was elected president, Anderson played a crucial role in making budgetary considerations, rather than political and national security considerations, central to Reagan's maritime policy.

When President Reagan took office Anderson was appointed assistant to the president for Policy Development; his wife, Anelise, was appointed to the Office of Management and Budget (OMB), where she was assigned oversight responsibility for the Maritime Administration. Ms. Anderson and her staff soon released a poorly researched document attacking the national security argument for supporting the U.S.-flag merchant marine. This one-sided report was presented to the president's Cabinet Council for Commerce and Trade on May 3, 1982. No record of its reception has survived. The shipowners, however, quickly let it be known what they thought of this position paper. On the first of July Admiral James L. Holloway III (U.S. Navy, retired), who had served as chief of naval operations and was at that time president of the Council of American-Flag Ship Operators, fired off an effective rebuttal.[2] The White House sidestepped the issue by neither endorsing nor repudiating Ms. Anderson's recommendations. Nothing further was heard of them, but a signal had been sent.

The Maritime Administration's ability to speak forcefully in behalf of the industry was significantly weakened in August 1981 when it was transferred from the Department of Commerce to the Department of Transportation. During the transfer the maritime administrator was stripped of the additional title of assistant secretary for Maritime Affairs and responsibility for formulating maritime policy was assigned to the secretary of transportation and his staff. The new arrangement meant that maritime policy proposals would be constrained from the outset by the budgetary restrictions that OMB was imposing on the various branches of the administration. The legislative authorization for maritime subsidies passed in 1936 remained in effect, but the administration was not obliged to seek annual appropriations to fund the authorized programs. It became clear that a new administrative strategy for curbing maritime subsidies had gained the upper hand.

On the same day that Ms. Anderson was presenting her report to the president's cabinet council, Drew Lewis, the new secretary of transportation, gave the council a general description of the maritime industry. Lewis recalled the commitments that the president had made during the campaign and listed ways in which they could be honored. He also offered an overall endorsement for continuing government programs that supported shipbuilding and U.S.-flag operations. But Lewis also noted that because of instructions from OMB, he would not be seeking appropriations for any shipbuilding subsidies during the fiscal years 1982 or 1983, nor would any new Operating Differential Subsidy (ODS) contracts be signed. These interruptions of subsidy programs begun in 1936 were presented as temporary responses to budgetary shortfalls, but in fact they marked the beginning of the end for the direct subsidy programs President Roosevelt had put in place during the Great Depression to get American shipbuilders and shipowners back on their feet.

Lewis was more the bearer of bad news than its author, and he told the cabinet council of a way in which some consequences of eliminating shipbuilding subsidies could be mitigated. In the 1981 Reconciliation Act Congress had agreed that since there would be no money for shipbuilding subsidies, shipowners should be allowed for one year to acquire foreign-built ships that would be eligible for operating differential subsidies and all government preference cargoes. This hiatus in the long-standing policy that subsidized ships must be U.S. built was a tacit acknowledgment that unsubsidized U.S. shipyards could not build commercial ships at anything approaching competitive prices and that U.S. carriers could not survive if they could not replace their aging fleets. The 1981 Reconciliation Act also contained a provision saying that if a special $100 million subsidy was provided to the shipbuilders, the window for purchasing foreign-built ships would remain open a second year. Secretary Lewis decided not to pursue this quid pro quo, and the window was only open during fiscal year 1982. Nonetheless, American shipowners made good use of this unusual opportunity by ordering thirty ships from foreign yards and having several others converted abroad.

In May 1982 Secretary Lewis announced a proposed maritime program that looked good on paper but was politically nothing more than a wish list. Its elements included:

1. Extending foreign building rights to the ODS fleet.
2. Allowing reflagged vessels to immediately become eligible to carry government-impelled cargoes.
3. Allowing foreign investment in U.S.-flag shipping to be increased from 49 percent to 75 percent.
4. Making all seamen employed on U.S.-flag vessels employed in foreign trade eligible to receive the foreign-earned income exclusion of $75,000 presently available to U.S. citizens employed abroad under the 1981 Tax Act.
5. Providing a Mariner-type build-and-charter program where DOD would build defense-relevant multipurpose carriers to be leased to private ship operators.
6. Allowing use of capital construction funds for foreign-built ships.[3]

Many in the industry considered Lewis's list too good to be true, and so it proved to be, for it soon became apparent that it would not be implemented. Those items that required additional funding were effectively blocked by budgetary restrictions, while those that sought to free the industry from entrenched positions aroused animosity in Congress. The House Merchant Marine Committee heard President Reagan's two maritime administrators, Harold Shear and John Gaughan, propose that certain elements of the Lewis program be implemented, but since all their proposals included allowing foreign building, the committee was not interested. The administration had other fish to fry and did not insist. High on Drew Lewis's agenda were increasing the Highway Trust Fund, improving the nation's highway system, enhancing the productivity of trucks, and removing many archaic restrictions on transportation. These were the areas in which he invested his considerable talents and achieved notable success.

By the final year of President Reagan's first term the failure to deliver on campaign promises made to the maritime industry was widely recognized and reluctantly accepted. The *Journal of Commerce* provided an appropriate obituary in an editorial published on March 22, 1984:

Maritime people clearly recall Mr. Reagan in mid-1980 unveiling a bold, seven-point plan to reverse the decline of the American merchant marine and shipbuilding industry. The cornerstone of that plan was a promise that Mr. Reagan would provide a "coherent, focused and consistent national maritime policy."

That was in 1980. Mr. Reagan took office in January 1981. It is now March 1984 and it's time to pose a few questions:

Has Mr. Reagan provided the nation a "coherent, focused and consistent national maritime policy," as he promised? Has Mr. Reagan targeted "a greater market share of exports and imports for U.S.-flag shipping," as he promised? Has Mr. Reagan revitalized the nation's domestic water transportation system, as he promised?

Mr. Reagan is ordering more Navy ships, and that does benefit the six U.S. shipyards that specialize in warships. Does he care about the dozen yards that build only merchant ships?

Has Mr. Reagan monitored all his maritime programs "from the top level of government," as he promised?[4]

The questions were rhetorical, no answers were forthcoming.

The Commission on Merchant Marine and Defense, 1986–1988

When faced with problems for which there are no obvious remedies, politicians frequently create commissions or special committees to study the issues involved and recommend responses. This is not an unreasonable course of action, for commissions of this sort can be useful when it is widely believed that the problems addressed require legislative action and that the members of the commission are prepared to make recommendations that can be implemented. There have been several commissions of this sort in the long history of federal maritime policy. In 1914 the Alexander committee brought to light serious malpractice in ocean shipping and laid the groundwork for the Shipping Act of 1916. In 1935 the Black committee exposed fraud in the mail subsidy system and set standards that were incorporated into the Merchant Marine Act of 1936. And more recently the Packard Commission made recommendations that were included in the 1986 Goldwater-Nichols Act that created USTRANSCOM.

Unfortunately, however, there is a parallel history of committees and commissions that, while launched with the best intentions, became so burdened with political considerations that they produced nothing of real value. A notable example is the post–Civil War Lynch committee, which did nothing to halt the rapid decline of the U.S. merchant marine; the same judgment can be applied to another commission formed over a century later, the 1986 Commission for the Merchant Marine and Defense. Both of these committees were created to find ways to halt and reverse the decline of the merchant marine; both, for political reasons, devoted most of their effort to attempting to find ways to support the ailing shipbuilding industry. In the end neither committee served the cause it was created to address or helped solve the problem on which it actually chose to focus.

The seeds of the 1986 commission's destruction were sown at the time of its creation. The idea of forming this commission emerged from hearings on military sealift that Congressman Charles Bennett, chairman of the Seapower Subcommittee of the House Armed Services Committee, held in 1983. Shocked by evidence that the merchant marine was rapidly declining, Bennett was determined to form a commission to investigate what could be done to ensure that the U.S. shipping industry was capable of meeting its sealift commitments in time of national emergency. It was not long, however, before the shipbuilders learned of Bennett's initiative. Realizing the commission might recommend that American shipowners be allowed to acquire ships in the world market, the shipbuilders were determined to at least extract a large increase in government funding for themselves. Their supporters in Congress therefore dutifully insisted that the commission's charge be expanded to include shipbuilding as well. Overburdened and divided at the outset, the commission never had a chance.

The commission's members were chosen in a carefully staged ritual that presumed disinterestedness. In the hearings prior to the commission's creation, it was strongly urged that no one with direct ties to the maritime industry, either as shipowner, shipbuilder, or labor leader, be made a commissioner. This was to be a truly blue-ribbon panel made up of public-spirited leaders in the public and private sectors. As it turned out, only two of the seven commissioners could be so characterized: Jeremiah Denton, the commission's chairman and former U.S. senator from Alabama, and Edward Carson, the retired chairman of United Airlines. The other members of the commission included the president of one of the nation's largest shipyards (the shipowners were not comparably represented), the president of a large maritime trade union, the former head of a Washington-based shipowners' trade association, a trustee of a state maritime academy, and the federal maritime administrator. Given this membership, a special law had to be passed exempting the commissioners from certain prohibitions governing conflict of interest before the commission could be sworn in. It was not an auspicious beginning.

The legislation establishing the commission, Public Law 98-525, was signed in October 1984 and defined its charge as follows: "The Commission shall study problems relating to transportation of cargo and personnel for national defense purposes in time of war or national emergency, the capability of the United States merchant marine to meet the need for such transportation, and the adequacy of the shipbuilding mobilization base of the United States to meet the needs of naval and merchant ship construction in time of war or national emergency. Based on the results of the study, the Commission shall make . . . specific recommendation[s], including recommendations for legislative action, action by the executive branch, and action by the private sector."[5] This charge and the fact that none of the commissioners had any experience as a shipowner or operator made it exceedingly unlikely that the roles of private investment and commercial operation would be seriously explored, and indeed they were not. After deciding by default to leave in place all the existing regulations and subsidy programs, the commission had no choice but to recommend a vast increase in government funding.

Although the commission's recommendations were foreordained, the testimony it collected during two years of hearings is instructive. All aspects of the industry were heard from, and the evidence presented was collected into four reports that total almost three thousand pages. In the first of these reports, issued in September 1987, the commissioners defined the central problem as they saw it: "The Commission has found clear and growing danger to the national security in the deteriorating condition of America's maritime industries. The United States cannot consider its own interests or freedoms secure, much less retain a position of leadership in the Free World, without reversing the decline of the maritime industries of this nation, which would depend so heavily in a protracted war upon adequate use of oceans for its military defense and for its economic survival."[6]

This note of alarm was sounded throughout the commission's reports, but it could only motivate action if all those involved agreed that the danger that alarmed the commission was credible. Did the United States have to be ready to wage a protracted, nonnuclear, foreign war, one that, like World War II, would require extensive and sustained sealift? The commission took this as a given and concluded, in its first report, that the United States was inadequately prepared: "There is today insufficient strategic sealift, both ships and trained personnel, for the United States, using only its own resources as required by defense planning assumptions, to execute a major deployment in a contingency operation in a single distant theater such as Southwest Asia[;] and maintaining the shipbuilding and repair segment of the industrial base required to sustain a protracted general war is essential to deterring or winning such a war. The base of shipyards and repair facilities, and their industrial suppliers, is currently inadequate in that sense and is continuing to deteriorate at a alarmingly progressive rate."[7]

The public hearings held between February and July of 1987 made it obvious that the commissioners, for all their alarm, were far from open-minded. Their response to the two leadoff witnesses clearly revealed that they had a pretty clear idea how the problem they were addressing should be solved. The first of these witnesses was Ran Hettena, president of Maritime Overseas Corporation, a company that owned a large fleet of bulk carriers. Hettena asked the commissioners to first answer the central question: "Does the country really need a merchant marine?" When they told him he was there to answer questions, not to ask them, he persisted in questioning the commission's basic assumptions. "The historic justification for the American merchant marine has been national defense, not commercial necessity," Hettena said. "The question is whether the justification has, after two centuries, disappeared. This is the question which must be addressed. Up to now, the movement to do away with the fleet has been tacit, without explicit public discussion. That is why perhaps you are hearing these things for the first time. The principal public service, I believe, that this Commission can render is to decide whether the long-established relationship between the fleet and defense has indeed vanished, or whether the military plans have been unconsciously modified to match diminishing resources."[8]

Only after being prodded by John Gaughan, maritime administrator and one of the commissioners, was Hettena willing to address the issue of cargo preference for oil imports. He readily agreed that if a U.S.-flag requirement were imposed on this trade, the order books of U.S. shipbuilders would swell and the number of U.S.-flag ships would increase. Having heard what they wanted to hear, the commissioners declined to explore the political difficulties that would be created by imposing such a policy.

The second witness, Joseph Klausner, was counsel for the Independent U.S. Tanker Committee; he, too, tried to get the commissioners to answer Hettena's question. He noted that "the one unavoidable conclusion" reached in a paper produced by the Maritime Administration was "that we do not need an American merchant marine in war." Klausner went on to say,

> I think that it is a mistake to start looking for the solutions when, for the first time, you have a statement that there is no problem.
>
> Mr. Gaughan's Department, which has been a leader in this movement, responsive by the way to strong intellectual pressure from the OMB—I say "intellectual pressure," I am not talking now about merely political command, but intellectual pressure—has produced a document which says that our actual needs in war-time are so small that we all but do not need a shipping instrument at all and, such as it is, it can quite safely be reposed not in the fleets of our allies, but in the world fleet, the third flag fleet.[9]

Klausner was probably the first shipping official to publicly acknowledge that maritime policy was now being directed by OMB. The chairman of the commission tried to wave this aside by suggesting that OMB's position was a result of the brief time it had had to study the problem and the pressures of the current national budget, but Klausner would have none of it. He pointed out that OMB had studied the issue in 1982 at the direction of Anelise Anderson, and hence one could hardly say their position was arrived at in haste. By this time one of the commissioners had become so upset with the way things were going that he said that if the commission accepted this conclusion, he would not "set aside four days a month to come back here and listen to these presentations."[10]

The storm soon passed, however, and the commissioners regained control of the hearings. Witnesses were found who spoke in favor of preserving the Jones Act and protecting and extending cargo preference. Two liner operators, representing Lykes and APL, strongly advocated being allowed to purchase foreign-built ships while retaining their subsidy contracts and access to government cargoes. The president of Lykes, James Amoss, also strongly criticized the Military Sealift Command (MSC) for its system of awarding contracts solely on the basis of competitive bidding. He demonstrated that in the existing conditions of excess tonnage this system yielded freight rates so low that they did not cover costs. Amoss urged that MSC return to the earlier system of military-cargo allocation based on negotiated bidding.

American owners of ships sailing under flags of convenience were represented by Philip Loree, chairman of the Federation of American Controlled Shipping. He took exception to the claim, made by an earlier witness, "that he had to be guaranteed cargoes." Loree went on to remind the commissioners that shipping is a business, not a federal work program: "In our business, we do not have guaranteed cargoes. We have to go out and compete for them and earn them. That is what shipping is. It is a high risk, capital intensive business, and anyone who says 'I cannot go there unless you give me the business,' you had better stop and think. That is not the way it should work."[11] If the commissioners understood what Loree was telling them, they did not let it divert them from their chosen path.

After the initial hearings concluded in March, the commission scheduled four additional sessions, one during each of the next four months. The five labor leaders who testified during the first of these sessions emphasized the need to guarantee more cargo for U.S.-flag ships. Specific recommendations included continuing the Jones Act, extending its offshore limit from twelve miles to two hundred miles, canceling the Jones Act exemption enjoyed by the U.S. Virgin Islands, and adopting the UNCTAD 40-40-20 bilateral cargo-sharing formula. For the labor leaders, the key to saving the U.S. merchant marine was more protection.

The shipbuilders presented their case in May. John Stocker, president of the Shipbuilders Council of America, summarized the position of the major builders:

> We need to sustain Navy new construction and ship repair as a business base. If you're going to maintain a 600 ship Navy, you need to order roughly on the basis of 20 ships a year—$11 billion or $12 billion a year in (naval ship construction). . . .
>
> We would ask for a restatement from this commission of the sanctity of the Jones Act and follow it up with efforts to insure that those markets that are potential markets for shipbuilders will be reaffirmed so that things like the tankers that are required for the movement of coastal products would, in fact, be ordered from U.S. shipyards.[12]

As Stocker's language indicates, his view of the world depended completely on legislative protection. New naval construction and ship repair work of $11 to $12 billion a year constituted merely "a business base," and he counted on additional government-mandated commercial building and repair work to sustain his industry. For Stocker, as for the union representatives, the Jones Act, which he asked the commission to "sanctify," was the mother of all protection programs. But the forms of protection Stocker called for did not exhaust his appetite for federal assistance. When asked by John Gaughan what his priority project would be, Stocker replied, "[I]f money is no object, obviously I would like to see the development of the build-and-charter program for tankers."[13] Since this program was one of those ultimately recommended by the commissioners, they evidently considered money no object.

In the months that followed, organizations representing the views of shippers, whose cargoes were carried by the shipowners, made their case to the commission. George Miller, executive director of Shippers for Competitive Ocean Transportation, and Stanley Smith, export coordinator for Cargill, argued persuasively that government cargo-reservation programs were detrimental to the nation's exports.[14] But the case they made, like the testimony of others who failed to call for increased government support, left no lasting imprint on the commission.

The commission issued its second report, containing its recommendations for meeting the problems faced by the maritime industry, on December 30, 1987. It began by projecting the ocean tonnage that the nation would need: "The Commission believes that, as an initial approximation of what is needed, the United States must have and should work toward a total oceangoing fleet of at least 650 modern cargo ships in the active commercial fleet, Military Sealift Command fleet, and Ready Reserve Force in order to meet wartime military sealift and economic support requirements. In comparison, the total oceangoing fleet in 1987 includes approximately 530 ships, and it is projected to decline to about 350 by the year 2000."[15] To ensure that in the next thirteen years the nation would not suffer the 300-ship shortfall that the commission projected, more subsidies and more protection were called for.

The commission recommended that shipowners be granted Operating Differential Subsidy (ODS) Reform, but not exactly in the form that shipowners had requested. The owners had asked for "reform" that would extend operating differential subsidies to all qualified operators and allow them to buy foreign-built ships without restriction. The reform the commission called for was more limited: "The Commission . . . generally supports the following ODS reform provisions commonly included in draft ODS reform legislation: (a) permission for a controlled number of foreign-built United States flag ships to be eligible for ODS; (b) allowance for a limited number of foreign-built United States flag ships to be immediately eligible to carry preference cargo under the cargo preference laws of the

United States; (c) allowance for a 'window' for any company to enter the ODS program."[16] The commission recommended only a one-year "window of eligibility" for buying foreign-built ships because it concluded that approval of unlimited foreign building was "not now desirable." It's longer-range hope was that "domestic shipbuilding programs recommended elsewhere in this report" would make buying foreign-built ships less attractive.

The shipbuilding program the commission was counting on was called Procure and Charter. The navy, the Maritime Administration, and the commercial operators were to cooperate in developing a design for a new ship that would be commercially viable and militarily useful. Once the government had produced the necessary ship's plans, U.S. shipbuilders would produce the ships at the government's expense. The ships would then be chartered to operators at whatever rate they were willing to pay. Here was a bounteous subsidy plan indeed, for since no foreign cost comparisons were required, as they had been under the Construction Differential Subsidy (CDS) program, the shipyards could charge whatever the government was willing to pay. Had such a program been funded, the campaign to buy foreign-built ships well might have drowned in a sea of subsidies.

The commission also addressed the provisions of the Tariff Act of 1930 that imposed a 50 percent ad valorem tax on repairs to U.S.-registered ships done in foreign yards. It seemed, the commission reported, that because of "overly liberal interpretations of the tariff . . . , the tariff appears no longer to function as intended under the original legislation."[17] The commission was concerned that despite the stiff tariff, some U.S. shipowners were getting their ships repaired abroad and paying the tax. What the commission ignored was that U.S. repair prices had risen so high that in some cases the net cost of having the work done abroad and paying the tax was less than having the work done in a U.S. yard. The commission recommended that the tax be raised to 100 percent.

The commission was also persuaded, if indeed persuasion was needed, that more cargo protection was necessary. It accepted the union officials' proposal to extend the Jones Act so as to require "the use of United States flag tankers for any voyage beginning and ending within the United States 200 nautical mile Economic Exclusion Zone," and it recommended that the U.S. Virgin Islands be brought under the Jones Act.[18] It also recommended that existing cargo preference programs be expanded to include all government or government-impelled cargoes. In 1986 the percentage of cargoes subject to cargo preference laws had been increased from 50 to 60 percent; the percentage for agricultural commodities was scheduled to increase further to 75 percent by April 1988. The commission was not satisfied and boldly called for 100 percent cargo preference protection.

The commissioners also considered the three-year waiting period that was imposed on foreign-built ships after they were transferred to U.S. registry before they could qualify to carry preference cargoes. They acknowledged that bringing such ships under U.S. registry would add modern, efficient ships to the U.S.-flag fleet, but they were worried that it would do so at the expense of the U.S. shipbuilding industry. They therefore proposed a linking formula that would make their new construction program more palatable: "The Commission suggests that eligibility to carry preference cargo be given to new foreign-built ships in proportion to the total number of oceangoing merchant ships under construction in any year in United States shipyards, including those ships built under the 'Procure and Charter' program."[19] Recommendations such as this made quite obvious where the commission's primary concerns lay.

The commission kept working for another year. It held additional hearings, and many witnesses who had appeared previously returned for a second round of testimony. Having heard a variety of opinions, the commission was now primarily interested in gathering information that would support its final recommendations.

The $13 billion Procure and Charter program was the centerpiece of the commission's recommendations; it was also the shipbuilders' number one priority. The program obligated the government to design, finance, and finally purchase the ships that were to be built. The program might have been justified commercially had it been possible to show that the federal funds would prime the shipbuilding pump by giving shipbuilders the capital and initial orders they needed to get up to speed for commercial production. But this would not have been an easy argument to sustain and the commissioners avoided it. Unconcerned with commercial justifications, they viewed additional subsidies and protection as essential to the continued existence of the U.S. merchant marine. The proposed building program was to them just one more public expenditure justified by national need.

The commissioners did include a set of revenue projections that indicated there would be no net cost to the government for the proposed program, but their estimates were unconvincing. They asserted, without any substantiating economic testimony, that the building program would add $81 billion to the gross national product over the course of eleven years and that the federal tax revenues on this increase would be some $13 billion. Suddenly, as if by magic, it appeared that a massive new subsidy for the maritime industry would cost the government nothing! There are occasions when government investment in industry is justified, but the facts of each case must be carefully examined. There is good reason to think that an investment of $13 billion in other industries, rather than the merchant marine, would have produced a far greater increase in national economic activity and tax revenue, but the commission did not examine this possibility. Nor did they look at the cost per job for subsidies to shipbuilding, as opposed to other industries. Given the narrowness of the maritime industry's focus and the extravagance of its claims, it is hardly surprising that OMB insisted on a more active role in setting federal maritime policy.

Early in 1989 the *Journal of Commerce* published the commission's obituary:

> The U.S. flag merchant fleet has continued to shrink in recent years, due to high costs, stiff foreign competition and the trend toward operating fewer but larger vessels. Past government policies, which subsidized inefficiencies and left carriers that had accepted federal construction aid bound by bureaucratic rules long after the aid ceased to be available, have contributed to the fleet's decline.
>
> The commission's report ignores these and other areas in which relatively inexpensive changes in government policy could make a difference. By failing to address the commercial viability of U.S. shipping and offering no politically feasible program, the Commission on Merchant Marine and Defense has ensured that its report, like the many that have come before it, will gather dust.[20]

In April Congressman Bennett, author of the legislation that created the commission, held hearings on a bill that would turn the commission's recommendations into law. The shipbuilders gave the bill strong support, but they were alone in doing so. Representatives of shipowners opposed the legislation and pointed out its many flaws. How would foreign shipbuilders react, they asked, when they saw U.S. shipbuilders simultaneously criticizing the modest subsidies they were receiving while seeking massive new government subsidies for themselves? The delusions of the shipbuilders found little support in Congress, and the commission and its recommendations soon expired. Having been captured early on by vested interests, it failed to resolve in any way the problems it was created to address.

The Bush Administration's Maritime Initiative, 1988–1992

By the time the *Journal of Commerce* announced the commission's demise, George Bush had been inaugurated as the forty-first president. Unlike his predecessors, Nixon and Reagan, Bush had made no promises to the maritime industry while campaigning for election, and the new administration gave no indication that it would reverse earlier decisions to cancel construction differential subsidies and phase out operating differential subsidies. This put the shipbuilders in a bind. The cost of building a ship in the United States was fast approaching twice the world price, and there was little likelihood of reviving direct subsidies large enough to cover this difference. Although the proposed Procure and Charter legislation had not yet died, its prospects were also dim. The shipbuilders therefore needed a new strategy; they found it in the subsidies that foreign nations provided to their shipyards. Acting through their lobbying organization, the Shipbuilders Council of America (SCA), they filed a petition under section 301 of the amended 1974 Trade Act, charging that Norway, Germany, Japan, and Korea were excluding the America shipbuilder from world competition by providing subsidies to their shipyards.

This charge of discrimination came with a long pedigree. After the Civil War, when U.S. shipbuilders could no longer find foreign buyers for their wooden ships, they attributed their inability to compete to the high cost of iron and steel in America. Their claim had some merit at the time, but the long-term effects of U.S. withdrawal from the world market were far more significant. Even after the price of American steel became competitive in the early years of the twentieth century, American ships were still priced above the world market. The reason given at that time, again with some validity, was the high cost of American labor. By the 1980s, however, it was evident that the hourly wage rates in American shipyards, but not labor productivity, were essentially the same as those in Great Britain and Japan and were well below those in the Scandinavian countries, in the Netherlands, and in Germany. Needing a new explanation for their lack of competitiveness, the shipbuilders seized on foreign subsidies. Congress, pressed by the political muscle of the big shipyards and happy to attribute the source of the problem to agents beyond their control, never seriously examined the shipbuilders' complaints of unfairness. The shipbuilders' case was soon being powerfully argued in Congress by their champion, Representative Sam Gibbons, member of the House Ways and Means Committee and chairman of its Sub-Committee on Trade.

The Federal Trade Commission, a nonpolitical arm of the federal government, hears claims of unfair international competition; if it finds them valid it can order punitive action. When it examined the shipbuilders' claims, the commission determined that the difference between U.S. ship-building prices and average foreign prices was approximately 96 percent. How much of this difference could be attributed to foreign subsidies? The commission concluded that if all foreign subsidies were eliminated, world shipbuilding costs would rise only 5.9 percent. In short, American shipyard costs were so out of line that "the elimination of subsidies was unlikely to make U.S. shipyards competitive."[21] Undaunted by the commission's findings, the shipbuilders pressed on and eventually gained House approval of Congressman Gibbons's bill to retaliate against any country that provided subsidies to its shipyards. Although the Senate later declined to pass the bill, it remained alive and would reappear again.

The ship operators, like the builders, also developed new subsidy proposals. In February 1992, Sea-Land and American President Lines (APL) jointly announced that they would apply to transfer most of their U.S.-flag fleet to open registry unless existing government restrictions were lifted and funds were provided to equalize U.S. crew costs with foreign competition. They also outlined a plan that would enable them to remain competitive as American operators. Their plan called for the following changes:

1) Taxation
   a. Allow use of the Capital Construction Fund (CCF), as provided by the Merchant Marine Act of 1936, for acquisition of foreign vessels and equipment for operations under the U.S. flag.
   b. Provide a five-year write-off (depreciation) of vessels on an accelerated basis. Eliminate treatment of both depreciation and CCF deposits as tax preference items for alternative minimum tax purposes.
   c. Eliminate the fifty percent ad valorem duty on foreign repairs of U.S. vessels.
   d. Exempt from income taxes crew wages on voyages in foreign commerce.
2) Government-Impelled Cargo
   a. Develop a pre-negotiated procurement system for use in contingencies. Ensure that compensation adequately covers actual costs on an ongoing basis (for existing military cargo).
   b. Eliminate the provision . . . that requires a three year waiting period before foreign-built U.S.-flag liner ships are qualified to transport preference cargo. . . .

      c. Broaden the definition of U.S.-flag service to include foreign flag feeders. . . .

3)   Vessel Design Standards

Adopt U.S. Coast Guard design and operation standards that impose no greater cost and compliance burdens on U.S.-flag vessels than are applied to other vessels that call at U.S. ports.

4)   Measuring Productivity

      a. Adopt standards for crew size that are competitive with foreign-flag service.

      b. Develop additional approaches to productivity, such as extending the number of months a mariner will be employed actively each year.

5)   Merchant Marine Ready Reserve

Institute a federally funded program to train American seamen for service on U.S. vessels to ensure the availability of seamen . . . in time of emergency.

6)   Re-flagging Requirements

If the integrated package of reforms . . . is not adopted, then re-flagging of U.S. owned and operated fleet should be allowed beginning in January 1995, upon six months notice. During the notice period, the government would have the option to purchase (the companies') ships.[22]

The companies had fired off a policy broadside that finally caught the administration's attention.

Although the Bush administration did little for the merchant marine during its first three years, Andrew Card, President Bush's last secretary of transportation, made a serious effort to develop a new program and establish a consensus within the administration. His was the most impressive effort undertaken on behalf of the American merchant marine in recent years.[23] He drew attention to the serious decline in the country's shipping assets. The United States, he noted, ranked sixteenth in the world, with only 393 privately owned merchant vessels. He projected that by the turn of the century this fleet would shrink to 117 ships if no further action was taken. The only ships that would remain under the U.S. flag would be overage and sailing with Jones Act protection. In 1960 there were 100,000 active merchant seamen; by 1992 there were only 27,000, and a further drop of a third, to 18,000, was projected for the year 2000. Card hoped to halt this decline with a program containing a number of the items proposed by Sea-Land and APL. Getting agreement within the administration

	Total		Privately Owned		Government Owned	
	Number of Ships	DWT*	Number of Ships	DWT*	Number of Ships	DWT*
Active Fleet						
Passenger	5	48	1	7	4	41
General Cargo	23	366	18	313	5	53
Intermodal	117	3,503	117	3,503	–	–
Bulk Carriers	12	536	12	536	–	–
Tankers	134	8,439	133	8,422	1	17
Total	291	12,892	281	12,781	10	111
Inactive Fleet						
Passenger	10	91	2	20	8	71
General Cargo	109	1,513	4	45	105	1,468
Intermodal	45	1,097	2	49	43	1,048
Bulk Carriers	3	76	3	76	–	–
Tankers	44	2,513	17	1,644	27	869
Total	211	5,290	28	1,834	183	3,456
Total Fleet						
Passenger	15	139	3	27	12	112
General Cargo	132	1,879	22	358	110	1,521
Intermodal	162	4,600	119	3,552	43	1,048
Bulk Carriers	15	612	15	612	–	–
Tankers	178	10,952	150	10,066	28	886
Total U.S. Flag	502	18,182	309	14,615	193	3,567

Table 13.1: U.S. Oceangoing Merchant Marine (July 1, 1996).
*Tonnage in thousands of deadweight tons. *Source:* Data from the Maritime Administration, "U.S. Merchant Marine: Data Sheet," July 1, 1996.

required compromise, however, and the bill that emerged fell far short of Sea-Land's and APL's needs.

Card's proposed program addressed the fact that many foreign shipowners pay no taxes, since their laws allow them to place profits earned from ship operations in tax-deferred accounts that can be used to improve existing ships or acquire new tonnage. Although there is a similar tax shelter in American law, the funds can only be used for domestic building. But since U.S. building costs are so high, this provision is essentially worthless, unless a building subsidy is provided, except for the occasional ship built for the coastal trades. Card's program would have allowed these funds to be used for building U.S.-flag ships anywhere in the world.

Card's program also would have phased out the ad valorem duty on foreign repairs over a five-year period, but this was a tax item, and the Treasury Department would have to concur. The Treasury, however, objected to providing any general tax relief to the shipping industry.

Hoping to avoid raising other issues that would require negotiation, Card did not include any proposals bearing directly on military matters. There appeared to be little interest in creating a Merchant Marine Ready Reserve to ensure there would be enough trained personnel to man the reserve fleet. Later, in 1993, when MARAD included the establishment of such a reserve in its budget request, it was eliminated by OMB. Card also steered clear of the Coast Guard, although it was a part of the Department of Transportation, and made no mention of eliminating the unnecessary and costly regulations imposed on U.S.-flag vessels.

Card also sought to remove the requirement that owners had to obtain government approval to transfer a U.S.-registered ship to a foreign flag. He tempered this change by stipulating that it would only apply to ships that "are not militarily useful." Even more controversial in the Card program was the most radical proposal, eliminating the requirement that ships must be built in American shipyards to qualify for most government subsidy and cargo preference programs.

Once again the shipbuilders demonstrated that if their interests were not taken care of, maritime legislation was going nowhere. However reasonable Card's proposals may have seemed, legislative approval would be blocked unless the provisions of Gibbons's bill were included in the administration's package. When Card's bill was sent to the Senate, the shipbuilders had it revised so that their complaints about foreign subsidies could be heard by the Federal Maritime Commission (FMC), which they hoped would be more supportive than the Federal Trade Commission. This was how Gibbons's bill, which had failed to obtain Senate approval at an earlier date, was brought back to life.

The final provision of the Bush administration's maritime initiative called for creating a contingency retainer program designed to offset the wage differential between U.S. crews and foreign crews. The new maritime subsidy was patterned on the air force's Civilian Reserve Airlift Fleet (CRAF) program, which provided an annual payment for a certain number of U.S. aircraft in return for a commitment to make them available to the government in time of emergency. The attempt to copy a successful program was laudable, but the proposed maritime program had serious flaws. Years earlier the Military Sealift Command (MSC) had insisted that a willingness to make ships available to the government in time of emergency was a condition for bidding on military contracts, so most of the U.S.-flag operators concerned had already made the central commitment called for by the new program. The cost of the new program also appeared excessive. The initial payment was to be $2.5 million per ship for the first two years, after which it was scaled back to $1.6 million in the final year.

The congressional reaction to Card's proposals revealed that support for maritime subsidies was approaching a terminal state. Unwilling to authorized open-ended programs, Congress insisted that the total cost of the program be funded up-front. This required DOD support, but DOD was unwilling to commit to what appeared to them to be a subsidy to private industry, and Secretary Card found himself unable to earmark the necessary funds in the time available. The program was expensive—it was estimated that providing subsidies for seventy-four ships over the seven-year life of the program would cost $1.1 billion—and the linkage to national security was not established in the public's mind. There was considerable talk of sustaining the manpower base needed to provide crews for the Ready Reserve Fleet, but there was no sense of urgency. The program was further damaged while being debated within the administration when a memorandum signed by the assistant secretary of defense for Production and Logistics was circulated. According to this official, even "in the most demanding scenario" the armed forces did not need the APL and Sea-Land ships for "surge shipping" in any future contingency. That being the case, he continued, "the issue of the two major U.S.-flag containership operators disposing of their U.S.-flag fleets is primarily an economic issue rather than a national security issue, and should be treated accordingly."[24] The divorce between maritime subsidies and military preparedness now appeared to be final, eliminating in a stroke the most persuasive and most frequently cited reason for using public funds to maintain a U.S.-flag merchant marine.

During the twelve years of the Reagan and Bush administrations nothing was done to halt or reverse the continued decline of U.S. merchant shipping. There had in fact been little significant legislative reform since 1970. While congressional committees were willing to hold hearings that provided forums for those wishing to express their concern for the plight of the industry, nothing was accomplished. In the end both administrations made it quite clear that they had little or no interest in providing support for what they considered to be a largely discredited system.

When Secretary Card attempted to formulate a semblance of a meaningful maritime program, the White House offered no serious support and most of the departments within the Administration, including OMB, were either hostile or showed no interest. Having failed to find allies outside the Department of Transportation, Card's proposals quickly disappeared in the rough-and-tumble of the legislative process.

The Commission on Shipping and Defense produced a catalogue of proposals that included just about every protectionist and subsidy scheme that had been proposed in recent decades. If they omitted any it was inadvertent and not readily apparent. What the commission's recommendations revealed most clearly, however, was that its members and the industry spokesmen they represented had utterly failed to appreciate the extent to which the political context in which the U.S. merchant marine operated had changed. The country was becoming more conservative, the political beliefs that had sustained the New Deal were fading, and "subsidy" had become a dirty word. Since the commission remained blind to these changes, it gave little thought to the political alliance building that would have been necessary to make their program acceptable and to see it through to implementation. The almost complete lack of a meaningful response to their efforts sent an unambiguous signal that peacetime support for sustaining the maritime industry was rapidly fading.

14

THE FINAL PUSH
TO EXTEND MARITIME SUBSIDIES,
1992–1999

Builders and operators of U.S.-flag ships had finally begun to appreciate the depth of opposition to traditional subsidy programs. As they reflected on the failed promise of the Reagan administration, the almost farcical results of the Commission on Merchant Marine and Defense, and Secretary Card's inability to carry forward his program, advocates of federal subsidies concluded that their only hope lay in working directly with appropriate congressional committees.

The builders decided to revive the argument that their problems arose entirely from the shipbuilding subsidies provided by foreign governments. It was these foreign government funds, they claimed, a source of funding no longer available to American builders, that made it impossible for them to compete in the world market. The American shipbuilders did not really expect foreign governments to yield to American pressure to eliminate their shipbuilding subsidies, rather they expected that a general unwillingness to eliminate subsidies would provide a basis for reviving shipbuilding subsidies in the United States. To their great surprise and considerable dismay, however, the Clinton administration succeeded in convincing foreign shipbuilders covered by the OECD to agree to eliminate direct subsidies for shipbuilding. This unexpected turn of events forced the American shipbuilders to reverse field suddenly and campaign against ratification of the commercial agreement that they had urged the administration to negotiate. This desperate campaign was a success, and worldwide shipbuilding subsidies continued as before.

The shipowners also continued their campaign for operating subsidies to cover higher U.S. crew costs. Although previous arguments for operating subsidies had been ineluctably linked to a requirement that subsidized vessels be U.S. built, it now began to appear that this ancient and burdensome linkage could be broken. The shipbuilders' flip-flop on the whole subsidy question had left them with little credibility. Thus, remarkably and without any explicit debate, it became possible to propose that operating subsidies be provided for American crews sailing on foreign-built ships. The elimination of this venerable impediment to a rational maritime subsidy program was indeed a step forward.

The Clinton administration made several attempts to develop a new and quite modest operating subsidy program. The effort was ultimately successful largely because of the relentless insistence shown by the shipowners and their maritime unions while working with congressional committees to obtain legislation that provided at least some stopgap assistance.

THE FIRST YEARS OF THE CLINTON ADMINISTRATION

Frederico Peña, a former mayor of Denver and President Clinton's choice for secretary of transportation, showed the unguarded enthusiasm of a "new boy" at his Senate confirmation hearing by promising to produce a comprehensive maritime program by the following month. His obvious unfamiliarity with maritime issues and with Washington politics was excusable, but his failure to retract ill-advised commitments made in public was unwise. Not realizing how little he knew about the subject, Peña pressed on.

In early May 1993 Peña presented his merchant marine program to the recently formed National Economic Council (NEC). The proposals developed by his staff, although not fully fleshed out, built upon those that Andrew Card had developed in the previous administration. The shipowners had already spoken with Peña, telling him they wanted a fifteen-year program covering a greater number of ships than would have been covered under the seven-year program Card had proposed. They were apparently unfazed by asking for a program that would cost $5 billion immediately after Congress had refused to fund the $1.1 billion program Card had submitted. When asked to comment on Peña's proposals, the Department of Defense requested more time before deciding whether his program was required for national security. When it was realized that Peña's maritime program was actively opposed by OMB and the Department of Agriculture and lacked support from any branch of government other than the Department of Transportation, its fate was clear. His maritime proposals expired without ever being presented to the president.[1]

By midsummer the administration finally got around to sending the name of the new maritime administrator to the Senate for confirmation. He was Vice Admiral Albert J. Herberger (U.S. Navy, retired), and he later proved to be unusually effective in persuading various elements in the administration to support, or at least not oppose, the maritime legislation being drafted by congressional committees. Secretary Peña, having been bruised while defending the ill-advised maritime proposals generated by his own staff, was increasingly willing to turn management of maritime issues over to his experienced maritime administrator.

Long-smoldering agricultural resentment toward PL 480 cargo preference requirements broke into flame about the time that Pena was proposing new legislation. When the president decided to send millions of tons of grain as aid to Russia, $150 to $200 million of the funds allocated for the program was set aside to pay the added cost of shipping the grain on American carriers. A number of midwestern farm-state senators cosponsored a bill that would have limited U.S.-flag cargo preference whenever the rates charged exceeded market rates. Senator Henry Brown of Colorado gave vent to the anger and exaggeration that swirled around this issue when he spoke on the Russian grain proposal:

> Greedy shipowners faced with the ability to corner the market because of this (U.S. preference) law have not only demanded 50% more than the world market rates but have been bidding 200% more, and 300% more, and 400% more.
>
> Most Senators are going to be shocked to find that the bids on the Russian grain exports are now almost five times the world market rate. This is the kind of greed and corruption the American people are demanding to be changed.[2]

Brown greatly overstated the case, but clearly the PL 480 program was generating intense frustration. Even senators who were well disposed toward the maritime industry wondered why a 50 percent freight-rate differential would not suffice, and the carrier industry once again had a serious public relations problem on its hands.

When it became clear that Secretary Peña's initial maritime program was going nowhere, Sea-Land and American President Lines (APL) again announced they were determined to transfer their U.S.-flag ships to foreign registry. Sea-Land, which already had forty-three ships under foreign flag, applied immediately to transfer thirteen more of its most modern U.S.-flag vessels to the Marshall Islands Republic registry. The company claimed that if required to employ U.S. seamen, it would cost them an extra $2.5 to $3 million per year for each ship to make them competitive. In the absence of a subsidy the company could not ask its shareholders to continue to assume that burden. APL filed a similar application to transfer seven ships to foreign registry, even though it continued to operate as a subsidized carrier. APL also announced that it had signed contracts with foreign shipyards to build six ships having a total container capacity of 28,000 TEUs and that it planned to place these ships under foreign registry.

Sea-Land's announcement incensed the maritime labor unions, and they turned immediately to their friends in Congress for an appropriate response. The chairman of the House Merchant Marine and Fisheries Committee was persuaded to amend the Maritime Administration's Appropriation Bill to include a two-year prohibition on the transfer of any ship from U.S. to foreign registry. The House passed the amended bill by a large margin on July 29. The Senate, however, was not prepared to go along. John Breaux, chairman of the Senate Merchant Marine Sub-Committee, let it be known that his committee would not support such a measure. The fault lay with the administration, he said, not the companies, and the legislative remedy proposed was probably unconstitutional.

In mid-August the Defense Department announced its decision on the military need for merchant shipping. The department's position was described by William Lynn, director of DOD's Office of Program Analysis and Evaluation, in testimony before Senator Breaux's committee. DOD had concluded that the funds it had available for logistic support could best be spent obtaining ships capable of quickly moving supplies and equipment in time of war. The existing ships in the merchant marine, they said, did not have that capability. Senator Breaux was not pleased and spoke with a frankness seldom heard in maritime-military policy discussions. "It seems to me," he told Lynn, "that you want ships available to you to use but you don't want to pay for them." Breaux later told reporters that the military had had a "free ride for years," but that it was about to end because soon there would not be any ships.[3]

Maritime policy seemed to be entering a meltdown. In the absence of effective Department of Transportation leadership and with U.S. companies preparing to transfer their ships to foreign flag now that there were no more subsidies to be had, individual congressmen began drafting and submitting bills to extend operating differential subsidies and provide other benefits, such as tax relief. Maritime policy was still dominated by a small group of players, however, and none of them suggested changing U.S. building requirements, altering or eliminating the Jones Act, or revising the cargo preference laws. Legislators moved in directions they believed would please their constituents, but without backing from the administration, the programs they proposed would remain unfunded.

SHIPBUILDING SUBSIDY NEGOTIATIONS WITH THE OECD

The shipbuilders found themselves in a similar situation. For a while it seemed that Gibbons's bill, which would penalize foreign shipowners who purchased ships built with subsidies, was on track. It had passed the House in the previous administration and was being prepared for a second vote. Rufus Yerxa, the new deputy U.S. trade representative, told Sam Gibbons's committee that President Clinton, like President Bush, preferred to seek elimination of foreign shipbuilding subsidies through negotiation within the OECD. He assured the committee, however, that if negotiations failed, he was prepared to work with Congress to achieve its objective by other means.

That was in 1993. The following year, much to the surprise of many and the obvious chagrin of the largest shipbuilders, a milestone agreement was reached with the OECD. While it was not perfect, the agreement would end most, if not all, direct government support for shipbuilding. It would require all government loan programs to conform to the existing international standard of twelve-year, 80 percent guarantees. The U.S. Title XI program of twenty-five-year, 87.5 percent loans, which was obviously more attractive, would have to be scaled back. In an effort to accommodate political considerations, the United States was allowed to retain the provision in its cabotage laws that requires all ships in those trades to be not only U.S. flagged and U.S. crewed, but also U.S. built. No other country was afforded this protection. It should be noted that the U.S. government regularly takes other nations to task when they provide this kind of protection for any of their industries.

Before the OECD agreement was reached, the Clinton administration had implemented a transitional shipbuilding program to help the industry manage major cutbacks in naval orders and adjust to the commercial market. The new OECD agreement would require that the generous subsidies the shipbuilders were receiving be modified. The shipbuilders could no longer have it both ways. For years they had portrayed themselves as champions of free enterprise, claiming that the only reason they could not participate in the international commercial market was the existence of foreign government subsidies. But the playing field had suddenly been leveled, and when they found themselves faced with the possible loss of their government support, they rapidly decided that subsidies were not so bad after all. The important question turned out to be who was getting them.

The pending global agreement was to go into effect on January 1, 1997, providing all the countries involved approved the accord. By 1996 all nations except the United States had ratified the agreement. Early in the year the Clinton administration pushed Congress for ratification, but the administration's interest drained away as the presidential elections drew nearer and opposition by the major shipbuilders and their labor unions grew more intense. The shipbuilders themselves had spilt over the issue. The smaller yards, which were generally far more productive, favored adoption of the OECD agreement; the six largest yards, which built primarily for the navy, split off to form their own lobbying organization to oppose ratification.

The six big yards wield considerable political clout. Avondale Shipyards is the largest industrial employer in Louisiana. Bath Iron Works is the largest employer in Maine. General Dynamics owns both Bath Iron Works and Electric Boat in Connecticut and was prepared to deploy its major lobbying organization to reinforce the shipbuilder's effort. Ingalls Shipbuilding, Mississippi's largest employer, is a division of Litton Industries, another powerful defense contractor. National Steel and Shipbuilding is located in San Diego. Because of the importance of California's electoral votes, it has a substantial voice in any election year. Newport News Shipbuilding of Virginia is the largest shipyard in the country. The senators from these states had little difficulty persuading most of their colleagues to avoid bringing the question of ratifying the OECD agreement to a vote, and it died in the waning days of the 104th Congress.

When he introduced legislation to implement the international agreement, Senator John Breaux of Louisiana, the only senator from a shipbuilding state to champion the smaller, more efficient builders, said: "The new shipbuilding agreement negotiated through the OECD, while not perfect, appears to be our last best chance to eliminate unfair subsidies, to counter injurious pricing policies, to rein in trade distorting export financing and to institute an effective binding dispute settlement system for shipbuilding controversies." Senator Breaux concluded by saying: "We stand before a window of opportunity for the U.S. commercial shipbuilders. The $265 billion commercial shipbuilding market is fast approaching its cyclical peak. I hope we will seize this moment and implement this agreement. It may be our best and only chance to end foreign shipbuilding subsidies and finally give our workers and yards the level playing field for which they have asked for too long."[4]

The spokesman for the American Shipbuilding Association, the big yards' new lobbying organization, made the group's priorities quite plain. The first was to kill the OECD agreement. Their other initiative was to convince Congress and the public that the naval shipbuilding base should be preserved. It seemed unlikely, however, that the level of funding would be enough to maintain the base as they defined it. A related effort was to try to prevent the navy from buying any foreign-built ships for future logistic support. Noticeably missing from this game plan was any mention of the previously much-touted commercial shipbuilding effort.

Forced to acknowledge the unlikelihood of reestablishing a direct subsidy program for shipbuilding, several of the major companies began looking for ways to build for the commercial market without such assistance. The key to this new initiative was to be modification of Title XI of the 1936 Merchant Marine Act, which authorizes federal mortgage guarantees for shipbuilding. The provisions that rendered Title XI particularly useful at this time were contained in the National Shipbuilding and Shipyard Conversion Act, which became law in November 1993. As the title of this law indicates, it was designed to aid the builders at a time when naval contracts were being cut back. One of its provisions allowed Title XI mortgage guarantees to be used, for the first time, to insure the debt on vessels being built for export. Newport News, the nation's largest shipyard, moved quickly to take advantage of this new source of financing and began planning a forty-thousand-DWT product tanker that could be built in series production. Financing of this initiative was also aided by a $3 million grant from the navy.

When planning for these new tankers had been completed, the Eletson Corporation, a Greek company, placed an order for the first four. The price of $38 million for each ship was only a few million above the world price, a difference which was partially offset by the Title XI financing, which was considerably better than that available on the open market. To help get this project underway, Tenneco, the parent company of the shipyard, invested $118 million in capital improvements.

Newport News had not built a ship for the commercial market for twenty years, a fact whose consequences were not long in becoming apparent. Costs soon began to exceed original estimates, and Tenneco noted in its 1996 year-end financial statement that it would take a $31 million reserve "to cover estimated contract losses related to construction of commercial product tankers."[5] Early in the construction process it was found that the prefabricated sections of the ships did not fit together precisely. This was not unprecedented. Years earlier Litton Industries had encountered a similar problem after it purchased the Pascagoula Yard and announced it would show the world how to build ships using modern aircraft assembly systems. Their prefabricated sections did not fit on the first ships built for Farrell Lines, and before long Litton gave up the effort and returned exclusively to naval building.

Newport News obtained a second order for five ships from the Van Ommeren/Hvide Group. These ships were designed for the Jones Act trade and will meet the double-hull requirements of the federal Oil Pollution Act (OPA) of 1990. These ships will be much more sophisticated than those built for Eletson and are expected to cost in excess of $50 million each. The federal Title XI program is guaranteeing 87.5 percent of the debt on this project, but to secure the order the Newport News Shipyard had to take an equity position in the venture as well.

In December 1996 Tenneco spun off the Newport News Shipyard to its stockholders, a move that was probably motivated by the end of the glory days of naval construction. The shipyard's ability to build good ships was never in doubt, but its ability to do so and make a profit was less certain. For many years a bronze plaque at the entrance to the Newport News Shipyard proudly announced, "We will build good ships here; at a profit if we can; at a loss if we must, but we will build good ships." Thirty years ago, when Tenneco bought the shipyard, they had this plaque moved to the nearby Mariners Museum. While it is unlikely that the plaque will be returned, the commitment to building good ships remains firm. In March 1998 the company's chairman, William Fricks, announced that the shipyard was withdrawing from further commercial building. The four-ship order to Eletson Holdings was reduced to three, with the first delivery being to Mobil Oil for domestic use. The other order, for five ships, was reduced to two vessels to be taken by Hvide Marine Incorporated. This decision was made, Mr. Fricks explained, because the shipyard would have lost $150 million on the original venture.[6]

The failure of the United States to ratify the agreement that would have outlawed commercial shipbuilding subsidies had not killed this eminently sensible proposal. The European Union, Norway, Japan, and South Korea continued their endorsement and were determined to go ahead without United States agreement. After all, America represented no more than 1% of the world's commercial shipbuilding production. As orders for U.S. built ships continued to decline, existing yards proceeded to consolidate.

RENEWED ATTEMPTS TO ENACT OPERATING SUBSIDIES

Early in the new administration Vice President Gore headed a National Performance Review formed to identify ways to make government more efficient. The committee's preliminary report on maritime issues was deliberately leaked and immediately aroused hostile public debate. The report called for eliminating just about every law or regulation that had provided subsidies and protection for the previous fifty years. It recommended Congress deregulate the maritime industry by

1. Fostering competition by not extending operational differential subsidies beyond their current contract expiration dates.
2. Disallowing the proposed stimulus package for Title XI and shipbuilding.
3. Opening the U.S. flag to foreign built and repaired ships [and] foreign investors and foreign crews.
4. Increasing foreign competition on U.S. domestic trade by eliminating the Jones Act and the Passenger Act of 1986.
5. Reducing the costs of sending government shipments overseas by eliminating cargo preference.
6. Eliminating the shipping conference exemptions under antitrust immunity managed by the FMC.
7. Eliminating the tariff filing requirements of ocean liners with the FMC.[7]

Such recommendations provoked the shipping industry into an immediate counterattack. A labor leader was reported to have said, "[I]t is so absurd on its face, and so inimical to U.S. interests, that it cannot be taken seriously." A shipping company executive was quoted as saying, "[T]he authors have no sense of the economic and political realities of shipping. Just look at those crazy recommendations."[8] But not everyone found the recommendations "absurd" or "crazy." Karla Corcoran, the report's author, was a deputy assistant inspector general for Audit serving on loan from the Treasury Department to Vice President Gore's committee, and she had had the fortitude to think the unthinkable and give voice to the unspeakable. It appeared that mere dismissal of critical observations made by outsiders no longer constituted an adequate defense of badly outdated maritime policies.

The loudest cries came from maritime labor leaders who had given President Clinton considerable support during the recent electoral campaign. Their voices could not be ignored. They requested a meeting with the president to discuss the recommendations that they found so offensive and were assured that that part of the Gore report would not be implemented. Shortly thereafter the president proposed a new operating subsidy program called the Maritime Security and Competitive Act. It was to be funded by a tax on imports and exports, a dubious proposition at a time when the U.S. trade deficit was soaring to all-time heights. The proposed program's only merits were political, and after lingering in the wings for a while it disappeared.

ATTEMPTS TO REDUCE SHIPPING RESTRICTIONS

The Reagan and Bush administrations had tried to modify certain aspects of existing maritime regulation, but their attempts to eliminate some of these restrictive laws and regulations had not been well received. Although both presidents were prepared to continue some form of operating subsidy, opposition to the changes they proposed was so intense that their initiatives died in Congress. In the program proposed by the Clinton administration, national defense was to justify new operating subsidies. But the Clinton proposal also contained three important regulatory changes.

The first was to eliminate the requirement that a company receiving subsidy be committed to serve a particular trade route. Containerization had rendered the concept of established trade routes irrelevant. The two major U.S. carriers, APL and Sea-Land, were already serving a global market with feeder ships that connected their hub ports to secondary ports and, through a growing number of interchange agreements, with other carriers. The second change was of much greater significance: the Clinton program would allow foreign-built ships to be registered under the U.S. flag and qualify for subsidy. The shipbuilders, following the abortive efforts of the Commission on Merchant Marine and Defense, had reluctantly concluded there was no hope of commanding the level of funding necessary to enable them to produce ships at competitive prices. The shipowners in turn had made it quite plain that they could no longer afford to subsidize the U.S. shipbuilder.

The ultimate failure of the Construction Differential Subsidy (CDS) system introduced in the Merchant Marine Act of 1936 deserves further examination. Even when the maximum allowable subsidy of 50 percent was paid, the shipowner was the loser. Why this was so can be illustrated by briefly examining the costs incurred by APL in the early 1980s when the company acquired the last three ships built in the United States for foreign trade. Construction had been held up for almost a year while the navy tried to force the company to have steam turbines installed instead of the slow-speed diesel engines that had been specified. The navy could do this because to qualify for CDS, the ship's design had to be approved by the navy. At the time this dispute occurred ship's bunkers cost $25 to $30 per barrel. It would have added approximately $6 million more in fuel cost per ship each year if the ships had steam rather than diesel propulsion. While APL had agreed to upgrade their fleet when signing earlier subsidy contracts, they could not be obliged to commit commercial suicide. The navy eventually relented, but, as the following table indicates, that was not the end of the trouble associated with building these ships in the United States.

	United States	Foreign
	(in thousands of dollars)	
Shipyard price	120,000	42,000
Construction Differential Subsidy (CDS) (50%)	(60,000)	–
Construction interest (net)	6,000	1,000
Net capital investment	66,000	43,000
Financed investment	49,500	34,400
Equity investment	16,500	8,600
Interest rate	14%	8.75%
Financing term	25 years	8.5 years
Annual depreciation	3,300	2,150
First year interest	6,860	2,920
First year capital cost, book basis	10,160	5,070
Present value of equity investment plus debt payments, cash basis (14% discount)	66,000	37,510

Table 14.1: Comparative Capital Cost of New 2,500-TEU Containership: United States and Foreign (1980).
Source: Data from Andrew E. Gibson, "Elements of a U.S. Shipping Policy," in The U.S. Merchant Marine in Search of an Enduring Maritime Policy, ed. Clinton H. Whitehurst (Annapolis: Naval Institute Press, 1983), 95.

The cost differential between foreign and U.S. shipbuilding prices had risen so high because U.S. builders had to charge enormously high prices to make even a modest profit. When the construction differential subsidy came within a few million dollars of covering the entire difference in construction costs, and when U.S. interest rates were close to foreign rates, the shipowner could afford to absorb the uncovered cost difference. But these conditions no longer pertained in the 1980s, and the owners could no longer afford to make up the difference. If APL had tried to replace its entire fleet at the construction costs that prevailed in the 1980s, the company would have gone bankrupt.

Indeed, the situation was even worse than table 14.1 indicates. Since it took twice as long to construct a ship in a U.S. yard, the capital that APL had tied up in progress payments during this extended period was providing no return. After the ships were delivered, hull insurance on a vessel that cost $120 million to build in a U.S. yard would be considerably higher than on a comparable vessel that cost $42 million to build in a foreign yard. APL survived in these adverse conditions by taking full advantage of the 1981 Reconciliation Act window for foreign building and by making extensive use of foreign-flag charters and space-sharing arrangements on the ships of their foreign-flag associates.

The third change, one that drew little attention during the hearing on the new legislation, may prove to be the most important of all. The legislation stipulated that once a company had offered a ship for the Maritime Security Program (MSP) and it had been rejected, the company had the immediate right to dispose of that vessel as it saw fit. One of the available options was registering the ship under a foreign flag. This new flexibility meant that a company's decision to divest itself of its American-flag tonnage would no longer be vulnerable to political reversal.

The story of how the Operating Differential Subsidy (ODS) system collapsed is also instructive. The subsidized liner companies had accepted crew costs that were completely out of line not only with comparable foreign-flag ships, but with nonsubsidized American companies that chartered their ships to MSC or competed for PL 480 cargoes as well. Comparative data was readily available, but it seemed the Maritime Administration officials who approved ODS payments had never applied the "fair and reasonable" standard to these payments.

Sea-Land's situation was unusual. Although a nonsubsidized company, Sea-Land had long operated with the same crew costs as the subsidized operators. When the company was young it was highly profitable and could afford to pay high wages, and doing so seemed preferable to fighting the unions and risking a possible strike. In later years the company had been able to operate profitably because it had a large foreign-flag fleet; well-established Jones Act operations to Puerto Rico, Alaska, and Hawaii; and considerable involvement in carrying military cargoes. Its determination to shift many of its U.S.-flag ships to foreign registry was in part a response to the decline in military cargoes.

While George Bush was president the Maritime Administration commissioned two professors at the Massachusetts Institute of Technology to undertake a study of comparative crewing costs. A subsidized major U.S. carrier was compared to a nonsubsidized U.S. company, a German company employing a German crew, and a ship flying the Bahamian flag with an international crew. The results were so startling that the Maritime Administration refused to release the report. An error had been made in distinguishing between the direct wages and fringe benefits for the American captain when listing the crew costs for the U.S. subsidized operator. Although this figure was subsequently revised and the error had not invalidated the figure given for total monthly crew cost, those opposed to releasing the report seized on this flaw and successfully stifled objective consideration of the report itself. Citizens Against Government Waste, a taxpayer watchdog group, finally obtained copies of the report and publicized it. Although neither the public nor the Congress showed any particular interest, the cost differentials developed in the report are of such magnitude that they should not be ignored.

Tables 14.2 and 14.3 make possible some instructive comparisons. The base wage on a nonsubsidized U.S.-flag ship (table 14.2) is comparable to that on the European (German) ship (table 14.3), but the European wage includes overtime. American wages are inflated by overtime and benefits. This is the case because in U.S. maritime labor contracts rules relating to watch standing and work rules, while reasonable for shoreside factories, are unrealistic for seafaring. When one compares the wages paid on the Bahamian-flag containership and the large U.S. subsidized ship, the enormous disparities involved are evident. Many FOC vessels have higher costs than the ships that are shown but are still far below the lowest crew costs on American vessels.

Meanwhile, attacks on the bastions of maritime political power continued. In the fall of 1994, when the Republicans took control of the Congress, one of the first acts of the new Speaker was to eliminate the Merchant Marine and Fisheries Committee in the House of Representatives. For years that committee had been the cornerstone of the maritime industry's iron triangle. The other two corners of this controlling structure were the shipowning and shipbuilding industries, together with their labor unions, and the Maritime Administration. This combination had achieved much in the past. There were few members serving on the committee who were not there to take care of their particular maritime constituency; the same could be said for the Maritime Administration. The elimination of this committee indicated how much the industry's political influence had declined.

The Clinton administration, undeterred by the failure of its original subsidy program or the loss of the Democratic majority in both houses of Congress, reintroduced a renamed but essentially identical program called the Maritime Security Program (MSP). As already noted, foreign-built ships would again be eligible and restrictive trade route requirements would be eliminated; all other existing restrictions were left in place. The new proposal called for annual appropriations in the order of $100 million per year. The goal was to support a fleet of some forty ships by providing payments of $2.5 million per ship per year, dropping to $2.1 million after the first year. Shipowners supported by this subsidy would commit to making their vessels available to the government in time of emergency.

Crew	slots	U.S.-Flag Ro-Ro (unsubsidized)				slots	Large U.S.-Flag Containership (subsidized)			
		Total	Base	Overtime	Benefits		Total	Base	Overtime[a]	Benefits[b]
Master	1	16,963	7,042	1,492	8,429	1	34,116	10,989	1,858	21,269
Chief Officer	1	14,005	3,210	4,848	5,947	1	24,563	5,183	4,204	12,176
2d Officer	1	11,555	2,909	3,287	5,359	1	21,119	4,382	6,091	10,646
3d Officer	1	9,492	2,360	2,627	4,498	1	17,675	3,581	4,978	9,116
Radio Officer	1	10,480	2,913	2,621	4,946	1	18,281	4,064	5,649	8,568
Boatswain	1	7,129	1,952	2,206	2,971	1	11,416	3,237	4,4499	3,679
Able-bodied Seamen	3	17,613	4,032	5,607	7,974	5	44,915	12,192	16,947	15,777
A.B. Gen. Pur.	2	12,672	3,004	4,174	5,494					
Chief Engineer	1	16,297	6,810	1,492	7,995	1	26,831	10,249	1,728	14,854
1st Engineer	1	15,490	4,109	5,794	5,587	1	22,276	5,183	7,204	9,890
2d Engineer	1	14,184	3,723	5,249	5,212	1	18,512	4,382	6,091	8,039
3d Engineer	1	11,839	3,030	4,272	4,537	1	15,422	3,581	4,978	6,863
Electrician						1	10,553	2,790	3,878	2,885
Reefer						1	10,181	2,658	3,695	3,828
Oiler	2	12,255	2,958	3,846	5,451	1	9,427	2,419	3,362	3,646
Chief Steward	1	7,607	1,952	2,655	3,000	1	9,487	2,687	3,734	3,067
Steward #2	1	6,307	1,530	2,020	2,757	1	8,631	2,401	3,338	2,892
Steward #3	1	4,744	1,042	1,230	2,472	1	7,459	2,011	2,795	2,653
Totals	20	188,632	52,583	53,420	82,629	21	310,867	81,989	88,030	140,847

Table 14.2: Comparative U.S. Crew Costs: Monthly Subsidized versus Nonsubsidized (in U.S. dollars).
[a]Overtime is calculated using 139% of base wage with exception of the Master and Chief Engineer, who receive a payment in lieu of time off in port.
[b]Includes the cost of crew travel and relief officers.
Source: Data from Henry S. Marcus and Peter T. Weber, *Competitive Manning of U.S. Flag Vessels*, rev. ed. (Washington, D.C.: Maritime Administration, 1994), 56–59.

At the end of September 1996, during the final days of the 104th Congress, the Senate passed the administration's subsidy bill with overwhelming support. Despite this support, it seems inevitable that the program will at best slow the decline that is already well advanced. The program will make it possible for some American shipowners to continue absorbing the high labor costs associated with American operations, along

Crew	European-Flag Containership[a]					Bahamian-Flag Container[b]				
	slots	Total	Base	Overtime	Benefits	slots	Total	Base	Overtime	Benefits
Master	1	9,697	5,945	0	3,752	1	5,500			
Chief Officer	1	8,247	4,153	1,485	2,609	1	1,925	1,100	495	330
2d Officer	1	7,036	3,567	1,258	2,211	1	1,488	850	383	255
3d Officer	1	6,607	3,401	1,094	2,112	1	1,400	800	360	240
Radio Officer						1	1,488	850	383	255
Boatswain						1	1,114	600	33	180
Able-bodied Seaman	3	13,530	10,533	1,760	1,238	3	2,787	1,500	837	450
A.B. Gen. Pur.	1	3,259	2,317	323	619	1	669	360	201	108
Deck Cadet	2	3,704	2,467	0	1,238					
Chief Engineer	1	8,912	5,458	0	3,454	1	5,200			
1st Engineer	1	8,425	4,153	1,664	2,609					
2d Engineer	1	7,845	3,630	1,679	2,536	1	1,925	1,100	495	330
3d Engineer						1	1,488	850	383	255
4th Engineer						1	1,400	800	360	240
Electrician	1	7,586	3,914	1,237	2,435	1	1,488	850	383	255
Mechanic G.P.						3	2,786	1,500	836	450
Engine Cadet	2	5,601	3,027	1,474	1,100					
Chief Steward	1	7,619	3,660	1,673	2,286	1	1,114	600	334	180
Steward #2	1	1,775	796	358	621	2	1,337	720	401	216
Steward #3	1	1,456	652	296	508					
Totals	19	101,299	57,672	14,301	29,325	21	33,109	12,480	6,185	3,744

Table 14.3: Comparative Crew Costs: Monthly European versus FOC (in U.S. dollars).
[a]Total crew includes 4 Cadets.
[b]2 U.K. Officers, PHL Crew.
Source: Data from Henry S. Marcus and Peter T. Weber, *Competitive Manning of U.S. Flag Vessels*, rev. ed. (Washington, D.C.: Maritime Administration, 1994), 56–59.

with the costs of the government restrictions, but it contains few incentives for long term American-flag operations. Once the present, rapidly aging fleet has been retired, the owners may well continue the current trend toward foreign-flag replacement.

The Maritime Security Program authorized expenditure of $100 million per year for ten years, subject to annual appropriations becoming available. All the liner companies currently receiving Operating Differential Subsidy payments are included in the program and four new companies were added. The new companies are Sea-Land, Central Gulf, Crowley Maritime, and an American subsidiary of a Danish company, Maersk Line. A total of 47 ships are to be supported.

Early in 1997 Lykes Lines, which was in bankruptcy, announced that the company had been sold to CP Ships, the maritime division of Canadian Pacific Limited, a container line operating from Canada to Europe. Lykes, having been awarded a subsidy for three ships under MSP, attempted to retain the grant by forming an American company that would own and operate the subsidized ships, which would be time chartered to the Canadian company. Sea-Land vigorously opposed the proposal, which was finally denied when the Maritime Administration concluded that the ships would be subject to undue foreign control.[9] Thus Lykes, which was once America's largest subsidized line, was soon to disappear as a U.S. flag carrier, while Sea-Land, formerly the most vociferous opponent of subsidies, took its place as the new leader; the fifteen Sea-Land ships covered by the Maritime Subsidy Program would receive a total of $31.5 million per year. In spite of the large subsidy, Sea-Land was unable to earn a satisfactory return on the capital invested. Two years later, in 1999, CSX, its parent, announced that the international division together with most of the ocean terminals had been sold to a Danish shipping company, A. P. Moller. All that remain of this once proud, pioneering company are the vessels sailing in a protected Jones Act trades.

In April 1997 the barely launched MSP was rocked by the announcement that American President Lines (APL) had been acquired by Singapore's Neptune Orient Line (NOL). APL had nine ships in the new subsidy program and now had a problem with foreign ownership, but its situation differed from Lykes's in one significant respect. APL, like Lykes, had been operating ships under an expiring ODS contract. Early in 1997, before the sale to NOL was announced, the Maritime Administration had agreed to allow APL to transfer its subsidy contracts if certain conditions were met. APL announced that it intended to transfer those ships to trusts that would register the ships under the U.S. flag. The trusts would provide bareboat charters to the newly formed American Ship Management Company (ASM), which would manage the ships and collect the subsidies. ASM would have no common ownership with APL.[10] It is assumed that the Sea-Land ships covered by the MSP program will be similarly operated.

With the sale of Lykes and APL, only one of the twelve liner companies running ships when the 1936 Merchant Marine Act was originally imple-

ntinues to operate. Farrell Line, under the able direction of its George Lowman, maintained a viable Mediterranean service.

During the year that the Senate was considering the MSP, the Defense Department's position underwent a marked change. In an April 9, 1996, letter that was sent to Senator Larry Pressler, then chairman of the Senate Commerce, Science, and Transportation Committee, John P. White, deputy defense secretary, stated that the department gave its full support to the Maritime Security Act (H. Rept. 1350). This reinforced the supportive testimony previously given by U.S. Air Force general Robert L. Rutherford, head of the U.S. Transportation Command.

In an act that would have gladdened the heart of any true mercantilist, Congress in 1996 passed, and the president signed, legislation that lifted the twenty-three-year-old ban on the sale of Alaskan oil to foreign buyers, but only if it was transported in ships built and registered in the United States. The legislation had the support of the U.S. shipowners and their maritime unions, while the OECD and the European shipowners were outraged by the continued protectionist nature of the U.S.-flag shipping requirements.

A few months later U.S. trade negotiators working toward an international agreement on maritime services made it clear that they would continue to oppose any greater access to the U.S. shipping market. Their position was a response to the widespread opposition of maritime labor interests to any agreement that might provide access to U.S. domestic trades. It was a position that penalized U.S. exporters and importers, who had been hoping for more open, multilateral competition.

The message that these actions sent to our trading partners was quite clear—the United States is willing to support free trade in principle and engage in endless talk in the OECD, in the World Trade Organization, and in G7 meetings of the world's leading industrial nations, but when any special interest is involved, particularly any that impinges on the U.S. maritime industry, no substantial movement should be expected.

As the century draws to a close, most of America's maritime policies, many of which contain restrictions that extend back to the mid-nineteenth century, are firmly in place. It is true, however, that the funds, both direct and indirect, needed to support these policies are diminishing. The United States has long believed it could continue to rely on subsidies and protection to preserve industries that were in obvious decline. It was quite willing to continue spending hundreds of millions of dollars annually to maintain the status quo rather than face up to the reality that much of this expenditure could be put to better use and should be subjected to much more realistic examination.

15

THE AMERICAN MARITIME INDUSTRY
IN A TIME OF TRANSITION

A time of transition can be defined as an interval between earlier and later periods of transition, and few American enterprises have seen a greater number of transitions than the maritime industry. An industry that in the first half of the nineteenth century successfully carried American free enterprise throughout the world had by the end of the twentieth century declined to the point where it was almost totally dependent on government protection and support for its survival. While this decline was driven in part by great technological and economic changes, it was even more powerfully shaped by attitudes that persisted and dominated in the political environment in which the industry operated.

If the American merchant marine is defined as the commercial fleet flying the national flag and employed in carrying its international trade, it now seems entirely possible that its history will come to an end in the early years of the twenty-first century. As those familiar with this industry know, the cost of operating U.S.-flag merchant vessels today greatly exceeds the cost of operating under other flags. This higher cost is the result of excessive wage rates and a host of restrictions and penalties imposed by federal regulations. Today the U.S.-flag fleet can only compete with the aid of government subsidy, cargo preference, or some other form of protection from international competition. But the political convictions that sustained such regulatory support in the past are losing credibility, and the number of U.S.-flag ships in foreign trade is rapidly declining.

It must be remembered, however, that every regulatory restriction that today makes the U.S. merchant marine uncompetitive was introduced at some earlier moment as a specific remedy to a problem that demanded attention. The gradual accumulation of these protections and constraints therefore does not pose a problem that needs to be explained; it was a natural outcome of the political process as it operates in the United States.

The fact that does require explanation, and the predisposition that has prevented timely reform, is persistent political opposition to reassessing the utility of these inherited constraints and eliminating those that no longer serve any reasonable purpose. It is this unwillingness to engage and reform U.S. maritime policy that is causing the industry to wither away.

It would be difficult to overstate the extent to which American maritime policies, since the middle of the nineteenth century, have fallen out of step with the policies of leading maritime nations. In the second half of the twentieth century only former colonies that gained their independence following World War II and the former Soviet Union adopted maritime policies as backward-looking as those of the United States. Other nations that wished to capture a share of the world's maritime trade realized that they would have to compete in the world market, stressful as that competition was bound to be. Like the United States during its first half-century of independence, the nations that were prepared to compete accepted the ground rules of commercial enterprise and did what had to be done to succeed.

Because the Soviet Union believed that communism would inevitably triumph over capitalism, its maritime policies ignored commercial considerations and reflected the dominance of central planning and state power. The newly independent nations, desperate to avoid the economic dependency of neocolonialism, sought to use political control over their own foreign trade to secure for themselves a place in international shipping. The maritime policies of the United States were shaped by different assumptions and concerns, yet during the decades following World War II American maritime policy resembled the policies of the Soviet Union and the ex-colonial nations more closely than it did the policies of the European and Asian nations that came to dominate international shipping.

In the second half of the nineteenth century the United States was only marginally involved in global imperialism, and at the beginning of the twentieth century its merchant fleet engaged in international trade was small and second-rate. But in the decades between the Civil War and World War I the United States became a mighty continental nation as it expanded westward, industrialized, and consolidated its economy. This rise to world prominence was accompanied by a widely held sense of American exceptionalism and a belief that the twentieth century would be the American century. Woodrow Wilson was only expressing his nation's buoyant moralizing self-confidence when he sought to subordinate traditional imperial commercial competition to the nobler ambitions of national self-determination and democratic regulation.

As part of his campaign to achieve this reordering of the worl[ ]
ical economy, Wilson, in 1916, launched a federal program that
lished the United States as one of the world's foremost maritime powers.
This historic moment remains relevant to America's maritime policy problems today, for the convictions that informed Wilson's initiative continued
to validate U.S. maritime policy until the end of the cold war. But in the
final decade of the twentieth century the circumstances that made the
Wilsonian dream and commitment credible no longer exist, and new realities must be acknowledged. Today many nations are carriers in international trade, and the thoroughly reinvigorated global maritime industry is
highly competitive. Yet in the United States the industry's fundamental
dependence on the federal government remains basically unaltered.

AMERICA'S POLICIES OUT OF STEP WITH THE MARITIME WORLD

Attitudes toward the way Americans engage in international commerce
have changed considerably in the closing decades of the twentieth century,
yet there is still strong resistance to bringing U.S. maritime regulations
into harmony with world standards. As a result there remains an enormous
variance between U.S. maritime practices and world practice:

1. No other modern nation demands that ships engaged in its cab-
    otage (coastal) trade be built domestically.
2. No other nation virtually insists that ships be built domestically
    if they are to carry much of government-sponsored foreign aid
    cargo. (After legislation creating the MSP had been passed, an
    amendment was added to a Defense Department appropriation
    act providing that ships included in the new MSP are to be
    exempted from the three-year waiting period for foreign-built
    ships operating under U.S. registry.)
3. No other first-class nation insists on levying custom duties on
    foreign repairs to its ships.
4. No other up-to-date nation insists on the retention of out-
    moded laws restricting the efficient use of shipboard personnel.
5. No other advanced nation insists on taxing the earnings of ships
    owned by its citizens and engaged in international trade as
    though they were a domestic operation.

The legislative changes made subsequent to the establishment of the Maritime Security Program have significantly improved the ability of the U.S.
owners of the forty-seven ships operating under the MSP to compete inter-

nationally. U.S. tax laws continue to be a serious impediment, however. The Clinton administration, represented by the Treasury Department, is unmoved by the fact that most companies engaged in international maritime trade either pay no taxes or can defer income-tax payments by setting funds aside for future vessel improvement or new construction. The 1936 Merchant Marine Act authorized a Capital Construction Fund (CCF) for this purpose, but because it requires that any new construction paid for with these funds must be built in U.S. shipyards, the program is worthless to all but those engaged in Jones Act trades. Many countries also exempt part or all of seamen's wages from taxation, but the United States has always preferred to compensate for this competitive disadvantage by subsidizing wages or limiting competition for U.S. cargoes. The Treasury Department, along with the U.S. shipbuilders, also continues to zealously defend the 50 percent ad valorem tax on repairs to U.S.-flag vessels made in foreign yards. It is indisputable that these tax policies force U.S. shipowners to seek additional operating subsidies or to operate their ships under foreign flags.

American maritime policy is also unique in depending heavily on the purported role of the merchant marine in national defense when justifying federal support. Only the former Soviet Union attached comparable importance to this rationale for its maritime policies. The defense justification has been especially important to the shipbuilding industry and has been used for more than a century to shield it from international competition. It is noteworthy, however, that the nations that build most of the world's commercial tonnage, such as Japan, South Korea, and Germany, provide no such shield. And, indeed, if history teaches any lessons, one of them must be that the supposed connection between naval building and commercial building is nothing more than a rhetorical linkage insisted upon by advocates of federal support for shipyards. The skills and organization required to build modern naval vessels are very different from those needed to build competitively priced commercial ships. A management and labor force proficient in building the finest combatant ships in the world has to be far more concerned with following voluminous, detailed building specifications and satisfying a vast number of naval inspectors during all phases of construction than with increasing productivity.

Significant legislative reform can only be effected after there have been fundamental changes in perception, understanding, and attitude. There are some recent indications that in the maritime policy arena new views are slowly taking hold. In Vice President Gore's report on the merchant marine, for instance, it was noted that "what is left of the U.S. maritime industry stands as a stark reminder of how protectionist, economic regulatory policies by our government literally razed the economic underpinnings of our once mighty and proud merchant marine fleet."[1]

The dysfunctional features of American maritime policy only became undeniably obvious when the cold war ended. So long as the United States was actually or potentially involved in war overseas, federal support for the merchant marine could be politically sustained. But with the end of the cold war and the pattern of international relations it created, the danger that there might be a renewed world war dissolved. It also became highly improbable that the global economy could be so disrupted that the United States could not find ships to carry its goods to and from foreign markets. The demise of the Soviet Union left the United States as the one remaining superpower, but in its commercial relations with other nations America became just one more competitor in an increasingly extended and diverse world market.

SUBSIDIES AS TAX SHIFTING

A number of economists have noted that while in the long run subsidies contribute little to the nation's economic welfare, they provide even less for their direct recipients. Would Sea-Land have been worse off if they had obtained and maintained their ships in the world marketplace? Could they possibly be disadvantaged if they obtained first-class crews, which undoubtedly would include some American officers, at wages determined primarily by economic considerations? It should be clear that the main beneficiaries of subsidies have been the seamen that the shipowners employ. Other beneficiaries include workers in U.S. shipyards and repair yards and marine equipment manufacturers. There are other expenses that are unique to the American merchant marine: the cost of endless litigation over interpretation of U.S. maritime laws and the cost of lobbying organizations that protect and seek to increase existing subsidies. The cost of insurance required to cover personal injury claims in the U.S. shipping industry vastly exceeds that encountered in any other industrialized country. Franz Eversheim, one of the most perceptive students of the economics of subsidies, rightly called the process "tax shifting."[2] The shipping company is treated as a vehicle for transferring subsidies provided by the taxpayers to other beneficiaries rather than to itself. At the same time the company retains the stigma of having been subsidized. In all maritime nations except the United States it is accepted that the sole purpose of a merchant ship is to make a reasonable net return on invested capital. In the United States political considerations tend to dominate because there are many beneficiaries other than the investors. The government is impelled to continue payments to make sure that the expectations of these many other recipients are satisfied.

There is a growing recognition that most shipbuilders and maritime labor unions have priced themselves completely out of the commercial market. This has little to do with the American standard of living; it has everything to do with the existing low levels of productivity that have been tolerated only because the U.S. taxpayer has been willing to provide the subsidies required to make up for the industry's failure to develop as a world competitor. The tired response that Americans are only doing what everybody else does simply begs the question.

Recently passed legislation authorizes payments that will keep a few ships engaged in foreign commerce operating under the national flag for a few more years, yet it is unlikely that funds will be appropriated to support a fleet that is in any way commensurate with America's position as the world's leading trading nation. In the absence of large operating subsidies, most American-owned ships engaged in foreign commerce will have to operate under an open registry. This problem could, of course, be addressed institutionally by creating a U.S. registry that meets the commercial needs of American shipowners. The Norwegians did this, as was shown in an earlier chapter, when they created a separate international registry. Because many Norwegian owners chose to operate their ships under this registry, a large number of seagoing jobs for senior Norwegian officers were preserved. Having ships sail under Norwegian registry also played a crucial role in preserving the vital institutional infrastructure of the Norwegian maritime industry. Maritime nations must have citizens who can provide such essential services as ship brokerage, financing, equipment manufacture, and naval architecture. But these services can only be sustained if there is a fleet to serve. Few Americans seem to realize that roughly half the world's merchant tonnage sails under flags that do not reflect the nationality of the owners.

While the debate over direct subsidies continues, increasing national attention is also being directed to the large indirect subsidies provided by the cargo preference laws. As was noted earlier, the $700 million worth of American grain that President Clinton promised as aid to the Russians early in his first term focused attention on this issue. Congress objected to the high rates required by U.S. shipowners to carry these cargoes. It was also noted that few if any of the vessels they would use have any military utility; indeed many would have been sent to the ship breakers years earlier had they not met the special requirements of the cargo preference laws. These legal requirements make for highly inefficient shipping operations. The ships that qualify to carry PL 480 cargoes for the most part serve only that market and lie idle between cargoes. When they do carry a cargo outbound, they usually return empty. Few of these ships are suitable for carrying other bulk cargoes.

Nationality of Owner	Percentage of Tonnage under Nonnational Flags (%)
American[a]	66
Belgian	100
British[b]	73
Chinese (PRC)	28
Danish[c]	41
Dutch	48
French[d]	41
German	74
Greek	65
Hong Kong Chinese	73
Japanese	76
Korean (South)	56
Norwegian[e]	46
Swedish	85
Taiwanese (Republic of China)	47

Table 15.1: Percentage of World Tonnage under Nonnational Flags.
[a]Including Great Lakes and Jones Act Fleets.
[b]Including Bermuda, Cayman Islands, Hong Kong, Isle of Man, and St. Vincent.
[c]Including Danish International Registry.
[d]Including Kerguelan Island.
[e]Including Norwegian International Registry.
Source: Data from *Lloyd's Register*, November 15, 1996.

Those few that could are discouraged from doing so by the "fair and reasonable" rate program of the Maritime Administration; the rate they receive for outbound PL 480 cargoes would be reduced by any profits realized carrying a commercial cargo on the back haul.

The Jones Act requirement that ships engaged in coastal trade be U.S. built, U.S. owned, and U.S. crewed is also being subjected to increasing consumer scrutiny. The insistence that such ships be U.S. built is the Jones Act's most constraining feature; no other nation imposes a similar requirement. U.S.-built ships are so expensive that ships in the Jones Act trades are kept in service long past their normal retirement age, the result being that the fleet is very old, indeed much of it virtually obsolete.

The very high cost of building new ships for the Jones Act trades is illustrated by Matson's experience building *R. J. Pfeiffer*. While undoubtedly a fine ship, its $140 million cost made it the most expensive container carrier in the world. Built to serve the U.S. West Coast–Honolulu route, the vessel was designed to carry nineteen hundred twenty-four-foot containers. Construction contracts for the ship were signed late in 1989. Less than four years later, Neptune Orient Lines Limited (NOL) of Singapore signed a contract with a Japanese shipyard for the same total price for two ships, each having a 4,300-TEU capacity.[3] In other words, four years after Matson contracted with an American yard for a new ship, NOL was able to get roughly twice the capacity on the world market at half Matson's cost. Matson must charge freight rates that cover the higher cost of its ship. In this instance it has been estimated that shippers sending goods between California and Hawaii are being charged an extra $10 million per year in freight rates to subsidize the building of that one ship in a San Diego shipyard.[4]

American shipping companies have been world leaders in creating cost-effective new systems and technologies. They have developed the modern cargo container and established the standards that made containerized shipping an innovation of international significance. They led the way in adapting Car Identification Systems (CIS), which were originally developed for railway use, to the identifying and tracking of containers. They also developed sophisticated computer programs for continuous on-line monitoring of container movements. They invented the double-stack railcar and unit train, and more recently they pioneered in building wide-beam, post-panamax containerships. The list could be extended considerably. But technological innovation is not by itself enough to ensure the continued existence of a U.S.-flag merchant fleet. International carriers are alert to the value of technological change, and such improvements are quickly copied. The originator at best enjoys a two- to three-year lead, after which the new technology rapidly follows the market. If American carriers lose their share of the international shipping market, they will no longer be able to maintain their distinguished record of technological leadership. If U.S. maritime laws are not changed so as to make owning ships much more attractive to American investors, the incentive to innovate and invest will be lost and those with strong market positions will take over as leaders of the innovation parade.

THE NEW REALITIES

If at least a nucleus of an American-owned merchant marine is to be retained, the relevant new realities in maritime commerce must be recognized and accepted, no matter how great the political pain incurred. Among these realities are the following:

1. Until recently political support for the U.S. maritime industry depended on war or the threat of foreign aggression. The nation no longer feels this pressure. There will be threats and wars in the future, but in the competition for federal funding the maritime industry will no longer command the privileged position it has occupied in the past.

2. The national identity of the various modes of transportation involved in international cargo movements has become blurred. More and more, the few successful U.S. companies were members of large consortia, and their containers were readily interchanged with those of other firms regardless of flag. Torn between unrealistic federal restrictions and union demands on the one hand and the competitive need to minimize costs on the other, Americans operating in international trade will continue to migrate to owning or chartering ships flying flags other than that of the United States. No combination of politically acceptable subsidies or protection will prevent this from happening once the present fleet has outlived its useful life.

3. One of the most significant new realities is the increased skill and sophistication of the managers who direct the leading shipping companies. The managers of these vast and complicated enterprises now realize that the costs associated with accepting government "assistance" almost always outweigh whatever benefits are provided.

4. While containerized cargo systems are incredibly productive when linked to extensive and sophisticated land-based transportation systems, they break down rapidly when shoreside facilities do not exist or are less than adequate. The Department of Defense (DOD), which must send weapons and supplies where they are needed rather than where it is most convenient, has recognized this limitation for some time. The armed forces' ship of choice for sealift is the Ro-Ro, which is far less dependent on shoreside support than the containership. It can operate virtually anywhere and rapidly load and discharge almost any type of cargo. Unfortunately, few Ro-Ros are American owned.

This last "reality" led the DOD, acting through the navy, to acquire a large number of foreign-built Ro-Ros, which have been reflagged for the Ready

Reserve Force (RRF) and placed in layup for rapid activation when needed. It was these ships, together with foreign-flag charters, that provided sealift for the initial force buildup in Saudi Arabia during the Gulf War. The navy has also embarked on a $6 billion program to build its own fleet of specialized sealift ships. With these programs in place, the armed forces will, in the future, rely much less on the U.S. commercial fleet than they did in the past. Containerships will still be used in major overseas operations to provide sustainment support when suitable shoreside handling and storage facilities are available, but such limited involvement is much less than the industry-military relationship anticipated by the authors of the preamble to the 1920 Merchant Marine Act.

While containerships are of limited usefulness to the military, the merchant marine remains the source of the officers and crews that man the ships in the RRF when they are activated to support military operations. This manpower need has therefore emerged as a second argument for continued federal support for a U.S.-flag merchant marine, and it does indeed have some merit. Its weakness, however, is the large cost involved in subsidizing ship operations to provide employment for a relatively few seafarers. Taxpayers will certainly be opposed to providing millions of dollars every year as operating subsidies when there are other and less expensive ways to train and maintain an adequate pool of qualified mariners. And it is unlikely that over the long run Congress will be prepared to provide $2 million per ship per year to provide employment for a twenty-one-man crew, most of whom will have no military service obligation in any case.

As the long and often glorious history of the American merchant marine comes to a close, it is well to be reminded that few of its fundamental problems are of recent vintage. As the authors of the 1937 *Economic Survey of the American Merchant Marine* observed,

> The history of the United States in subsidized shipping has been most unfortunate. We have generally acted under the lash of necessity which precluded the planning necessary to a sound commercial policy. Our present attempt to compete in the international carrying trades was born of the World War. Previous to the war we were a negligible factor in the overseas trade. Then, goaded by necessity, we requisitioned, bought, seized, and built vessels at a rate never equaled in the history of the world. We blanketed the oceans with ship services in a frenzied effort to make up for our negligence of the preceding half century. Today

we have gone back a long way from the ambitious program of the early 1920s. We are about to start again, not in a riot of enthusiasm, not with an expenditure of billions, but with a carefully planned program that gives due regard to the factors of need, method, and cost. Therein, we believe, lies our hope for the future of the American merchant marine.[5]

Maritime policy, according to the authors of the survey, should consist of "a carefully planned program that gives due regard to the factors of need, method, and cost." Unfortunately, recent maritime programs have generally lacked clearly defined objectives and have generated almost no public support. This absence of broader public involvement should not occasion surprise, however, for when measured against other sectors of American industry, the merchant marine is a very small part of the national economy. And when not linked to the national interest by emergencies that require its participation, the political interests of the merchant marine are largely sustained by narrowly based union-management support.

The collapse of the Soviet empire and the end of the cold war have essentially eliminated the traditional "national need" argument for subsidizing the merchant marine. The Department of Defense has previously said it does not need or want support from the commercial fleet, although this position has recently been reversed.

Once the national defense issue has been set aside, "cost" becomes the central concern in deciding what "method" should be used to maintain an American fleet. Here it must be acknowledged that it would be hard to improve on the 1936 Merchant Marine Act's intended arrangements for setting and limiting subsidy costs. The costs of implementing this act were at the outset modest, and prior to the frenzy of World War II, there was no reason to anticipate that an explosive rise in wartime production and labor costs would overwhelm the act's provisions and render them ineffective. But there was a war of massive proportions, and by its end it was obvious that the American merchant marine had become fundamentally a federal program rather than a commercial enterprise.

If the Jones Act requirement that the domestic fleet be U.S. built were eliminated, U.S. policy would suddenly conform much more closely than it now does to the policies of other developed countries. Eliminating the domestic building requirement would expand the fleet engaged in coastal trade and make it more competitive. While many nations protect their coastal fleets through laws similar to the Jones Act, none require that their ships be domestically built. It should also be noted that while domestic air transportation in the United States is reserved for American airlines, there is no requirement that the planes that serve these routes be built domestically.

The U.S. maritime industry has its sacred texts, two of which are the Jones Act and the requirements for U.S. registry. The political cost of revoking the Jones Act or opening a U.S. international registry would probably be so great as to be unjustified at this time, but some improvement could be achieved by undertaking more modest changes. It is highly likely, for instance, that market pressures will continue to force U.S. shipowners to change to open registries. While the U.S. container fleet is relatively new by U.S. standards, it is old by international standards, and foreign building for fleet replacement is well under way. When foreign registry of the U.S.-owned liner fleet is an accomplished fact and no longer subject to condemnation, it may be possible to reconsider the possibility of creating a U.S. international registry that would once again enable American shipowners to fly their country's flag with pride.

Resistance to changing the established political order in the maritime industry is deeply entrenched and will not be easily rooted out. Patriotic assertions abound and special interest and the national interest are tightly coiled around one another. The repeated failure of established policies to produce the intended effects provides endless opportunities for attributing bad faith and blaming opponents, while economic and commercial considerations get shunted aside as irrelevant. Heat and dust predominate; light and vision are obscured.

The commercial advantages of developing a well-designed maritime policy for the world's largest trading nation are rarely made part of the debate. Other nations, such as Japan and several European countries, maintain substantial commercial fleets even though they have little or no defense obligation to do so. While many of their ships are in open registries, they are still under effective control of their citizens. These nations support their merchant marine rather than burdening it excessively. Commercial realities dominate as the industry concentrates on the business of shipping. One rarely hears in these countries the kind of partisan wrangling that is so characteristic of disputes over maritime policy in the United States. Governments in these nations have other concerns and leave the merchant marine largely to itself.

American maritime policy has failed and a fresh approach is needed. The accepted practices of the international market must be taken as the new point of departure. The resources available are limited and, should they continue to be squandered, the end will come sooner rather than later.

Anyone familiar with America's proud record of maritime preeminence must be saddened by this prospect of final decline. The men whose skill and daring did so much to build the nation's wealth and commerce and later enabled it and its Allies to win the great global conflict that ended in 1945 earned their nation's undying gratitude. But their achievements adorn a past that can never be recalled, a time when this once great industry was sustained by levels of public and private expenditure that, in proportional terms, will never be seen again. Yet there is no reason to think that the transition the merchant marine is now going through must inevitably end in extinction. But in the absence of a truly new departure, of strong leadership and collective commitment to fundamental renovation, extinction is the most likely outcome.

We close as we began, by asking Admiral Mahan's profound question. He questioned whether "the national character of Americans is fitted to develop a great seapower." Like Mahan, we would wish to answer in the affirmative, but only after acknowledging his qualification, which was that "legislative hindrances [be] removed" and shipping be able to compete effectively with other opportunities for investment. In other words, the United States will only be able to develop a merchant marine commensurate with its position as the world's leading trading nation if national considerations are able to overcome the self-serving restrictions that have constrained the industry for over a century.

Appendix

The Nixon Maritime Program: A Personal Memoir

Andrew Gibson, the senior author, has been actively involved in various aspects of the American maritime industry since World War II and served as the federal maritime administrator from 1969 to 1972. Many of the observations and conclusions included in chapters 9 through 13 are based on his firsthand experience. The following memoir, written in the first person, describes some highlights of Gibson's long career.

Memoir

My seagoing career began during World War II when I sailed in North Atlantic convoys; it culminated in entering Japan in command of my own ship some sixty days after V-J Day. In the 1960s, I participated in most phases of the container revolution while serving as operations vice president and senior vice president of Grace Line. At that time I strongly urged the company to enter the South American trade with full containerships (see above, page 214). Although Grace Line's first venture in containerization failed, it paved the way for a successful ship design that was more suited to the trade at that time. Since these early days I have maintained contact with Malcom McLean and have watched with admiration as he developed the worldwide containerized cargo-carrying system that we know today (see above, pages 209–19). Years later, while serving as president of Delta Line, I played a part in bringing that company's LASH ships to their full profit-making potential (see above, pages 220–21).

In early February 1969, as the new Nixon administration was being assembled, I found myself being interviewed by the incoming secretary of commerce, Maurice H. Stans, for the position of maritime administrator. The interview was not a matter of chance. A close friend, Franklin Lincoln, was a senior partner in the president's former law firm, and through him I had let it be known that I was interested in joining the

307

government. Where chance did play a role was in the person of the new under secretary Rocco Siciliano, whom I had known when he was president of the Pacific Maritime Association. When my letter to Lincoln was forwarded to Stans, it was Siciliano who recommended that he see me.

On September 25, 1968, in Seattle, Washington, the president had given a far-ranging speech, called "Restoring the U.S. to the Role of a First-Rate Maritime Power," which outlined his maritime plan. Like many in the maritime industry, I was quite familiar with that speech; Stans was obviously interested in determining if I was the best qualified person to develop a program that would fulfill the commitments made in that address. I never learned who else was being considered, but with Siciliano's support I was chosen. Before the end of February I was on board on a temporary basis, and by early March my appointment, along with those of the assistant secretaries that Stans had selected, was confirmed by the Senate. This turned out to be the role for which I had been preparing myself for almost thirty years.

At this point it might be appropriate to consider some aspects of the Seattle speech that became the foundation of the future maritime program. Nixon started with his definition of seapower:

> Seapower is the ability of a nation to project into the oceans, in times of peace, its economic strength; in times of emergency, its defense mobility.
>
> Seapower is composed of all those elements enabling a nation to use the world ocean advantageously for either trade or defense—its navy, its merchant shipping, its shipbuilding, its fishing, its oceanographic research, and its port facilities.

He then went on to outline one of his main concerns, the buildup of the Soviet Union's maritime strength:

> Almost every day a ship leaves the Soviet Port of Odessa with cargoes for North Vietnam. An estimated 80 percent of the materials used by the enemy in Vietnam arrives in Soviet merchant ships. More than 97 percent of all supplies used by the Allied troops in South Vietnam also moves by water, most of it aboard old ships flying the U.S. flag but which are no match for the modern Soviet merchantman.

Two-thirds of our merchant ships are beyond their economically useful age. By contrast, half the Soviet fleet is less that 5 years old.

By contrast, the United States now has an active privately-owned merchant marine of fewer than 1,000 American flag ships. We are producing less than 15 ships a year. . . . If we permit this decay to continue we will find that we have abdicated our maritime position to none other than the Soviet Union.

The president then outlined specific areas to be addressed in his new program:

Commerce

Only 5.6 percent of the U.S. trade is carried on U.S. flag ships. This is the lowest since 1921.

Soviet flag ships already carry more than 50 percent of Soviet cargoes; Sweden, 30 percent of her own commerce; Norway, 43 percent; Great Britain, 37 percent; France, 48 percent; and Greece, 53 percent. Japan is carrying 46 percent. . . .

These nations have determined that a high degree of reliance on their own shipping resources is important to their self-interest. We have not.

To state it bluntly, our trade is predominately in the hands of foreign carriers, some of whom may be our trading competitors. We must have more control over the movement of our own cargoes not only for competitive reasons, but also because of the contributions our ships make to our balance of payments.

Shipbuilding

The new Administration's maritime policy will seek a higher level of coordination between naval and merchant shipbuilding.

In that way we can create a climate in which shipbuilding can attract the capital, as well as the stable labor force, needed to make it competitive with foreign yards and to provide an expansion base for national emergencies.

In turn I would expect initiative and cooperation from both industry and labor. Throughout the maritime industry, a new outlook must be encouraged to replace the current divisiveness and short sightedness.

Until such time as American yards can be independently competitive, I recognize that shipbuilding subsidies are necessary to enable shipyards to build ships and deliver them to operators at competitive world prices.

Operating Subsidy

Since the Merchant Marine Act of 1936 was passed, we have been living with an operating subsidy system. The system has been aimed primarily at removing the wage-cost disadvantage of the American operators who must pay seamen under U.S. working standards and levels of living.

The subsidy system has had its shortcomings. It has been extended exclusively to liner operators in the foreign trade; it has grown more costly; it has not created a modern merchant fleet even among its recipients nor has it had as a basic ingredient enough reward for increasing efficiency.

I propose, therefore, an immediately re-evaluation of this program, in consultation with industry members and labor representatives, with the goal of providing more incentives for productivity.

The unsubsidized sector of our merchant fleet must be given attention, so that it, too, can replace its deteriorating fleet in the immediate future. Included in this category are those who carry farm products to the underdeveloped nations, and the Great Lakes operators who daily face competition from their government-assisted Canadian counterparts.[1]

I had a number of advantages not shared by my predecessors as the new maritime administrator. I was the first with extensive experience in the maritime industry. This is in no way intended to depreciate the great achievements of Admiral Emory S. Land, who was chairman of the Maritime Commission during World War II, and his successor Admiral Cochrane, the first maritime administrator. Their talents lay in building ships for the government account, however; it was now intended that commercial considerations be given a higher priority.

I had entered the Massachusetts Maritime Academy in the spring of 1940 and was graduated two years later in time to be among the first Americans to sail in the North Atlantic convoys. As the vast fleet produced under Admiral Land's direction grew, it soon outstripped the supply of trained officers. As a result promotions came rapidly. By January 1945 I was placed in command of a Liberty ship at age twenty-two.

In the summer of 1946 I came ashore and two years later entered Brown University in Providence, Rhode Island, from which I graduated with honors in economics in 1951. During this time I retained a reserve commission in the navy, and as soon as I had graduated I was called to active duty as a lieutenant to participate in the Korean War.

Here I obtained another advantage. I was assigned to the newly created Military Sea Transport Service (MSTS). Although I had been mobilized for sea duty, I eventually became assistant comptroller for Budget in the Atlantic headquarters of MSTS. This position provided me with a working knowledge of the federal budget system. Since I had some 150 civilian employees working under my direction, I was able to master the intricacies of civil service procedures as well. When I joined the government I knew how the system worked, and when I wanted to effect changes in organization and staff I knew how to do it.

After leaving the navy I joined Grace Line and eventually rose to be senior vice president. Grace Line was a subsidized line serving the Caribbean and South America. As I was to learn later, when I was in a position to compare performance, Grace Line was one of the best managed of the subsidized lines. While at Grace I got to know most of the senior executives in the U.S. shipping industry as well as many labor union officials and key personnel in the Maritime Administration. This gave me a running start in the new job.

These advantages notwithstanding, the one that counted most was that I served under a dedicated president, ably assisted by a highly experienced secretary of commerce. Stans and his deputy each had eight years of senior service in the Eisenhower administration, Stans as director of the Bureau of the Budget, and Siciliano as a member of the White House staff for four years and then as assistant secretary of labor for four years. They were both outstanding mentors.

Stans insisted on a thorough housecleaning of the agency. Although in some instances politically motivated, his primary purpose was to recruit talent and bring in new blood. We were to try to hire Republicans, but the emphasis was on talent, and I also hired Democrats. I chose as my deputy Robert J. Blackwell, a lawyer who had extensive experience at the Federal Maritime Commission. I was most fortunate in being able to engage Roy G. Bowman as my general counsel. He had been planning to return to the New York law firm from which he had been recruited by the previous secretary of commerce, who needed him to establish the Office of Foreign Direct Investment (OFDI). Blackwell and Bowman became the first members of a highly talented team that joined the Maritime Administration (MARAD) over the next few months. I was able to generate the necessary vacancies and salaries for the new staff by persuading a number of individuals that it was time to retire. A few were convinced that voluntary departure was preferable to having a major contest with the new administrator who brought a "no nonsense" reputation with him. Easing these people out was facilitated by the recent creation of the Department of Transportation, where jobs were readily available; a number of MARAD personnel took advantage of the opportunity.

With a nucleus of experienced staff that included many who continued on in the MARAD organization, we began to develop a program that would implement the promises the president had made. The program was completed by May, and after being approved by Secretary Stans, it was reviewed by others in the Nixon administration in a series of meetings chaired by a White House assistant. These meetings were anything but friendly. It appeared that the Department of Commerce was the only part of the administration that wanted the new maritime program.

Opposition was so adamant during the summer of 1969 that I went to the secretary and told him that if the president really wanted the program, it was the best kept secret in Washington. If he had in fact changed his mind, I wanted to leave, since I was receiving half my former salary and had a large family to support. At that time Stans was scheduled to meet with the president in San Clemente, California, on some other business, but he also voiced my concern. Stans later told me that the president had said, "Tell Gibson to keep his shirt on," and that he still intended to keep his commitment.

I had expected that while promoting the new program I would receive help from the navy and the Department of Defense, but in this I was disappointed; their support vacillated at best. I also expected the Departments of State and Agriculture to oppose the new program, and this proved to be the case. The representative of the Council of Economic Advisors, a Harvard University professor, was in a class by himself. At one of the final meetings he was so upset by the proposal to extend the subsidy system that he became virtually incoherent.

The positions taken by the Department of Transportation were generally not helpful. Its attempts to take over the Maritime Administration did not end with the Johnson administration (see above, pages 195–97). The new secretary, John Volpe, a former governor of Massachusetts, brought along Paul Cherington, a Harvard Business School professor, to be his assistant secretary for Policy and Plans. Cherington sought to bring all U.S. transportation policy considerations into his office. Since the railroad, highway, and aviation administrators were subordinate to him, they had limited say in the development of policy in their areas of responsibility. Cherington obviously hoped to control maritime policy issues as well and thus was not particularly anxious to have me launch a successful program that would develop a constituency that could frustrate his intentions.

While the program was being developed, no attempt was made to include groups of shipping company executives, shipbuilders, or labor union officials in the planning process, although a number of meetings were held with individuals representing these interests. Prior experience indicated that large-scale meetings with these groups, who rarely agreed on anything, would probably be counterproductive; I therefore proceeded on the assumption that I generally knew what was needed. I was certain of one thing: each group, had they been involved, would have gone to the press to air its particular point of view, with the result that by the time a program was announced it could have been partially discredited.

Eventually the reviews within the administration were completed and the program was sent to the president with comments furnished by the departments and agencies having an interest in maritime matters. Secretary of Commerce Stans provided the only unqualified endorsement. Secretary of Labor George Schultz took a neutral although in no way negative position. Everyone else took exception to some aspect of the program. The chairman of the Council of Economic Advisors advocated complete rejection. But as President Nixon had assured Stans, he intended to fulfill his promise; unlike some of his successors, he did.

In spite of all the disagreement within the administration while the program was under review, there was no leak to the press, as now seems to be commonplace. Even more remarkable, in light of today's experience, once the president had made his decision the bickering stopped. It was now a presidential program, not merely one sponsored by the maritime administrator. All members of his administration were therefore obliged to give it full support, and they did.

Representatives of the various maritime organizations were called together on the evening of October 22 and given copies of the program. There was little discussion. The next day the program was formally announced, and shortly thereafter I presented it at a special hearing convened by Warren G. Magnuson, the powerful chairman of the Senate Commerce Committee. The main features of Nixon's maritime program have been described in chapter 10 (see above, pages 197–203).

All segments of the U.S. maritime industry gave the program a generally positive reception. Nothing like it had been developed since 1936. It clearly addressed problems that had become apparent with the passage of time, and it was designed to once again make the American merchant marine a major player on the world scene.

I had initially assumed that my work would be finished once the president adopted the program. I soon found out that it was only half done. A number of amendments to the existing law were required, and my general counsel was responsible for preparing much of the enabling legislation. He completed this task, which could easily have occupied half a year, in three months. In addition to writing the legislation, he coordinated his efforts with House and Senate staffs so that controversy would be minimized when the draft legislation was presented. The great value of having an experienced staff became especially apparent while we were preparing for hearings on the bill. It was essential that the chairmen of committees conducting hearings and the key members of those committees understand and support the proposed legislative changes. Largely because Bowman had earned the respect of everyone with whom he had worked since becoming MARAD's general counsel, our positions on the proposed bill were generally accepted.

During the spring and summer of 1970 both houses of Congress held hearings on various aspects of the bill, generally covering one part per week. I was always the lead witness for the administration's position, and as the program's author I had certain advantages. I could, with considerable credibility, explain why certain amendments had been made; I did not have to defend or justify conditions others sought to impose but that I opposed. The industry representatives who spoke when I was finished for the most part supported the program. There was, however, some controversy over a detail we had overlooked. Most of the owners of the bulk vessels carrying PL 480 cargoes had labor contracts with the SIU. The subsidized liner companies' labor contracts were with the NMU. This made the two sides natural enemies (see above, pages 183–84). The situation was further complicated because several of the owners of nonsubsidized SIU companies also owned foreign ships, which the subsidized lines were prohibited from doing. Some of the owners of bulk vessels indicated that they intended to build and operate tankers under the new subsidy program, but the NMU argued that they should be obliged to conform to existing rules for subsidized lines and divest themselves of their foreign holdings immediately. This they were not about to do. After extended debate it was agreed that the owners could keep their present foreign-flag tonnage but would have to get rid of it over a ten-year period. Since most of the ships in question were quite old, this last requirement did not raise a real problem.

Several owners of bulk carriers also had reservations about certain terms of the new law that were patterned on the liner provisions in the 1936 act, but since nobody knew exactly how these provisions would work, it was agreed that the administration would make whatever changes were required should problems arise. One such problem did arise in 1973 when oil prices escalated rapidly. The navy specified that all ships constructed under the building program be powered by steam turbines; most of the ships operated by foreign competitors were diesel-driven. The greater fuel efficiency of the diesels made little difference when ships' bunkers were one to two dollars per barrel. But when bunker costs rose to thirty dollars per barrel, American ships could not compete. The subsidy law contained no provision to offset cost differentials caused by different fuel efficiencies. American-flag bulk carriers therefore had no choice but to retreat into competition for PL 480 cargoes or enter the Jones Act trades, where their only competitors would be other Americans operating under the same handicap.

Hearings were concluded by October 1970. Only minor changes were made to the administration's bill and the House and Senate versions were easily reconciled.[2] As has been noted, the bill passed with only two dissenting votes and was sent to the White House for the president's signature.

Because Nixon, unlike Franklin Roosevelt, did not bring a long history of involvement with maritime affairs to the presidency, questions have been raised about his commitment to the industry. He was undoubtedly aware of the maritime problems that arose while he was vice president in the Eisenhower administration, and he must have privately considered how he would have handled them had he been in charge. He was also genuinely concerned about the rapid rise of the Soviet Union's merchant fleet. Yet his commitment to his maritime program was sustained more by political considerations than national security. Nixon appreciated the political potential of maritime industry and labor support following the presentation of his maritime proposals. He was comfortable with many of the maritime labor leaders and saw the advantage of fulfilling his campaign promises. His special counsel, Charles W. Colson, who was responsible for maintaining liaison with the unions, also valued this connection.

This union support paid off handsomely in the fall of 1971 when union cooperation made possible huge grain sales to the Soviet Union.[3] While much of this story has been told (see above, pages 203–4), my particular involvement may warrant further elaboration. For a number of years I had been one of the principal negotiators of labor contracts with both the longshoremen and maritime unions. Since the unions and companies bargained as groups, I had come to know the heads of many of these unions. While relationships during negotiation were often adversarial, a degree of mutual understanding and respect developed as well. After years of negotiation I had clearly demonstrated that once I had made a commitment, it would be fulfilled.

To obtain union cooperation in shipping grain to the Soviet Union, I had assured the union leaders that a substantial part of the trade would be carried by the ships of both nations. I had in mind something like the 40-40-20 cargo sharing agreement I had previously negotiated with Brazil. Once I began to focus on the details, however, I realized this agreement would be much more complicated. The longshoremen, with tacit approval of the Department of Defense, had essentially closed all U.S. Atlantic and Gulf Coast ports to Soviet and Communist bloc ships. The Department of Defense was involved because certain policy makers had convinced themselves that the Soviets might smuggle atomic bombs into the United States in the holds of merchant ships and then threaten to blow up key coastal cities, a threat that did not seem quite so far-fetched in the years before the Soviet Union had developed intercontinental ballistic missiles (ICBMs) or missile-launching submarines. Thus, even after the longshoremen had finally agreed to load grain bound for the Soviet Union, American ports were still closed to Communist vessels. During the first year of the grain agreement, therefore, all cargo had to be carried in third-flag ships while the details of the maritime agreement were being worked out between the two governments.

When Secretary Stans visited the Soviet Union in November 1971 he was the first senior member of the Nixon administration to do so. I accompanied the secretary. While he was engaged in his ceremonial duties, I worked with Lewis Bowden, the embassy's able, Russian-speaking commercial attaché, in discussions with members of the Soviet Ministry of Merchant Marine. Before going to Moscow I had had my staff obtain copies of the shipping agreements that the Soviets had previously negotiated with other Western countries. We analyzed these documents carefully and determined how they had been implemented in practice. We found that they had worked to the advantage of the Soviet Union because they had exploited many loopholes and conditions embedded in these agreements that had not been obvious when they were signed. I was determined that the agreement we negotiated would not be similarly flawed.

The experience I gained while negotiating with American labor unions proved to be a good preparation for my negotiations with the Russians. One union representative in particular, a former Communist who was counsel for the Marine Engineers Beneficial Association, taught me a great deal about difficult contract negotiations. Learning the lessons as he taught them had been unpleasant at the time, but I did learn how to cope successfully with duplicitous bargaining.

It soon became apparent that my Russian counterpart had been instructed to obtain an agreement. I, too, had been told to reach an agreement, but I was under much less pressure. I therefore had the advantage of proceeding according to the well-known advice of Dr. Henry Kissinger, Nixon's national security advisor at the time: "Better no agreement than a bad agreement." It took only a few days to forge a tentative agreement of the sort I was seeking, and I signed a protocol before leaving Moscow. I later learned that the commercial attaché evidently had not kept his superiors in Washington informed of our activities, the result being that the State Department knew nothing of our negotiations. State Department officials responsible for Soviet Affairs were extremely upset and tried to have me dismissed for exceeding my authority, but the president's continuing support saved me. My discussions with union leaders before going to Russia had been held in Colson's office, and he had assured me that the agreements we reached had the president's blessing. I therefore escaped unscathed from this first foray into international diplomacy.

We next turned to the problem of opening U.S. ports to Soviet ships. The U.S. Army was not happy with the prospect and at one meeting argued vigorously that Soviet ships entering the Great Lakes might blow up the locks. When it was pointed out that Soviet vessels had been passing through these same locks for years as they proceeded to Canadian ports, a pained silence followed.

The State Department was obsessed with the niceties of reciprocity and insisted that each nation open ports of the same size and general configuration to ships of the other power. This made sense in the abstract, but in fact the United States had many large, efficient ports while the Soviets had few. Somehow it had been decided that approximately forty ports in each country would be opened. Implementing this decision involved pairing such American ports as New Bedford and Wood's Hole, Massachusetts, with Soviet Arctic ports having little commercial value. The ridiculousness of this exercise soon became obvious and it was dropped. It was then agreed that all ports in both countries would be opened except for those having military significance. The excluded ports, all of which had major naval facilities, were Norfolk, Virginia, and San Diego, California, while in the Soviet Union they included Sevastapol in the Black Sea and Nahodka in the Pacific.

The last problem to be resolved was the freight rates to be charged for shipping grain to Russia. The Brazilian model offered no help on this issue since it covered only liner traffic, not bulk, and the American liners in foreign trade were already subsidized. Furthermore, liner rates between Brazil and the United States were set by a conference agreement signed by all companies in the trade, while the rates for bulk shipment to the Soviet Union were not covered by a comparable conference agreement. The agreement with the Soviets, rather than following the 40-40-20 terms of the Brazilian agreement, eventually stipulated that one-third of the grain trade would be allotted to ships from each country and one-third to third-flag carriers, with freight rates being set essentially by the market. Since American ships could not operate at world market rates, the agreement to employ market rates meant no American ships could participate. This posed a problem, for without American participation there could be no agreement.

Although it was again obvious that the Soviet negotiators had been instructed to reach an agreement at whatever cost, the rate problem was only settled after extended and extremely difficult negotiations. These were brought to a conclusion by Robert Blackwell, my former deputy, who became maritime administrator in July 1972.

At that time the then assistant secretary for Domestic and International Business, the Commerce Department's agency for international trade, had been forced to resign after publicly criticizing another member of the administration. On the basis of my success in the maritime negotiations, I was offered the trade assignment, which I readily accepted. This shift occurred just as U.S.-U.S.S.R. trade negotiations were getting underway. These negotiations were conducted at a much higher level than had been required for the maritime agreement and participating in them turned out to be a most stimulating experience.

The full story of the maritime agreement is instructive. I negotiated the initial protocol with the Soviets in two days; it took eight months of hard bargaining within the administration and with the Soviets to reach a satisfactory agreement. Setting broad policy is only the beginning; implementation is often far more difficult. If policies are vague or less than carefully crafted, disappointment is bound to follow and blame is often assigned to the implementer.

The rate problem was solved when the Soviets agreed to provide built-in subsidies in the form of higher charter rates that would enable the Americans to carry their one-third share. When the unions tried to force the Soviets to subsidize American ships during the Kennedy administration, they were rebuffed. But during the Nixon administration Soviet leader Brezhnev was determined to achieve detente with the West, and especially with the United States. Subsidizing the U.S. merchant marine became politically acceptable. But the Soviets also had a larger game plan that served them well.

When the final agreement was signed in October 1972, the Russians set about recouping their subsidy payments by moving into the European-American trade with a vengeance. They cut shipping rates aggressively and quickly took over a substantial portion of North Atlantic trade. Between 1971 and 1976 Soviet annual participation in all U.S. trade increased from 261,000 tons to 5.7 million tons.[4] Many of their ships discharged full cargoes in the ports of their satellite Cuba before heading for the U.S. Gulf Coast to load American grain for the return trip to Russia.

The Nixon administration's lasting achievement was reopening relations with China and the Soviet Union. While the objective in both instances was to improve the political environment, trade paved the way. Had the American maritime unions not been cooperative, no maritime agreement with the Soviets could have been reached. Without that agreement, further trade agreements probably would not have followed. Although it could not have been foreseen at the time, Nixon's determination to honor a commitment made during the presidential campaign in 1968 was the seed that produced a mighty oak.

NOTES

INTRODUCTION
    1. Mahan, *Influence of Seapower on History*, 57.

PART I: FREE TRADE AND AMERICAN ENTERPRISE

1: FROM COLONY TO NEW NATION, 1600-1790
    1. Marvin, *American Merchant Marine*, 2.
    2. Bauer, *Maritime History of the United States*, 31.
    3. John Holroyd, Earl of Sheffield (1735–1821), wrote extensively on the commerce of Great Britain and its American colonies.
    4. Bates, *American Navigation*, 31.
    5. Faulkner, *American Economic History*, 117.
    6. Bauer, *Maritime History of the United States*, 48.
    7. Faulkner, *American Economic History*, 121.
    8. Hope, *New History of British Shipping*, 43–44.
    9. Marvin, *American Merchant Marine*, 12–18.
    10. Ibid., 30–31.
    11. Bates, *American Navigation*, 40–41.
    12. Beard and Beard, *Economic Interpretation*, chap. 5.
    13. Bates, *American Navigation*, 61–66. See also Dunmore, *Ship Subsidies*, 14; where Dunmore notes that "between 1789 and 1828 Congress passed no less than fifty tariff laws intended directly or indirectly to protect American Shipowners or Shipbuilding."

2: MARITIME WARS AND RECIPROCITY, 1790-1830
    1. Marvin, *American Merchant Marine*, 51.
    2. Hagan, *This People's Navy*, 22.
    3. Ibid., 36.
    4. Ibid., 60.
    5. Ibid., 40.
    6. Ibid., 51.
    7. Marvin, *American Merchant Marine*, 55.
    8. Ibid, 54.
    9. Lauder, *American Commerce and the Wars*.
    10. Marvin, *American Merchant Marine*, 106.
    11. Hagan, *This People's Navy*, 67.
    12. Faulkner, *American Economic History*, 234.
    13. Lauder, *American Commerce and the Wars*, 146.
    14. Hagan, *This People's Navy*, 72.

15. Hickey, *War of 1812*, 190–93.
16. Marvin, *American Merchant Marine*, 129.
17. Hickey, *War of 1812*, 295.
18. Marvin, *American Merchant Marine*, 177.
19. Hutchins, *American Maritime Industries*, 264–65.

3: THE GOLDEN AGE, 1830–1860

1. Albion, *Square-Riggers on Schedule*, 106–19.
2. Smith, *Wealth of Nations*, 13.
3. Ibid., 31.
4. Ibid., 32.
5. Hutchins, *American Maritime Industries*, 261.
6. Ibid., 328–29.
7. Hunter, *Steamboats on the Western Rivers.*
8. Hope, *British Shipping*, 300.
9. Howarth and Howarth, *The Story of P&O*, 15.
10. Ibid.
11. Hope, *British Shipping*, 268.
12. Ibid., 270.
13. Hutchins, *American Maritime Industries*, 347.
14. Ibid., 351.
15. Saugstad, *Shipping and Shipbuilding Subsidies*, 49.
16. Ibid., 50–51.
17. S. Morison, *Old Bruin*, 256–59.
18. Marvin, *American Merchant Marine*, 247–50.
19. Ibid., 276.
20. Ricardo, *Anatomy of the Navigation Laws*, unpaginated preface.
21. Ibid., 43.
22. Ibid., 44.
23. Ibid., 52.
24. Ibid., 91–92.
25. Ibid., 170–74.
26. Palmer, *Repeal of the Navigation Laws*, 126.
27. Ibid., 143–55.
28. Ibid., 148.
29. Ibid., 168.
30. Marvin, *American Merchant Marine*, 259.
31. See Semmel, *Rise of Free Trade Imperialism.*
32. Smith, *Wealth of Nations*, 464–65.
33. Marvin, *American Merchant Marine*, 255–56.
34. Hutchins, *American Maritime Industries*, 266.
35. Ibid., 303.
36. S. Morison, Commager, and Leuchtenburg, *American Republic*, 1:579–80.
37. Hutchins, *American Maritime Industries*, 318–19.

38. Marvin, *American Merchant Marine*, 245.
39. Ibid., 262.

4: THE CIVIL WAR AND THE TURN TO THE WEST, 1860–1880

1. Catton, *The Coming Fury*, 187.
2. Hagan, *This People's Navy*, 162.
3. Marvin, *American Merchant Marine*, pp.339–40.
4. Hagan, *This People's Navy*, 162.
5. Dalzell, *Flight from the Flag*, 157.
6. Ibid., 239.
7. Ibid., 243.
8. Ibid., 165.
9. Beard and Beard, *American Civilization*, 53–54.
10. S. Morison, Commager, and Leuchtenburg, *American Republic*, 1:676.
11. Ibid., 725–26.
12. Beard and Beard, *American Civilization*, 129.
13. Ibid., 139.
14. Marvin, *American Merchant Marine*, 313.
15. U.S. House, *Report of the Select Committee on . . . Navigational Interests*, ix.
16. Dalzell, *Flight from the Flag*, 249.
17. U.S. House, *Report of the Select Committee on . . . Navigational Interests*, i.
18. Ibid., 2.
19. Ibid., 5.
20. Codman, "Ship Building versus Ship Owning," 482.
21. U.S. House, *Report of the Select Committee on . . . Navigational Interests*, 22.
22. Ibid., xi–xii. All of the above quotations are taken from this report.
23. Zeis, *American Shipping Policy*, 20.

5: MARITIME DECLINE, NAVAL REVIVAL, 1880–1914

1. Allin, "Period of Decline," 88.
2. Zeis, *American Shipping Policy*, 21–22.
3. Allin, "Period of Decline," 71.
4. Zeis, *American Shipping Policy*, 24.
5. Marvin, *American Merchant Marine*, 414–15.
6. Ibid., 419–20.
7. Bates, *American Navigation*, 417.
8. Ibid., 424.
9. Dunmore, *Ship Subsidies*,.21–31.
10. Marvin, *American Merchant Marine*, 359–60.
11. Zeis, *American Shipping Policy*, 54.
12. Bauer, *Maritime History of the United States*, 260.
13. Tuchman, *The Proud Tower*, 149.
14. Ibid.
15. Marvin, *American Merchant Marine*, 425.

16. Lawrence, *Shipping Policies*, 34.
17. Zeis, *American Shipping Policy*, 44.
18. Hutchins, *American Maritime Industries*, 465.
19. Allin, "Period of Decline," 77.
20. Hutchins, *American Maritime Industries*, 571–72.
21. See McCraw, *Prophets of Regulation*, 100, for a revealing analysis of Louis Brandeis's failure to appreciate the structural differences between Morgan's IMM and George Pullman's Palace Car Company.
22. Ibid., 142.

PART II: WAR-IMPELLED INDUSTRIES

6: THE VORTEX OF THE WAR AND ITS AFTERMATH, 1914–1930

1. Zeis, *American Shipping Policy*, 85.
2. Safford, "World War I Maritime Policy," 117.
3. Ibid.
4. Lawrence, *Shipping Policies*, 39.
5. Zeis, *American Shipping Policy*, 86.
6. Forester, *The General*, 186.
7. Stokesbury, *World War I*, 219.
8. Hagan, *This People's Navy*, 249.
9. Ibid., 245.
10. Stokesbury, *World War I*, 128.
11. Ibid., 222.
12. E. Morison, *Admiral Sims*, 342.
13. Ibid., 345.
14. Ibid., 350.
15. Zeis, *American Shipping Policy*, 95.
16. Ibid., 97.
17. S. Morison, Commager, and Leuchtenburg, *American Republic*, 2:378.
18. McDowell and Gibbs, *Ocean Transportation*, 52.
19. Lawrence, *Shipping Policies*, 36–37.
20. Kemble and Kendall, "The Years between the Wars," 152–53.
21. Lawrence, *Shipping Policies*, 41.
22. Hagan, *This People's Navy*, 255.
23. Zeis, *American Shipping Policy*, 136.
24. Kemble and Kendall, "Years between the Wars," 157.
25. U.S. Senate, *Investigation of Air Mail and Ocean Mail Contracts*, 39–40.
26. Ibid., 11–12.
27. McDowell and Gibbs, *Ocean Transportation*, 256.

7: MARITIME POLICY IN THE NEW DEAL, 1930–1939

1. Zeis, *American Shipping Policy*, 153.
2. U.S. Senate, *Investigation of Air Mail and Ocean Mail Contracts*, 2.

3. Ibid., 5–6.
4. Ibid., 10.
5. Ibid., 13.
6. Ibid., 14–15.
7. Ibid., 16.
8. Ibid., 19.
9. Ibid., 23.
10. Ibid., 36.
11. Ibid., 37.
12. Ibid.
13. Ibid., 4.
14. Ibid., 38–46.
15. Ibid., 47.
16. U.S. House, "Message from the President of the United States," 1.
17. Schlesinger, *Coming of the New Deal*, 37.
18. Ibid., 38.
19. Ibid., 40.
20. U.S. Public Law 835, sect. 201.
21. Ibid., sect. 212.
22. Ibid., sect. 301.
23. Ibid., sect. 404.
24. Ibid., sect. 501.
25. Ibid., sect. 502.
26. Ibid., sect. 503.
27. Ibid., sect. 507, 508.
28. Ibid., sect. 511.
29. Ibid., sect. 601, 603.
30. Ibid., sect. 606.
31. Ibid., sect. 701, 703, 704.
32. Ibid., sect. 803.
33. McCraw, *Prophets of Regulation*, 202–3.
34. Lawrence, *Shipping Policies*, 70–71.
35. Ibid., 71–72.
36. U.S. Maritime Commission, *Economic Survey*, 1.
37. Ibid., 4.
38. Ibid., 8.
39. Ibid., 10.
40. Ibid., 20.
41. Ibid., 28.
42. Ibid., 34.
43. Ibid., 36.
44. Ibid., 43.
45. Ibid., 46–47.

46. Ibid., 51.
47. Ibid., 53.
48. Ibid., 54.
49. Ibid., 60–61.
50. Ibid., 62–69.
51. Lawrence, *Shipping Policies*, 73.
52. Ibid., 63.
53. U.S. Public Law 705, sect. 1001.

8: THE MERCHANT MARINE IN WORLD WAR II, 1939–1945
  1. Churchill, *The Gathering Storm*, 138.
  2. Ibid.
  3. Albion and Pope, *Sea Lanes in Wartime*, 255–57.
  4. Hagen, *This People's Navy*, 294.
  5. S. Morison, Commager, and Leuchtenberg, *American Republic*, 547.
  6. Gannon, *Operation Drumbeat*, 389–40.
  7. Ibid. A copy of the British message is reproduced on an unnumbered page.
  8. Cohen and Gooch, *Military Misfortunes*, 60.
  9. Gannon, *Operation Drumbeat*, 378.
  10. Ibid., 176.
  11. S. Morison, *Naval Operations in World War II*, 1:200.
  12. Cohen and Gooch, *Military Misfortunes*.
  13. Albion and Pope, *Sea Lanes in Wartime*, 351.
  14. Behrens, *Merchant Shipping*, 437–38.
  15. Ibid., 287.
  16. Ibid., 288.
  17. Ibid., 328–30.
  18. Ibid., 364–65.
  19. Ibid., 429.
  20. Lawrence, *Shipping Policies*, 74.
  21. Levine and Platt, "Contribution of U.S. Shipbuilding," 180.
  22. Ibid., 185.
  23. Lawrence, *Shipping Policies*, 74.

9: DREAMS OF A NEW GOLDEN AGE, 1945–1960
  1. Lawrence, *Shipping Policies*, 84.
  2. Ibid., 98.
  3. The survey is untitled and was forwarded to Congress by the chairman in a letter of transmittal dated December 31, 1948.
  4. Ibid., 10–11.
  5. Lawrence, *Shipping Policies*, 169–70.
  6. Acheson, "Crisis in Asia."
  7. Uhlig, *How Navies Fight*, 297–300.
  8. Lawrence, *Shipping Policies*, 110.

9. Hagan, *This People's Navy*, 365.

10. Lawrence, *Shipping Policies*, 112.

11. Brief histories of maritime labor organizations and dates they were founded can be found in De La Pedraja, *Historical Dictionary of the Merchant Marine;* see the alphabetic entries and appendix C.

12. On the historical background to this strike, see Nelson, *Workers on the Waterfront.*

13. Kemble and Kendall, "Years between the Wars," 164–66.

14. Ibid., 166–68.

15. U.S. Maritime Commission, *Economic Survey*, 45.

16. Lawrence, *Shipping Policies*, 91.

17. Ibid., 151–52.

18. The section covering the 1965 strike was written from notes kept by the senior author, who was one of the principal negotiators for management while vice president of Grace Line.

PART III: THE APPROACHING END

10: ATTEMPTS TO AVOID A SECOND DECLINE, 1960–1980

1. Lawrence, *Shipping Policies*, 304.

2. The senior author discussed the substance of this meeting with Franc Nemec shortly after it had taken place.

3. U.S. House Merchant Marine and Fisheries Committee, *Hearings on Independent Maritime Agency*, 498.

4. Nixon, *Nixon Speaks Out*, 217. This book is a collection of speeches made during the 1968 election campaign.

5. For additional details see the appendix, "The Nixon Maritime Program: A Personal Memoir."

6. Richard M. Nixon, President's message, *To the Congress of the United States*, October 23, 1969, press release.

7. Lawrence, *Shipping Policies*, 168.

8. U.S. Department of Commerce, *MARAD 1978*, v.

9. *New York Times*, September 22, 1974, sec. 4, 2.

10. Ibid., October 12, 1974, 30.

11. O'Brien, "The Making of Maritime Policy,"52–54.

11: THE RAPIDLY CHANGING MARITIME WORLD

1. Gibson, "Current Conditions in the Transport Services."

2. Carlisle, *Sovereignty for Sale*, 14–17.

3. Lawrence, *Shipping Policies*, 102.

4. U.K. Committee of Inquiry into Shipping, *Rochdale Report*, 369–70.

5. U.K. Report of Working Group, *British Shipping.*

6. Sletmo and Holste, "Competitive Advantage," 249–51.

7. Ibid., 245.

8. Cecil, "The Shipping Act of 1984," 210. The cases discussed are summarized in this journal (*Dickinson Journal of International Law* 3 (1985): 197–232).
9. Ibid., 214.
10. Ibid., 213.
11. Statistics taken from Loree, "Some Thoughts on the Future of Open Registries."

12: MILITARY SEALIFT AFTER WORLD WAR II
1. Gibson and Calhoun, *The Evolution of USTRANSCOM.*
2. U.S. Department of Defense, "Memorandum of Agreement," *Instruction 5030.3.*
3. President, "National Sealift-Policy."
4. *Washington Post*, September 11, 1990, "Skinner May Seek Aid for Maritime Industry," A12.
5. Gibson and Shuford, "Desert Shield and Strategic Sealift," 6–19.

13: THE SEARCH FOR A WORKABLE MARITIME PROGRAM, 1980–1992
1. Reagan, "Effective Maritime Strategy."
2. Holloway, *Critique of OMB Review.*
3. Lewis, "Maritime Policy."
4. Editorial, "What Happened Mr. Reagan? *"Journal of Commerce*, March 22, 1984.
5. U.S. Commission on Merchant Marine and Defense, "First Report: Finding of Fact and Conclusions," (Washington, DC: GPO, September 30, 1987), 5.
6. Ibid., 1.
7. Ibid.
8. U.S. Commission on Merchant Marine and Defense, "Public Hearings, February–July 1987," 32–33.
9. Ibid., 107.
10. Ibid., 108.
11. Ibid., 110.
12. Ibid., 519–20.
13. Ibid., 573.
14. Ibid., 791–813.
15. U.S. Commission on Merchant Marine and Defense, "Second Report: Recommendations," (Washington, DC: GPO, December 30, 1987), 9.
16. Ibid., 13.
17. Ibid., 16.
18. Ibid., 20.
19. Ibid., 13.
20. Editorial, "Gathering Dust," *Journal of Commerce*, February 23, 1989.
21. U.S. International Trade Commission, "Shipbuilding Trade Reform Act of 1992," xx–xiv.

22. John Clancy, "APL/Sea-Land Maritime Policy Proposal," press release dated February 3, 1992.

23. Andrew H. Card, Jr., "Proposals for Maritime Reform," statement before the Subcommittee on Merchant Marine of the Committee of Science and Transportation, U.S. Senate, June 17, 1992.

24. Policy Coordinating Group (PCG) Working Group, "Decision Memorandum on Commercial Maritime Policy," Memorandum for Chairman, Policy Coordinating Group (DOD), signed June 8, 1992, by Colin McMillan, Assistant Secretary for Production and Logistics.

14: THE FINAL PUSH TO EXTEND MARITIME SUBSIDIES, 1992–1999
1. "Subsidy Decision May Sink U.S. Flag Merchant Fleet," *Washington Post*, May 15, 1993.

2. "U.S. Bill Would Curb 'Greedy' Shipowners," *Lloyd's List*, June 2, 1993.

3. "Two Senators Rip Clinton Over Lack of Plan . . . ," *Journal of Commerce*, August 6, 1993.

4. "Stop Shipbuilding Subsidies," *Journal of Commerce*, 30 October 1995.

5. *Seatrade Week Newsfront*, 1 November 1996.

6. *Journal of Commerce*, March 18, 1998.

7. Vice President Gore's Office, National Policy Review, "Reinventing Government," Report to the Vice President for public release, September 7, 1993.

8. *Journal of Commerce*, August 11, 1993.

9. Letter dated June 20, 1997, from the Maritime Subsidy Board to Joe B. Freeman, President of Lykes Bros. Steamship Co., Inc., signed by the secretary.

10. *American Shipper* (August 1997).

15: THE AMERICAN MARITIME INDUSTRY IN A TIME OF TRANSITION
1. Vice President Gore's Office, *Reinventing Government*, introduction (not paginated).

2. Eversheim, *Ship Subsidization*, 25.

3. *Journal of Commerce*, July 28, 1993.

4. *Lloyd's List International*, August 14, 1992.

5. U.S. Maritime Commission, *Economic Survey*, 85.

APPENDIX
1. Nixon, *Nixon Speaks Out*. Contains major speeches and statements by Richard M. Nixon in the presidential campaign of 1968.

2. U.S. House Committee on Merchant Marine and Fisheries, *Hearings on a Maritime Program*.

3. The story of the Soviet grain sales is quite accurately told in M. Marder and M. Berger, "U.S.-Soviet Grain Deal: Case History of a Gamble," *Washington Post*, December 7, 1971.

4. U.S. Maritime Administration, *Expansion of the Soviet-Merchant Marine*, 27.

BIBLIOGRAPHY

The following bibliography is organized into two parts. Part one, "Sources for Further Reading," lists works under three headings: (1) Maritime History: General Accounts and Bibliographies, (2) U.S. Maritime Industries: Special Studies, and (3) Maritime Policy Studies: Political, Industrial, and Economic. Part two, "Sources Cited in the Notes," lists its entries alphabetically in one list. Complete citations are given for works included in part one. Works included in part two that are also listed in part one are cited in short form in part two.

PART I: SOURCES FOR FURTHER READING

The literature on maritime history and policy is enormous and only some of the more useful items that pertain to the subjects addressed in this book are listed below. These works provide reliable points of entry into the subjects they cover. Since the older literature can most easily be identified by referring to the notes and bibliographies given in subsequent scholarly studies, a special effort has been made to list recent publications.

MARITIME HISTORY: GENERAL ACCOUNTS AND BIBLIOGRAPHIES

Albion, Robert G. *Naval and Maritime History: An Annotated Bibliography.* 4th ed. Mystic, CT: Marine Historical Association, 1972. Supplement by Benjamin W. Labaree. *A Supplement (1971–1986) to Robert G. Albion's Naval and Maritime History: An Annotated Bibliography.* Mystic, CT: Mystic Seaport Museum, 1988. The standard general bibliography.

Bauer, K. Jack. *A Maritime History of the United States.* Columbia: University of South Carolina Press, 1988. Lengthy, up-to-date, annotated bibliography.

De La Pedraja, Rene. *A Historical Dictionary of the U.S. Merchant Marine and Shipping Industry since the Introduction of Steam.* Westport, CT: Greenwood, 1994. Highly informative; excellent bibliographic essay on many topics, such as government publications, labor, business, and technology, that are often slighted.

Hagan, Kenneth J. *This People's Navy.* New York: Free Press, 1991. Reliable recent survey of U.S. Naval history.

Hugill, Peter J. *World Trade since 1453: Geography, Technology, and Capitalism.* Baltimore: Johns Hopkins University Press, 1993. Well-informed broad-view history of modern transportation.

Kennell, Susan K., and Suzanne R. Ontiveros. *American Maritime History: A Bibliography.* Santa Barbara, CA: ABC-CLIO, 1986. Convenient first source.

Labaree, Benjamin W., et al. *America and the Sea: A Maritime History.* Mystic, CT: Mystic Seaport, 1998. Handsomely illustrated and produced comprehensive narrative history; scholarly and readable.

Marvin, Winthrop L. *The American Merchant Marine: Its History and Romance from 1620 to 1902.* New York: Scribner's, 1902. Valuable older survey.
Morison, Samuel E. *History of United States Naval Operations in World War II.* 14 vols. Boston: Little Brown, 1947–60. Classic official history by a great maritime historian.
Rodger, N. A. M. *The Wooden World.* Annapolis: Naval Institute Press, 1986. Scholarly account of naval careers and life at sea in eighteenth-century Britain.
Runyan, Timothy J.,ed. *Ships, Seafaring, and Society: Essays in Maritime History.* Detroit: Wayne State University Press, 1987. Wide range of topics; some useful essays.
Schultz, Charles R. *Bibliography of Maritime and Naval History: Periodical Articles.* College Station: Texas A&M Press, 1972–present, alternate years. Provides abstracts of articles.
Uhlig, Frank, Jr. *How Navies Fight.* Annapolis: Naval Institute Press, 1994. Uhlig has written what may be the definitive study of the essential role of the wartime navy, which is to protect the sea-lanes, deny their use to the enemy, and land and sustain the army on a hostile shore.

U.S. MARITIME INDUSTRIES: SPECIAL STUDIES
Albion, Robert G. *Square-Riggers on Schedule.* Princeton: Princeton University Press, 1938. Origins of scheduled freight-liner service around 1820.
———. *Rise of the Port of New York: 1815–1860.* New York: Scribners, 1939. Classic port study; reprinted by Northeastern University Press in 1984.
Albion, Robert G., William A. Baker, and Benjamin W. Labaree. *New England and the Sea.* Middletown, CT: Wesleyan University Press, 1972. Standard regional history.
Baughman, James P. *The Mallorys of Mystic.* Middletown, CT: Wesleyan University Press, 1972. Detailed exemplary history of early maritime business.
Bolster, W. Jeffrey. *Black Jacks: African-American Seamen in the Age of Sail.* Cambridge: Harvard University Press, 1997. Award winning recent study.
Braynard, Frank O. *Famous American Ships.* 2nd ed. New York: Hastings House, 1978. One of many books by prolific maritime historian.
Bunker, John. *Heroes in Dungarees: The Story of the American Merchant Marine in World War II.* Annapolis: Naval Institute Press, 1995.
Clark, John G. *The Grain Trade of the Old Northwest.* Urbana: University of Illinois Press, 1966. Study of early transportation on Great Lakes.
Creighton, Margaret S., and Lisa Norling, eds. *Iron Men, Wooden Women: Gender and Seafaring in the Atlantic World, 1700–1920.* Baltimore: Johns Hopkins University Press, 1996.
Cutler, Carl C. *Greyhounds of the Sea.* 3rd ed. Annapolis, MD: Naval Institute Press, 1984. Definitive history of U.S. clipper ships.
Delgado, James P. *To California by Sea: A Maritime History of the California Gold Rush.* Columbia: University of South Carolina Press, 1990.

Bibliography

Gannon, Michael. *Operation Drumbeat*. New York: Harper & Row, 1990. Thorough study of German submarine warfare in 1942.

Goldberg, Mark H. *The "Hog Islanders."* Kings Point, NY: The American Merchant Marine Museum, 1991. Detailed history of these World War I era ships.

Goldenberg, Joseph A. *Shipbuilding in Colonial America*. Charlottesville: University Press of Virginia, 1976.

Gorter, Wytze, and George H. Hildebrand. *The Pacific Coast Maritime Shipping Industry, 1930–1948*. 2 vols. Berkeley: University of California Press, 1952–54.

Headrick, Daniel R. *The Invisible Weapon: Telecommunications and International Politics, 1851–1945*. New York: Oxford University Press, 1991. Detailed account by leading historian of technology and imperialism.

Heinrich, Thomas R. *Ships for the Seven Seas: Philadelphia Shipbuilding in the Age of Industrial Capitalism*. Baltimore: Johns Hopkins University Press, 1997.

Hunter, Louis C. *Steamboats on the Western Rivers*. 1949. Reprint, New York: Dover, 1993. Classic economic and technological history.

Kilmarx, Robert A., ed. *America's Maritime Legacy: A History of the U.S. Merchant Marine and Shipbuilding Industry*. Boulder, CO: Westview Press, 1979. Excellent topical essays cover key historical periods.

Marcus, Henry S., and Peter T. Weber. *Competitive Manning of U.S. Flag Vessels*. Rev. ed. Washington: Maritime Administration, 1994. Detailed study of comparative manning costs in contemporary merchant marine.

McCullough, David. *The Path between the Seas*. New York: Simon & Schuster, 1977. Prize-winning history of building of Panama Canal.

McCurdy, H. W., and Gordon Newell. *The Maritime History of the Pacific Northwest*. 3 vols. Seattle: University of Washington Press, 1966–77.

Morison, Samuel Eliot. *Maritime History of Massachusetts, 1783–1860*. Boston: Houghton Mifflin, 1921. A classic.

Nelson, Bruce. *Workers on the Waterfront: Seamen, Longshoremen, and Unionism in the 1930s*. Urbana: University of Illinois Press, 1988. Excellent labor history.

Niven, John. *The American President Lines and Its Forebears: 1848–1984*. Newark: University of Delaware Press, 1987.

Palmer, S. R. *Politics, Shipping and the Repeal of the Navigation Laws*. Manchester: Manchester University Press, 1990. Detailed study of Parliamentary decision that radically altered competitive context of world shipping.

Van der Vat, Dan. *The Atlantic Campaign: World War II's Great Struggle at Sea*. New York: Harper & Row, 1988.

STUDIES OF MARITIME POLICY ISSUES: POLITICAL, INDUSTRIAL, AND ECONOMIC

Albion, Robert G., and J. B. Pope. *Sea Lanes in Wartime*. 2nd ed. Hamden, CT: Archon Books, 1968.

Bess, David H., and M. T. Ferris. *U.S. Maritime Policy*. New York: Praeger, 1981.

Bowman, Roy G. "The Merchant Marine Act of 1970." *Journal of Maritime Law and Commerce* 2 (1970–71): 715–34.

Boyce, Gordon H. *Information, Mediation, and Institutional Development: The Rise of Large-Scale Enterprise in British Shipping, 1870–1919.* Manchester: Manchester University Press, 1995. Not about U.S. industry, but a relevant and challenging analysis of British shipping firms in age of cartels.

Carlisle, Rodney. *Sovereignty for Sale.* Annapolis, MD: Naval Institute Press, 1981. History of origins and use of flags of convenience.

Dalzell, George W. *The Flight from the Flag: The Continuing Effect of the Civil War upon the American Carrying Trade.* Chapel Hill: University of North Carolina Press, 1940.

De La Pedraja, Rene. *The Rise and Decline of U.S. Merchant Shipping in the Twentieth Century.* New York: Twayne, 1992. Focuses on rise and decline of operating companies (firms); excellent annotated bibliography.

Dunmore, Walter T. *Ship Subsidies: An Economic Study of the Policy of Subsidizing Merchant Marines.* New York: Houghton Mifflin, 1907.

Eversheim, Franz. *Effects of Shipping Subsidization.* Bremen: Institut für Schiffahrtsforschung, 1958. Translated from German.

Ferguson, Allen R., et al. *The Economic Value of the U.S. Merchant Marine.* Evanston, IL: Transportation Center at Northwestern University, 1961.

Fortune 16, no.3 (September 1937). Special issue on U.S. Shipping. Lively snapshot of U.S. industry just as 1936 Merchant Marine Act being implemented; excellent graphics.

Goldberg, Joseph H. *The Maritime Story: A Study in Labor-Management Relations.* Cambridge: Harvard University Press, 1958. A history of maritime labor organizations, 1900–1950s.

Gorshkov, Sergei G. *Seapower and the State.* Annapolis, MD: Naval Institute Press, 1976.

Gorter, Wytze. *United States Shipping Policy.* New York: Harper, 1956.

Goss, Richard O. "Sense and Shipping Policies." In *Shipping Policies for an Open World Economy*, edited by George N. Yannopoulos, 61–87. London: Routledge, 1989.

Goss, Richard O., and P. B. Marlow. "Internationalism, Protectionism and Interventionism in Shipping." In *Current Issues in Maritime Economics*, edited by K. M. Gwilliam, 45–67. Dordrecht: Kluwer Academic Publishers, 1993.

Graham, G. S. *Seapower and British North America, 1783–1812.* Cambridge: Harvard University Press, 1941.

Harper, Lawrence. *English Navigation Laws: A Seventeenth-Century Experiment in Social Engineering.* New York: Columbia University Press, 1939. Rationale and historic background on navigation acts.

Harvard Business School. *The Use and Disposition of Ships and Shipyards at the End of World War II.* Washington, DC: GPO, 1945.

Hutchins, John G. B. *The American Maritime Industries and Public Policy, 1789–1914.* Cambridge: Harvard University Press, 1941. Exhaustive and a classic.

Jantscher, Gerald R. *Bread upon the Waters: Federal Aids to the Maritime Industries.* Washington, DC: Brookings Institution, 1975. Rigorous economic analysis of intended and unintended consequences.

Bibliography

Kilgour, John G. *The U.S. Merchant Marine: National Maritime Policy and Industrial Relations*. New York: Praeger, 1975.

Lane, Tony. *The Merchant Seaman's War*. Manchester: Manchester University Press, 1990.

Lauder, A. C. *American Commerce and the Wars of French Revolution and Napoleon, 1792–1812*. Philadelphia: Augustus Kelly, 1932.

Lawrence, Samuel A. *U.S. Merchant Marine Shipping Policies and Politics*. Washington: Brookings Institution, 1966. First-rate scholarly treatment.

Lovett, William A., ed. *United States Shipping Policies and the World Market*. Westport, CT: Quorum Books, 1996.

Luce, Stephen B. "The Manning of Our Navy and Mercantile Marine." *The Record of the United States Naval Institute* 1 (1874): 17–37, in *The Papers and Proceedings of the United States Naval Institute* 1 (1874).

Mahan, Alfred T. *The Influence of Seapower on History, 1660–1783*. Boston: Little Brown, 1890. Classic text on maritime and naval power.

McCraw, Thomas. *Prophets of Regulation*. Cambridge: Harvard University Press, 1984. Excellent historical introduction to federal antitrust regulation.

Nimitz, Chester W. *The Great Sea War*. Edited by E. B. Potter. Englewood Cliffs, NJ: Prentice Hall, 1960.

Safford, Jeffrey J. *Wilsonian Maritime Diplomacy, 1913–1921*. New Brunswick, NJ: Rutgers University Press, 1978. Origins of federal revival of U.S. merchant marine.

Schlesinger, Arthur M., Jr. *The Coming of the New Deal*. Boston: Houghton Mifflin, 1958. Describes political and industrial context in which 1936 Merchant Marine Act was written and passed.

Sturmey, S. G. *British Shipping and World Competition*. London: Athlone Press, University of London, 1962.

Taylor, George Rogers. *The Transportation Revolution: 1815–1860*. New York: Holt, Rinehart and Winston, 1951. Classic in economic and transportation history.

Ubbelohde, Carl. *Vice-Admiralty Courts and the American Revolution*. Chapel Hill: University of North Carolina Press, 1960.

White, Lawrence. *International Trade in Ocean Shipping Services: The United States and the World*. Cambridge, MA: Ballinger, 1988.

Whitehurst, Clinton H., ed. *The U.S. Merchant Marine in Search of Enduring Policy*. Annapolis, MD: Naval Institute Press, 1983.

Zeis, Paul M. *American Shipping Policy*. Princeton: Princeton University Press, 1938.

PART II: SOURCES CITED IN THE NOTES

Complete citations of works cited below in short form can be found in part one of this bibliography.

Acheson, Dean. "Crisis in Asia: An Examination of U.S. Policy." Speech given at National Press Club, Washington, DC, January 12, 1950.

Albion. *Square-Riggers on Schedule*.

Bibliography

Albion and Pope. *Sea Lanes in Wartime.*

Allin, Lawrence C. "The Civil War and the Period of Decline." In *America's Maritime Legacy,* edited by Robert A. Kilmarx, 65–110. Boulder, CO: Westview Press, 1979.

Bates, William W. *American Navigation, the Political History of Its Rise and Ruin and the Proper Means for Its Encouragement.* Boston: Houghton Mifflin, 1902.

Bauer. *Maritime History of the United States.*

Beard, Charles A., and Mary R. Beard. *An Economic Interpretation of the Constitution of the United States.* 1913. Reprint, New York: Macmillian, 1972.

———. *The Rise of American Civilization (The Industrial Era).* New York: Macmillian, 1930.

Behrens, C. B. A. *Merchant Shipping and the Demands of War.* London: HMSO and Longmans, Green, 1955.

Carlisle. *Sovereignty for Sale.*

Catton, Bruce. *The Coming Fury.* Garden City, NY: Doubleday, 1961.

Cecil, Martha L. "The Shipping Act of 1984: Bringing the United States in Harmony with International Shipping Practices." *Dickinson Journal of International Law* 3 (1985): 197–232.

Churchill, Winston S. *The Gathering Storm.* Boston: Houghton Mifflin, 1948.

Codman, John. "Ship-Building versus Ship-Owning." *North American Review* 142 (1886): 478–84.

Cohen, Eliot A., and J. Gooch. *Military Misfortunes: The Anatomy of Failure in War.* New York: Collier Macmillian, 1990.

Dalzell. *Flight from the Flag.*

De La Pedraja. *Historical Dictionary of the Merchant Marine.*

Dunmore. *Ship Subsidies.*

Eversheim. *Shipping Subsidization.*

Faulkner, Harold U. *American Economic History.* 5th ed. New York: Harper, 1943.

Forester, C. S. *The General.* London: Michael Joseph, 1936.

Gannon. *Operation Drumbeat.*

Gibson, Andrew E., "Current Conditions in the Transport Services Involving Japanese Vehicle Imports to the United States." Brief filed by Automar International Car Carriers, Inc., in response to a U.S. International Trade Commission investigation into Japanese shipping practices, 1985.

———. "Elements of a U.S. Shipping Policy." In *The U.S. Merchant Marine in Search of an Enduring Maritime Policy,* edited by Clinton H. Whitehurst, 90–103. Annapolis: Naval Institute Press, 1983.

Gibson, Andrew E., and William M. Calhoun. *The Evolution of USTRANSCOM.* Newport, RI: Naval War College, 1990.

Gibson, Andrew E. and J. L. Shuford. "Desert Shield and Strategic Sealift." *Naval War College Review* 44 (1991): 6–19.

Hagan. *This People's Navy.*

Hickey, Donald R. *The War of 1812.* Urbana: University of Illinois Press, 1995.

Holloway, J. L. *Critique of OMB Review of Maritime Programs and Policy.* Washington: Council of American Flag Ship Operators, 1982.

Hope, Ronald. *A New History of British Shipping.* London: John Murray 1990.

Howarth, David, and S. Howarth. *The Story of P&O.* London: Weidenfeld and Nicholson, 1986.

Hunter. *Steamboats on the Western Rivers.*

Hutchins. *American Maritime Industries.*

Jantscher. *Bread upon the Waters.*

Kemble, John H., and Lane C. Kendall. "The Years between the Wars, 1919–1939." In *America's Maritime Legacy*, edited by Robert A. Kilmarx, 149–74. Boulder, CO: Westview Press, 1979.

Kilmarx. *America's Maritime Legacy.*

Lauder. *American Commerce and the Wars.*

Lawrence. *Merchant Marine Shipping Policies and Politics.*

Levine, David, and Sara Ann Platt. "The Contribution of U.S. Shipbuilding and the Merchant Marine in the Second World War." In *America's Maritime Legacy*, edited by Robert A. Kilmarx, 175–214. Boulder, CO: Westview Press, 1979.

Lewis, Drew. "Interim Report on Maritime Policy." *American Shipper* (July 1982), 51–56. Lewis's May 3 report to the Cabinet Council and his May 20 outline of his maritime program.

Loree, Philip J. "Some Thoughts on the Future of Open Registries." Paper prepared for Panama Maritime II: Second Maritime World Conference, 1994.

Mahan. *Influence of Seapower on History.*

Marcus and Weber, *Competitive Manning of U.S. Flag Vessels.*

Marvin. *American Merchant Marine.*

McCraw. *Prophets of Regulation.*

McDowell, Carl E., and Helen Gibbs. *Ocean Transportation.* New York: McGraw-Hill, 1954.

Morison, Elting E. *Admiral Sims and the Modern American Navy.* Boston: Houghton Mifflin, 1942.

Morison, Samuel Eliot. *History of United States Naval Operations in World War II*, vol.1., 1947.

———. *Maritime History of Massachusetts.*

———. *Old Bruin: Commodore Matthew C. Perry.* Boston: Little, Brown, 1967.

Morison, Samuel Eliot, Henry S. Commager, and William E. Leuchtenburg. *Growth of the American Republic.* 7th ed. 2 vols. New York: Oxford University Press, 1980.

Nelson. *Workers on the Waterfront.*

Nixon, Richard M. *Nixon Speaks Out.* New York: privately printed by the Nixon-Agnew Campaign Committee, 1968.

O'Brien, L. T. "The Making of Maritime Policy." In *The U.S. Merchant Marine in Search of an Enduring Maritime Policy*, edited by Clinton H. Whitehurst, 44–56. Annapolis: Naval Institute Press, 1983.

Palmer. *Repeal of the Navigation Laws.*

Reagan, Ronald. "A Program for the Development of an Effective Maritime Strategy." Released by the Reagan and Bush Headquarters, Arlington, VA, 1980.

Ricardo, J. L. *The Anatomy of the Navigation Laws.* London: Charles Gilpin, 1847.

Safford, Jeffrey J. "World War I Maritime Policy and the National Security, 1914–1919." In *America's Maritime Legacy,* edited by Robert A. Kilmarx, 111–48. Boulder, CO: Westview Press, 1979.

Saugstad, Jesse E. *Shipping and Shipbuilding Subsidies: A Study of State Aid to the Shipping and Shipbuilding Industries in Various Countries of the World.* Washington, DC: GPO, 1932.

Schlesinger. *Coming of the New Deal.*

Semmel, Bernard. *The Rise of Free Trade Imperialism: Classical Political Economy, the Empire of Free Trade, and Imperialism, 1750–1850.* Cambridge: Cambridge University Press, 1970.

Sletmo, Gunnar K., and Suzanne Holste. "Shipping and the Competitive Advantage of Nations: The Role of International Ship Registers." *Maritime Policy and Management* 20 (1993): 243–55.

Smith, Adam. *The Wealth of Nations.* Edited by Roy H. Campbell and Andrew S. Skinner. Oxford: Clarendon Press, 1979.

Stokesbury, James L. *A Short History of World War I.* New York: William Morrow, 1981.

Tuchman, Barbara W. *The Proud Tower.* New York: Macmillian, 1962.

U.K. Committee of Inquiry into Shipping. *Rochdale Report.* London: HMSO, 1970.

U.K. Report of Working Group. *British Shipping: Challenges and Opportunities.* London: HMSO, 1990.

U.S. Commission on Merchant Marine and Defense. Four Reports and Three Volumes of Hearings. Washington, DC: GPO, 1986–89.

U.S. Department of Commerce. Maritime Administration. *MARAD 1970: Year of Transition.* Washington, DC: Department of Commerce, 1971.

U.S. Department of Commerce. Maritime Administration. *MARAD 1978.* Washington, DC: Department of Commerce, 1979.

U.S. Department of Defense. "Memorandum of Agreement between the Department of Defense and the Department of Commerce Dealing with the Utilization, Transfer and Allocation of Merchant Ships." In *Department of Defense Instruction 5030.3,* 20 October 1954.

U.S. Department of Transportation. *Annual Reports of the Federal Maritime Commission.* Washington, DC: GPO, 1970–1992.

U.S. Department of Transportation. *Annual Reports of the Marine Administration.* Washington, DC: GPO, 1970–1992.

U.S. House. *Message from the President of the United States.* 73rd Cong., 1935 H. Doc. 118.

U.S. House. *Report of Select Committee on the Causes of the Reduction of American Tonnage and the Decline of the Navigational Interests, 1870*. 42nd Cong., 2nd sess., H. Rept. 28.

U.S. House Committee on Merchant Marine and Fisheries. *Hearings on Independent Maritime Agency, 1967*. 90th Cong., 1st sess.

U.S. House Committee on Merchant Marine and Fisheries, Senate Committee on Commerce. *Hearings on a Maritime Program*. 91st Cong., 2nd sess., 1970.

U.S. International Trade Commission. *Shipbuilding Trade Reform Act of 1992; Likely Economic Effects of Enactment*. Washington, DC: USITC Publication 2495, 1992.

U.S. Maritime Administration. *Expansion of the Soviet-Merchant Marine*. Washington, DC: GPO, 1978.

U.S. Maritime Commission. *Economic Survey of the American Merchant Marine*. Washington, DC: GPO, 1937.

U.S. National Advisory Committee on Oceans and Atmosphere. *Shipping, Shipyards and Sealift: Issues on National Security and Federal Support*. Washington, DC: GPO, 1985.

U.S. Public Law 835. 74th Cong., 2nd sess., 1936. *The Merchant Marine Act*.

U.S. Senate. *Investigation of Air Mail and Ocean Mail Contracts*. 74th Cong., 1st sess., 1935, S. Rept. 898.

Uhlig. *How Navies Fight*.

Whitehurst. *Merchant Marine in Search of an Enduring Policy*.

Zeis. *American Shipping Policy*.

INDEX

Commerce Department, 137, 173, 178, 185, 194–95, 198, 245, 256, 312, 313, 318
commercial carriers, 43–46, 177–78
Commission on Merchant Marine and Defense, 260–68, 275–76, 285
Committee of American Steamship Lines (CASL), 186, 196
Compact of Free Association, 233
Companie Generale Maritime (CGM), 230
Confederacy. *See* Civil War
Congress of Industrial Organizations (CIO), 182–83
Congress, U.S.: and Adams, 30; and antitrust law, 237; and Black committee, 134; and Boyd Report, 195; and "cargo equity" bill, 205; and commerce, 22–23, 174; and defense appropriations, 243; and Department of Defense (DOD), 274, 277; and Embargo Act, 34; and grain boycott of Soviet Union, 185; and Great Depression, 136; and Gulf of Tonkin Resolution, 180; and Jefferson, 103; Lynch committee of, 5, 73–74, 76–78, 80, 146, 259; Madison's address to, 35–36; and merchant marine policy, 72, 75, 80, 135–37, 199–200, 207, 232, 258, 303; and military readiness bills, 80, 113, 161, 246, 251, 260, 274, 277, 281; and Neutrality Act of 1935, 156–57; and New York Shipowners' Association, 75; and Organization for Economic Cooperation and Development, 281; and Panama Canal Act, 95; and Philippine Islands, 120; and registry policy, 84–85; and Republican Party, 27, 206, 288, 311; and Shipping Act of 1984, 238; and subsidies, 51, 53,

75–76, 80–86, 94, 103, 134, 150, 257, 268, 270, 285, 288, 303–4; and tariffs and taxes, 23, 238–39; and trade policy, 39–42, 269, 279; and treaty with Algeria, 28; and Volstead Act, 226; and War of 1812, 35; and World War I, 110. *See also* House of Representatives, U.S.; Senate, U.S.
Connecticut (battleship), *following page* 154
Connecticut (steamboat), *following page* 154
Connor, John, 194
Constellation, 30, 180
Constitution, 30, 36, *following page* 154
Constitution, U.S., 21–25, 47, 226
Construction Differential Subsidy (CDS), 138–39, 266, 285–86
containerization and containerships, 6, *following page* 154, 176, 208–20, 224, 232, 239, 241, 254–53, 274, 285, 286, 299–304
Continental Grain, 185
convoys, 111–12, 158, 161–62, 173, 310
Corcoran, Karla, 284
Corn Laws, 55–56
cotton gin, 49
cotton industry, 24, 46, 65, 69, 87, 104
Council of American-Flag Ship Operators, 256
Council of Economic Advisors, 198, 312–13
Cramp (William) and Sons shipyard, 85, 121–22
Crescent, 28
crews. *See* sailors
Crimean War, 59, 61
Cromwell, Oliver, 15–16
Crowe, William, 244
Crowley Maritime, 216–17, 221, 291
Cuba, 84, 91–92, 128, 318–19

competition to shipping, 62–63, 87–88, 120, 125, 171–72; and containerization, 211, 217, 300; development of, 2–3, 50–51, 300, 312; and intermodalism, 217–18; and land grants, 77; and opposition to Panama Canal, 95; transcontinental railroad, 58; and Western expansion, 70–72, 78
Railway Labor Act, 190
Railway Mediation Boards, 153
Ready Reserve Force (RRF), 221, 241, 247, 251–53, 265, 271–74, 301–2
Reagan, Ronald: and Bush, 268–69; and Maritime Administration, 258–59; maritime policy of, 255–68; on merchant marine, 274; and military readiness, 247; and subsidies, 233, 276; and U.S. Navy, 193, 207, 243
Reconciliation Act of 1981, 257, 286–87
Red D Line, 84
registry: during Civil War, 69, 75–77; and flags of convenience, 6, 80, 225–34; and foreign-built transports, 91, 150, 202, 206, 273, 278–79, 304; laws of, 84–86, 133, 175, 190, 207, 225–34, 255, 266–67, 279, 295, 304; open registry ships, 209, 226, 231, 270, 298; and Pacific Rim, 231; and subsidies, 138, 225–26; during World War I, 104
Reliance, 226
Republican Party: and Congress, 27, 206, 288, 311; and Eisenhower, 127, 174; and Great Depression, 123, 134; and Jefferson, 34; and Nixon, 198; platform of, 83; and tariffs and taxes, 98
reserve fleet. *See* National Defense Reserve Fleet
Resolute, 226
Reuben James, 158

Revenue Code, U.S., 232
revenue pooling agreements, 171
Reynolds, R. J., 215, 217
Ribbentrop, Joachim von, 156
Ricardo, David, 55–56
Ricardo, J. Lewis, 56–58
Robert E. Peary, following page 154
Rochdale Report, 229
Rockefeller, John D., 71, 95
roll-on/roll-off ships (Ro-Ros), *following page* 154, 208, 214–17, 220–22, 241, 247, 252–53, 301
Roosevelt, Franklin D.: and Black committee, 130, 133–36; compared to Nixon, 315; election of, 127; and Great Depression, 182, 257; and New Deal, 136, 141–44, 153, 182, 257, 275; on subsidies, 174; and U.S. Navy, 109; and World War II, 157–58, 161–66
Roosevelt, Theodore, 93–94, 125
Roper, 160
Ro-Ros (roll-on/roll-off) ships, *following page* 154, 208, 214–17, 220–22, 241, 247, 252–53, 301
Royal William, 51
Rutherford, Robert L., 292

sailors: Maritime Commission on, 148; training for, during World War II, 167–68; wages of, 117–18, 136, 138, 140, 147, 184–86, 287–90. *See also* labor unions; merchant marine; Navy, U.S.
Sailors' Union of the Pacific (SUP), 182
Sampson, William T., 92
San Diego, 281, 300, 317
San Francisco: and domestic trade, 88; labor strike in, 182–83; and New York City, 60–61, 94–95; and Pacific trade, 41, 61, 213; ports of call in, 71, 84, 212, 220
Santa Elena, 122